The Canadian Space Program

From Black Brant to the International Space Station

Andrew B. Godefroy

The Canadian Space Program

From Black Brant to the International Space Station

 Springer

Published in association with
Praxis Publishing
Chichester, UK

Andrew B. Godefroy
Kingston, Ontario
Canada

SPRINGER-PRAXIS BOOKS IN SPACE EXPLORATION

Springer Praxis Books
ISBN 978-3-319-40104-1   ISBN 978-3-319-40105-8   (eBook)
DOI 10.1007/978-3-319-40105-8

Library of Congress Control Number: 2017933959

Printed on acid-free paper

This Springer imprint is published by Springer Nature
The registered company is Springer International Publishing AG
The registered company address is: Gewerbestrasse 11, 6330 Cham, Switzerland

Contents

Acknowledgements... vi

Introduction.. viii

1 Northern Mariner: The Origins of Canadian Space Activities.................... 1

2 Forging a Space Nation: Policy and Program Development, 1957–1963...... 19

3 Challenge and Commitment: Canada's Space Program
  in Transition, 1964–1974... 76

4 Ad Astra: Establishing a Permanent Presence in Space, 1974–1984........... 121

5 Maple Leaf in Orbit: Institutionalizing the Canadian
  Space Program, 1984–1995 170

6 Build Up: Canada's Space Station Era Begins, 1996–2009.................. 215

7 Conclusion: Towards Space Station Expeditions 290

Appendix A: Canadian Space Agency Presidents, 1989–2013.................. 293

Appendix B: Biographical Notes on Canada's Astronauts 294

Select Bibliography 305

Index... 317

Acknowledgements

This was a book that I always knew I would eventually write. As a young boy I had the chance to make several trips to the Kennedy Space Center in Florida. Every year I marveled over the past of the Apollo and Skylab programs that I had been too young to witness unfold and also wondered at the future potential of the Space Shuttle program which I hoped someday to be a part of and fly on into space. It was a remarkable experience that unequivocally shaped my imagination and desire to learn as much as possible about the space program of my own country, Canada. At the time, however, before the age of the Internet and with few readily available resources at my local libraries, I was left to gather whatever scraps of information I could on my own.

What started as a single file folder of mostly newspaper clippings in the early 1980s eventually flourished over the following three decades into a rather sizable private archive and library. And while books on Canada's role in space began to appear with greater frequency, they often focused on single subjects. Much about the origin and evolution of Canada's space program remained obscure, overshadowed by more recent events in Canadian human spaceflight. Reflecting on the situation as it was, the time was right to finally sit down and craft the book that had been patiently sitting in my mind for a very long time.

As with any project of this nature, it simply could not have come to fruition without the inspiration and assistance of others for which I have much gratitude. Fellow Canadian space writers and historians Doris Jelly, Lydia Dotto, and Chris Gainor all provided me with different inspiration through their successful books. Gainor's own work has even made it out of this world, with his book *Arrows to the Moon* being flown in space by Canadian astronaut Robert Thirsk during his stay aboard the International Space Station. Other less well-known books were equally important to this project, from official government publications and archives to corporate publications and even enthusiast self-published works.

Beyond the literature, a number of other people provided valuable assistance. John Brebner, head of the Friends of Communications Research Center organization, was very kind in announcing this project to his membership as well as arranging for a number of interviews with people who worked on some of Canada's first satellites. In particular, Harold Raine, who worked on the Alouette satellite project as a launch coordinator among others, gave freely his time to answer the many questions I had about Canada's first space project. The staff at the Library and Archives of Canada, the Department of National Defence, the Canadian Aviation and Space Museum, and the Canadian Museum of Science and Technology all provided excellent assistance in responding to various queries. There is, in fact, a major, underappreciated treasure trove of Canadian space history at these locations just waiting for curious historians to research it. I also wish to thank in particular Clive Horwood and Rachael and Jim Wilkie at Praxis Publishing for their editorial and technical support in bringing this project to completion, as well as the anonymous editors and consultants who took time to review various parts of the manuscript. Last but not least, I must thank my family greatly for their patience with and tolerance for my affliction for wanting to write so many books and also for encouraging my never-ending interests in the exploration of outer space. It is always difficult to spend precious time away from them, and I am especially lucky to have such understanding people around me. Without them, books such as this one could have never been written.

Ad Astra

Andrew B. Godefroy
Kingston, Ontario, Canada
Summer 2016

Introduction

Long before Europeans first stepped onto its shores, the first nations of the land that became known as Canada used celestial observations to assist them in all aspects of their daily lives, from culture and religion to the determination of seasons and the necessary migration it created. The people of *Siksika* (Blackfoot) Nation, for example, traveled across the Great Plains of Alberta and Northern Montana guided by a widely recognized constellation, known to their people as *Ihkitsikammiksii*.

According to Blackfoot oral history, the constellation *Ihkitsikammiksi* evolved from a camp of ten lodges, one of which housed a family of seven boys and two girls. While the six eldest brothers were away, the eldest girl, Bear Skin Woman, married a grizzly bear. Her father was so angered by this that he, with the help of others in the village, later surrounded the grizzly bear's cave and killed him. When Bear Skin Woman learned of her husband's death, she changed into a huge grizzly bear herself and attacked the lodge, killing everyone including her father and mother.

Although Bear Skin Woman spared the lives of her youngest brother and sister, the two were greatly frightened when they later overheard her talking to herself, planning how she might kill them. One day, when the younger sister went to the river for water, she met her six brothers returning from the warpath. She told them what danger they were in, and together they made a plan to rescue their siblings from their predicament. The younger sister gathered many prickly pears and was instructed by her brothers to place them in front of the lodge in such a way that there would be a safe way for the children to escape. The children later left the lodge at midnight. When the older sister heard them leave, she followed, only to step on the prickly pears. Roaring with pain, she changed herself into a bear again and ran after her sister and brothers.

But the youngest brother had strong medicine powers so, when Bear Skin Woman overtook them, he shot an arrow into the air. Immediately, all the children found themselves just as far in advance of their sister as the arrow flew. Bear Skin Woman got close again, but the younger brother waved his medicine feather, which brought thick underbrush in her way. Then he made a lake come between them. Finally, in the last effort to escape, he made a large tree into which the seven brothers and their little sister climbed. But the grizzly knocked the four lowest from the tree. She was about to kill them when the younger brother waved his medicine feather once more and, singing a song, shot an arrow into the air. Immediately the little sister rose into the sky. He shot six more times, and a brother went up into the sky on each arrow. Finally, the younger brother followed, and together they formed the family of the Seven Brothers in the sky. They took the same position as they had in the tree, forming to more modern eyes what looked like a big dipper.[1]

Traditions and tales such as the story of *Ihkitsikammiksii* remind us of how the stars and outer space have captured the wonder and attention of humankind for thousands of years. And although humans have reached orbit, landed on the moon, and sent robotic probes to other worlds and even to the edges of the solar system and beyond, these efforts thus far have also demonstrated just how little we still know about our celestial surroundings. Imagine if just our own galaxy was the size of an 18-hole golf course. So far, humans have investigated perhaps half of a single blade of grass.

Humankind's desire to expand the collective knowledge of outer space has teased imaginations and fueled dreams since the days of classic civilization, but it was only during the twentieth century that the knowledge and technologies needed to turn these dreams into reality became accessible. Yet, like so many great technological innovations of the last century, rocketry was brought to full fruition in total war, and space flight, lunar landings, and space stations during the political and strategic uncertainty that followed it.

Perhaps no other scientific, technological, and cultural competition was a more popular icon of the Cold War era than the "space race" between the United States of America (USA) and the Union of Soviet Socialist Republics (USSR). Fueled by politics, fear, propaganda, and prestige, the two superpowers engaged in a progressively challenging and expensive technological contest above the Earth that changed the very nature of the Cold War itself and ultimately defined a new era in human civilization.[2]

---

[1] [FN]Recorded as "The Seven Stars" in Clark Wissler and David Duvall's papers *Mythology of the Blackfoot Indians* (1908). Anthropological Papers of the American Museum of Natural History, Vol. 11, Part 1. This story does not appear in Duvall's manuscripts in the Glenbow Institute Archives (Calgary); the original source is therefore unknown. Accessed online on 16 January 2016 at: https://www.blackfootdigitallibrary.com/publication/ihkitsikammiksi-seven-aka-big-dipper#sthash.QQhTOAF0.dpuf.

[2] General histories of the Cold War space race include T.A. Heppenheimer, *Countdown: A History of Spaceflight* (New York: John Wiley & Sons, 1997); W. McDougall, *...The Heavens and Earth: A Political History of the Space Age* (New York: Basic Books, 1985); Rip Bulkeley, *The Sputnik Crisis and Early United States Space Policy: A Critique of the Historiography of Space* (Bloomington: Indiana University Press, 1991); Andrew Chaikin, *A Man on the Moon* (New York: Penguin Books, 1994); and Asif A. Siddiqi, *Challenge to Apollo: The Soviet Union and the Space Race, 1945–1974* (Washington: NASA, 2000).

Though Canadian interest in the direct exploration of the upper atmosphere and outer space began at the end of the nineteenth century, official rocket and space programs only emerged after the Second World War.[3] For reasons of self-preservation and self-interest, the country took its first steps toward the development of a national rocket and space program that, although not large, had tremendous vision and a high degree of success. It was a program born out of scientific curiosity and shaped by technological and cultural innovation demanded by the security requirements of the Cold War, and it was something that in later decades fundamentally transformed Canada itself. Most importantly, however, Canada was able to parlay its own national technological competence in this field of endeavor into political, scientific, technological, and military saliency among its larger and more powerful allies, allowing the country to play a considerable part during what was arguably the greatest and perhaps most dangerous period of human discovery since the Renaissance.

Born out of privately driven research and development, dozens of university upper atmospheric and space-related ventures began in the late 1940s under the direction of Canada's two main government science and technology organizations. As policies and programs evolved and matured, the National Research Council (NRC) turned its attention toward those activities related to pure physical sciences research, while the newly created Defence Research Board (DRB) became largely responsible for Canada's rocketry and space technology engineering projects. This split, desired by the NRC and approved by the government, was fortuitous as it resulted in the Department of National Defence (DND) acting as the lead advocate for Canada's rocketry and space development at a time when these activities were seen as being only for defense-related purposes.

The DRB soon initiated an ambitious program in both rocketry technology and space research and development. The organization continued supporting upper atmospheric studies initiated by the armed services before the war, as well as the design and construction of, first, long-range missiles and, later, sophisticated upper atmospheric small launch vehicles for both military and civilian applications. In 1955, the Canadian Army, in cooperation with the United States Department of Defense (DoD), established a permanent launch facility at Fort Churchill, Manitoba, where the Canadians designed and built the *Black Brant* sounding rocket and other American-built launchers and missiles were later tested and employed. At the same time, the DRB initiated discussions with the DoD for Canadian participation in US ballistic missile defense research and development that ultimately led to considerable bilateral cooperation in a number of fields. Further leveraging this defense cooperation, scientists and engineers from Canada's Defence Research Telecommunications Establishment (DRTE) succeeded in gaining access to United States Air Force (USAF) launch services for its own satellite project. When the Soviet Union shocked the world in October 1957 with the launch of *Sputnik*, the first man-made object into space, plans for Canada's own first space satellite were already well advanced.

---

[3]The main subject of this book is Canada's postwar era rocketry and space exploration policy and programs. While acknowledging that European astronomy-related activities in Canada may be traced back to the sixteenth century, this topic has been dealt with elsewhere by others and therefore will not be repeated here. The best introduction remains Richard A. Jarrell, *The Cold Light of Dawn: A History of Canadian Astronomy* (Toronto: University of Toronto Press, 1988).

Canada's rocketry and space program activity first peaked during the 1960s. In late 1962, Canada launched into orbit its first spacecraft, an experimental defense communications satellite named *Alouette*, with plans for several more to follow.[4] The following year, the first real steps were taken to formalize a national Canadian space policy, and plans were also considered for the creation of a national space agency similar to the recently created American National Aeronautics and Space Administration (NASA). The Royal Canadian Air Force (RCAF) initiated its own comprehensive defense space program, which included projects such as ballistic missile reentry, upper atmospheric, orbital mechanic, and rendezvous studies; small payload launch vehicle design; satellite interception vehicle design and construction; satellite identification and tracking; and cooperative engineering personnel exchanges with American defense organizations and NASA. The National Research Council, meanwhile, expanded the activities of its recently created Associate Committee on Space Research (ACSR) to include high-altitude research projects as well as university-sponsored space research and development.

From this point, the focus of Canada's space program suddenly and rapidly changed course. An official "way ahead" for Canada's space activities was proposed by a government review commission in 1967, followed by a government white paper on domestic satellite telecommunications in 1968. In 1969, the entire Canadian rocketry and space sector was then reorganized to exploit these new agendas and opportunities. When Canada's space effort finally paused at the end of the decade, a legacy of achievement during the golden age of space exploration, largely based on security, defense, and national interest, was clearly identifiable, but that time had now passed.

The 1970s witnessed a much more narrowly focused Canadian space program that concentrated on leveraging just those capabilities where it had achieved the greatest political, strategic, technological, and economic success in the decade before. Costly and technologically unviable technologies such as indigenous launch capability were abandoned, whereas communications, robotics, and earth remote sensing received greater support. In 1974, an official national space policy was subsequently ratified that made satellite communications, the emerging field of remote and space-based sensing, as well as a number of engineering projects related to the ongoing human exploration of space the country's new top priorities. Large-scale defense space programs were significantly reduced, though the government continued to invest heavily in intelligence and early warning technologies in cooperation with the United States. Canada also maintained its own tremendous access to, and advantage within, America's space program through several other cooperative ventures, such as direct participation in the National Space Transportation System (space shuttle) and afterward various space station planning projects. During the same time frame, Canada also broadened its scope of cooperative space partnerships to other countries, including much of Europe, Japan, and, even its old political adversary, the Soviet Union.

---

[4] Canadian official documents have suggested that Canada was the third country to go into space after the Soviet Union and the Americans. This ranking has been openly challenged by other early spacefaring nations, especially the United Kingdom, and remains a matter of debate depending on how one frames the question.

In 1980, the Canadian government approved a new five-year plan for space activities that would ultimately result in the genesis of a modern national space program for the country. Projects for indigenous remote sensing capabilities, advanced telecommunications capabilities, space robotics, as well as direct Canadian participation in human space exploration were instituted. Canada was by this point a critical contributor to the US Space Shuttle program and was being courted to provide further advanced space robotics for a future US space station. The country leveraged this technological cooperation to its own advantage, creating its own astronaut corps in 1983 and successfully getting Canada's first astronaut, Marc Garneau, aboard an American space shuttle flight the following year. It was a tremendous accomplishment for Canada and clear recognition of a national space program that continued to evolve. Still, the government of the day appeared wary of the costs involved with space exploration and achievement, and the program's dispersed nature across several organizations and government departments did not allow the country to take full advantage of its potential. The future development of Canada's space program ultimately depended on a national unity of effort. Finally, in 1989, the Canadian Space Agency (CSA) was officially opened to rectify this problem as well as create a formal single government body responsible for all future Canadian civilian space exploration. DND, meanwhile, remained responsible for the country's defense space policy and programs. Together, the two organizations promoted a national space program that equaled – if not surpassed – the efforts of many others.

Throughout the 1990s, Canada's space agency continued to promote a vision of ambitious scientific exploration and technical discovery, which included the launch and operation of RADARSAT, a world-class synthetic aperture radar remote sensing satellite, the sustainment and expansion of the country's qualified astronaut corps, the launch of several new scientific experiments, and direct participation as a core partner in the construction of the new International Space Station (ISS). In 1999, Canadian Julie Payette was the first Canadian to board the ISS. Two years later, another Canadian, Chris Hadfield, performed the country's first spacewalks and led the installation of Canadarm2. This was a remarkable engineering achievement in and of itself, but it also opened the door for Hadfield to take on greater responsibilities including one day commanding the space station itself.

Canadian astronauts participated in three more shuttle missions during the early 2000s, conducting further spacewalks and participating in both the evolution of the space station and the science conducted aboard it. Eventually, Canadians would join the crew assignments that have permanently manned the ISS since the beginning. Robert Thirsk formed part of an ISS expedition in 2009, and in 2012, Chris Hadfield returned to space once more, this time to command the ISS. It was an incredible accomplishment for a space program that, for most of its existence, had struggled just to get by from fiscal year to fiscal year.

Today, Canada's space agency continues to serve as the primary advocate for the country's civilian exploration of outer space and is charged not only with preserving Canada's space legacy but also with conceiving and designing its future direction. This book tells the story of that legacy, with a view to inspire the next generation of activities that will see Canada return to the International Space Station and beyond.

Andrew B. Godefroy

1

Northern Mariner: The Origins of Canadian Space Activities

When European settlers began arriving in North America in the sixteenth century, they brought their somewhat limited knowledge of astronomy with them. Still, with rudimentary instruments and supplemented by information and observations from others such as the First Nations, slowly over time that knowledge grew both in volume and value. When the separate provinces of Canada entered into Confederation in 1867, a new government was formed that one day would begin to deliberately invest in the sciences and technology which would eventually bring the country to the very edge of space itself. It would be a slow and uncertain journey, but one that Canada and many other countries pursued in their quest to break free from the bonds of the Earth.

The genesis of Canadian government interest and, to a more limited degree, departmental support in the investigation of the upper atmosphere and outer space can be traced back to the end of the nineteenth century when investigators began coordinating special events associated with the advancement of the natural and physical sciences. Notable among these early activities was the declaration of an International Polar Year (IPY) in 1882–1883, when several European countries and their colonies came together to promote the discovery and exploration of the Arctic and Antarctic regions. As part of the overall effort, dominion scientists and investigators participated in internationally coordinated studies of meteorological, magnetic, and aurora phenomena in northern Canada. A second IPY was held in 1932, and another generation of Canadian scientists and explorers successfully conducted the country's first measurement of the ionosphere using newly-developed radio technologies, techniques, and procedures.[1]

Canadian scientists and researchers even continued to expand their knowledge of the upper atmosphere throughout the difficult economic depression of the 1930s, when most government funding had been severely cut out of such fields. Perhaps sadly, their efforts were eventually saved by the outbreak of the Second World War. Under the country's

---

[1] Doris Jelly, *Canada: 25 years in Space* (Ottawa: National Museum of Science and Technology, 1988), pp.13–17; and A.M. Pennie, *Defence Research Northern Laboratory, 1947–1965* (Ottawa: Defence Research Board Report No.DR179, April 1966), pp.5–11.

© Springer International Publishing AG 2017
A.B. Godefroy, *The Canadian Space Program*, Springer Praxis Books,
DOI 10.1007/978-3-319-40105-8_1

National Resources Mobilization Act (NRMA) of June 1940, Canada's scientific and engineering communities were widely mobilized to support the war effort and, as a result, previously strapped research organizations suddenly received a massive influx of funding and resources to develop their work. In addition to atmospheric studies already under way, chemistry, physics, ballistics, engineering, as well as other physical and biological sciences, all saw tremendous advances during the war years.

By the end of the war, sustained government support for the development of national level science and technology programs had paved the way for an ambitious postwar era foray into upper atmosphere/space technology research and development under the aegis of Cold War defence-related priorities.[2] Further supported by new cooperative arrangements with the United States of America, and challenged by a competing Europe and an unstable and potentially hostile Soviet adversary, the Cold War assured Canada an early entry into the exploration and exploitation of outer space.

Defence Space Research and Development: Beginnings

At the end of the First World War, Canada's national science program remained largely underdeveloped. The annual expenditure on government laboratory research was approximately $1 million; of nearly 2400 leading Canadian firms, fewer than forty had their own research laboratories. In Canadian industry writ large, less than $150,000 was spent annually on research and development.[3] Though the National Research Council (NRC) was officially created in 1916 to direct national scientific research and development, its early post-Great War years were plagued with internal rivalries as well as competition from Canadian universities for scarce government funding.[4] The situation ameliorated somewhat during the interwar years, with a notable but limited increase in federal funding and of scholarships tendered through Canadian universities by the NRC.[5] Scientific and

---

[2] Donald Avery, *The Science of War: Canadian Scientists and Allied Military Technology During the Second World War* (Toronto: University of Toronto Press, 1998); and Andrew B. Godefroy, *Defence & Discovery: Canada's Military Space Program, 1945–74* (Vancouver: University of British Columbia Press, 2011). See also Privy Council Office [PCO]. John H. Chapman, P.A. Forsyth, P.A. Lapp, and G.N. Patterson, *Upper Atmosphere and Space Programs in Canada: Special Study No.1* (Ottawa: Science Secretariat, February 1967); also published in abridged form as *A Space Program for Canada: Report No.1* (Ottawa: Science Council of Canada, July 1967).

[3] D.J. Goodspeed, *A History of the Defence Research Board of Canada* (Ottawa: Queen's Printer, 1958), p.6; see also M. Girard, 'The Commission of Conservation as a Forerunner to the National Research Council, 1909–1921', in Yves Gingras and Richard Jarrell (Eds), *Building Canadian Science: The Role of the National Research Council. Scientia Canadensis*, 15:2 (Ottawa: Canadian Science and Technology Historical Association, 1991), 19–40.

[4] P. Enros. 'The Onery Council of Scientific and Industrial Pretence: Universities and the Early NRC's Plans for Industrial Research', in Yves Gingras and Richard Jarrell (Eds), *Building Canadian Science: The Role of the National Research Council. Scientia Canadensis*, 15:2, 41–52.

[5] Yves Gingras, *Physics and the Rise of Scientific Research in Canada* (Montreal & Kingston: McGill-Queen's University Press, 1991); For this period see also M. Thistle, *The Inner Ring: The Early History of the National Research Council of Canada* (Toronto: University of Toronto Press, 1966); and W. Eggleston, *National Research in Canada: The NRC 1916–1966* (Toronto: Clarke Irwin, 1978).

technical manpower also increased slowly during this period, though a number of Canadian scientists and engineers continued to travel to the United States in search of gainful employment. In general, however, while the federal government was interested in scientific development it remained very reluctant to devote any resources to research beyond that already ascribed in the NRC Act of 1924.

The detached relationship between scientific research, engineering, and government in Canada drastically changed during the Second World War. The NRMA, designed to concentrate all of Canada's resources on the defeat of the Axis powers, essentially collected nearly all of Canada's scientists and engineers within the rapidly-expanding NRC and its ancillary departments. There, they were provided with funding and resources unlike any ever received previously from the federal government, and were able to quickly develop the country's scientific capabilities and post-war potential for more advanced research and development. Between 1938 and 1945, for example, government research and development expenditure increased sevenfold from $4.9 million to $34.5 million, or roughly 0.3 percent of gross national expenditure.[6] This expenditure decreased somewhat at the end of hostilities, yet the massive influx of funding and effort into centralized research and development forever altered, thanks to defence, the traditional relationship between science, engineering, and government in Canada.[7]

The war also created an environment that allowed prominent scientists and engineers, such as NRC president Dr. Chalmers J. Mackenzie, to join the central ranks of the political decision-makers in Ottawa. A close personal friend of the Minister of Trade and Commerce, the Honourable Clarence D. Howe, Dr. Mackenzie exerted an influence over the direction of Canada's national science research efforts during and after the war that his profession had never previously enjoyed.[8] While perhaps not equal to the relationship some have suggested that existed between British Prime Minister Winston Churchill and his wartime science advisor, Dr. Frederick Lindemann, there is little question that Dr. Mackenzie's personal access to C.D. Howe and the innermost circles of wartime government allowed him and his fellow scientists and engineers to enjoy an unprecedented position of power and status within Canada's scientific community.[9]

---

[6] A. King et al., *Reviews of National Science Policy: Canada* (Paris, Organization for Economic Cooperation and Development, 1969), p.43. See also Canada, Department of Reconstruction and Supply, *Research and Scientific Activity, Canadian Federal Expenditures, 1938–1946.* (Ottawa: King's Printer, 1947).

[7] Ibid. 42–45.

[8] Other notable scientists included John Cockcroft (1897–1967), scientific director of the Anglo-Canadian atomic projects at Montreal and Chalk River; E.G.D. Murray (1890–1964), director of Canada's biological warfare program between 1941–1945; Omond Solandt (1909–1993), a leading expert in Second World War operational research and the first director general of Canada's post-war Defence Research Board; and George Wright (1904–1976), a prominent influence in Canada's wartime explosives and propellants programs.

[9] The relationship between Churchill and Lindemann is described in C.P. Snow, *Science and Government.* (Cambridge: Harvard University Press, 1961); for details on the relationship between Howe and Mackenzie and the rise of scientific influence in wartime Canada see G. Bruce Doern, *Science and Politics in Canada* (Montreal and London: McGill-Queen's University Press, 1972), pp.4–49.

Fig. 1.1 Dr. Chalmers J. Mackenzie served as the wartime head of Canada's National Research Council and oversaw early government space science projects

A direct example of this may be found in the NRC's annual report for the years 1944–1945. Almost twenty years after its inception, the organization explicitly noted for the first time that one of its major functions was to act as "advisor to the various departments of government, particularly those of National Defence, Munitions and Supply, and Reconstruction."[10] Such a status was the result of Mackenzie and other leading scientists gaining access to those inner circles of government.

In mid-1944, senior scientists, ministers, and soldiers also began to consider the issue of post-war defence research in Canada. It was obvious to those in the know that, at the war's end, a large portion of the mobilized government scientific and engineering corps would request their release from military service or defence duties and seek civilian

---

[10] NRC, *Annual Report: 1944–45.* (Ottawa: King's Printer, 1945), p.7.

employment elsewhere. In order to ensure that some form of defence research program existed to both capture corporate knowledge and oversee ongoing projects in the post-war period, Air Vice-Marshal Ernest W. Stedman, the Director General of Air Research, recommended that a permanent cabinet committee on defence research be formed under a chairman nominated by the government. The idea was endorsed by the Cabinet War Committee and later approved by Cabinet on October 3, 1944.[11] The composition of the committee, which was to be chaired by the Honourable Minister of Munitions and Supply, C.D. Howe, would include the three armed service chiefs of staff, the President of the NRC, two representatives from industry, and two other civilian members. Cabinet again concurred and approved the terms of reference for the committee on August 10, 1945.[12] It was expected that, after an initial period to consider the issue, the committee would meet sometime near the end of the year.

The government's next step was to determine how Canada's post-war defence research capabilities would be organized. One suggestion to remove defence research from the control of the three armed services was the creation of a military division within the National Research Council, with one of the agency's vice-presidents appointed as a new Director of Defence Research. General Charles Foulkes, then serving as the army's Chief of the General Staff, discussed this option with Dr. Mackenzie, but he discovered that the NRC President was reluctant to accept responsibility for Canada's postwar defence-related scientific research. Although the NRC had assumed this responsibility in wartime, Dr. Mackenzie pointed out that it had been assigned to his organization and he was anxious to return the NRC to fundamental physical science research and the civilian sector.[13] He also argued that even though the NRC had substantial resources, the financial and administrative responsibility for postwar defence research would be large enough that it would ultimately require its own dedicated organization and staff.[14] Mackenzie's persistence in separating defence research out of the NRC ultimately proved to be a blessing in disguise. Unknown to him and other senior leaders at the time, the NRC laboratories had been infiltrated by a number of Soviet spies who sought to steal technological secrets including those associated with atomic weapons research. Had the NRC overseen Canada's postwar defence research, the impact might have proven disastrous for future force development. In the end, however, in 1947 a decision was made to create a separate body within the DND which became known as the Defence Research Board (DRB). A respected wartime operational research scientist, Dr. Omond Solandt, was appointed as its first director general soon after.[15]

---

[11] Privy Council Office Series A-5-a, Volume 2636, 2, Cabinet Conclusion – Committee on Research for Defence 3 October 1944. Record Group [RG] 2, Library and Archives Canada [LAC].

[12] D.J. Goodspeed, *A History of the Defence Research Board of Canada*, pp.3–16.

[13] Ibid., 21.

[14] Avery, *The Science of War*, pp.228–255.

[15] Jason Sean Ridler, *Maestro of Science: Omond McKillop Solandt and Government Science in War and Hostile Peace, 1939–1956* (Toronto: University of Toronto Press, 2105).

Fig. 1.2 Dr. Omond M. Solandt was a Colonel in the Canadian Army Operational Research Group before being chosen to head the new postwar Defence Research Board

Canada's access to the upper atmosphere and orbit so early on during the space race would have been impossible without organizations such as the DRB and the investment it made into Canadian science and engineering. The tools needed for space flight and research were essentially converted weapons of war. Missiles carrying warheads became rockets carrying payloads. In addition to these tools, the defence research community also had the political and financial means to exploit this technology and transform space research from a concept to reality. As in other developed countries, Canada's defence research organizations were at the time the only organizations in the country that could create the essential conditions for success; the DRB, for example, could advance Canada's own goal to explore and if necessary defend the upper atmosphere and outer space as well as make a meaningful contribution to Cold War missile, rocketry, and space programs in the United States and elsewhere.

The DRB served as the aegis for a number of research and development suborganizations, some of which would later conceive, design, build, and manage many of Canada's first rocketry and space technology projects. Existing facilities included: the Canadian Armament Research and Development Establishment (CARDE) at Valcartier, Quebec; the

Suffield Experimental Station (SES) in Alberta; the Defence Research Chemical Laboratories in Ottawa; the Kingston (biological warfare) Laboratory; the Radio Propagation Laboratory (RPL) in Ottawa; the Defence Research Establishment at Churchill (renamed the Defence Research Northern Laboratory in 1947); and the Naval Research Establishment at Halifax. Other research offices included: the Weapons Research Section; the Electrical Research Section; the Special Problems Research Section, the Biological Research Section; the Naval Research Section; and the Scientific Intelligence Section. Additionally, the DRB formed or co-chaired a number of advisory committees between its own armed services as well as cooperatively with other defence allies.

At its formation, the DRB's organizations employed roughly a thousand personnel including two hundred scientists and engineers and between thirty and forty technical officers. In 1947 and 1948, approximately CDN$4 million was expended on research and development, an amount that quadrupled in 1948–1949 when materials, supplies, equipment, and salaries were added to the budget. Defence research expenditures continued to increase between CDN$7 and $10 million a year through to 1956–1957, a definite indication of the strong commitment made to defence science and technology by Canada during the early years of the Cold War space race. More importantly, perhaps, the creation of the DRB clearly signaled Canada's intent to take its own defence research and development seriously and make a solid contribution to its own strategic interests as well as those of its main allies.[16]

The DRB initially sought to focus on nineteen specific fields of research, including guided missiles, rockets, meteorology, communications, and upper atmospheric studies. While the first two fields fell into the research domain of the army's armament research and development establishment, the latter fields included investigations into the nature and composition of the upper atmosphere such as those navy and air force projects already under way at Fort Churchill, Manitoba. Like its allies, Canada sought knowledge on both rocketry and upper atmospheric conditions early in the postwar period, realizing the importance of both in the development of advanced weapons and defence systems. Canada became particularly proficient in the latter field, so much so that the United States was satisfied to initially curb its own efforts and instead rely largely on Canada for raw data and information.[17]

A Focus on Physics

Among the many disciplines that led Canadian defence research towards the threshold of outer space, physics played the dominant role during those early years. Here Canada's expertise was considerable; the country's scientists and engineers had made major contributions to allied wartime technological developments in ballistics, radar, and even atomic energy. They were also considered world experts in the study of the effects of the

---

[16] A. King et al., *Reviews of National Science Policy:* Canada (Paris: Organization for Economic Cooperation and Development, 1969).

[17] H.E. Newell, *Beyond the Atmosphere: Early Years of Space Science* (Washington: NASA Publication SP-4211, 1974).

upper atmosphere on electronics and communication, a field that Canadian researchers had seriously pursued since the beginning of the twentieth century.

Canada's association with long-range 'over the horizon' communications has been no less lengthy. On December 12, 1901, for example, Guglielmo Marconi received a Morse signal in Newfoundland from a transmitter located 2900 kilometers away in England. At that time, no one could explain exactly how this was even possible, as no one yet understood how radio waves were able to bounce off the atmosphere and travel over the horizon. The following year, however, two scientists – an American named Arthur Kennelly and the Englishman Oliver Heaviside – suggested that the radio waves were able to travel great distances due to the existence of a conducting layer somewhere in the upper atmosphere which reflected the waves sent from one station to another. Their ideas started a long series of detailed studies and debates over the definition, origin, and nature of what for many years was known as the Kennelly–Heaviside layer; a part of the upper atmosphere which later was more appropriately named the ionosphere.

The ionosphere marks a region of the Earth's atmosphere where there is a significant concentration of ions. The region has no definite boundaries or limits; rather, it is a dynamic area anywhere from sixty kilometers upwards to approximately ten earth radii where neutral particles are ionized and then combined into neutral particles once again. The degree of ionization of that region of the atmosphere is also dependent on a number of factors such as solar radiation, atmospheric composition, pressure, and even temperature. The ion concentration maximum, also known as the electron number density maximum, is typically located at a height of approximately three hundred kilometers from the surface of the Earth.[18]

At first very little detailed study of the ionosphere was made in Canada, but after the introduction of wireless communications following the First World War, a greater comprehension of its effects became essential to understanding and predicting its impact on radio waves and, especially, wireless radio traffic. In the mid-1920s, American physicists Gregory Breit and Merle Anthony Tuve labored to construct what was known as a pulse ionosonde, a remarkable machine that fundamentally revolutionized ionospheric research in both the United States and Canada. A radio transmitter and receiver, the ionosonde allowed the user to deduce the height of a radio-wave reflecting level by illuminating the ionosphere with short pulses and noting the time difference between transmission and the return of an echo. Also, the particular frequency of a radio wave was determined by the degree of ionization at the reflecting level which, when combined with the pulse delay, provided the user with the height level in the ionosphere of that particular electron number density, and frequency required to propagate through it relatively unaffected.[19]

Sustained Canadian research and analysis of the ionosphere began with the massive influx of federal funding and support during the Second World War. In 1941, the National Research Council constructed its own ionosonde at Chelsea, Quebec, where it

---

[18] Theodore R. Hartz and Irvine Paghis, *Spacebound* (Ottawa: Department of Communications – Minister of Supply and Services Canada, 1982), p.31.

[19] Ibid., p.49.

was operated under the direction of Frank T. Davies, an experienced scientist then employed with the Operational Intelligence Center of the Royal Canadian Navy (RCN). Davies' tasks were to record constant measurements of the ionosphere and predict optimum operating frequencies for military communications. At the time, Canadian ships protecting conveys and chasing German submarines in the North Atlantic depended greatly on timely communications, making it essential that the link between RCN headquarters and ships and planes operating at sea remain uninterrupted as much as possible. Fortunately, the initial effort made by Mr. Davies was so successful that further ionosonde installations were built along the Canadian Atlantic coast, including one by the American Carnegie Institution at Clyde River, Baffin Island. In 1943, Mr. Davies along with Lieutenant Jack H. Meek, then serving as Superintendent of the Radio Propagation Laboratory (RPL), installed yet another ionospheric station at Cape Merry. By then Davies and Lieutenant Meek were responsible for overseeing the training and inspection of all ionosonde stations, including all those then manned by members of the United States military. By the end of the Second World War, a total of seven stations were located at various points across Canada.[20]

Fig. 1.3 The original DRB Radio Physics Laboratory located in Ottawa c.1956. Much of Canada's early ionospheric research was conducted here as a prelude to satellite operations

---

[20] Ibid., p.52.

Ionospheric research in Canada continued after the end of hostilities. Canada had attained considerable recognition and reputation amongst its allies as a leader in this field, and both the United States of America and the United Kingdom took a serious interest in supporting further efforts after the war. In 1946, the RPL took the lead in furthering Canada's research activities while the Department of Transport took over the operations and manning of the existing ionosonde stations.[21] Meanwhile, three additional stations were built, one each at Baker Lake, Resolute Bay, and Fort Chimo. With the transfer of the RPL and its cousin organization, the Radio Wave Propagation Committee, to the Defence Research Board in 1947, Dr. Solandt recommended that the time had come for a considerable expansion of Canada's defence research facilities, resources, and funding then devoted to Canadian ionospheric research. The decision was timely. In late 1947, the American Department of State also requested a technical conference with Canada to discuss options for the expansion of post-war ionospheric research, resulting in further pressure from the DRB to the Cabinet Defence Committee for new and improved research stations right across the Canadian arctic.[22]

In the early postwar era there were clear security reasons for encouraging such invest-ment, not the least of which was to provide Canada some degree of early warning against a possible Soviet attack. Beyond this, however, Canada had other strategic motives for expanding its expertise in atmospheric research as well. The physical attributes of the atmo-sphere in the Canadian north, which were affected by the proximity of the geographic North Pole, the situation of the North Magnetic Pole entirely within northern Canada, and the phenomenon of other occurrences such as Aurora Borealis, presented unique challenges for other technologies dependent on radio waves, in particular electronics and communica-tions. The large expanse of Canadian territory and its relatively small population – approxi-mately seventeen million people lived in Canada in the 1950s – implied that long-range communications would become essential to any defence of the country in the case of war, and any modern development of the country while peace ensued. Moreover, it was yet unknown what effects the upper atmosphere might have on advanced aerodynamics, jet or nuclear propulsion, or on unmanned objects such as guided missiles and rockets.[23] Obtaining answers to these and other questions about the unique environment of Canada would ulti-mately provide considerable value to the development of both security and sovereignty.[24]

Upper atmospheric studies in Canada were formalized in February 1951 with the creation of the Defence Research Telecommunications Establishment (DRTE) under the superintendence of Frank T. Davies. Essentially an amalgamation of the two other labora-tories, the RPL and the Defence Research Electronics Laboratory (DERL), the new orga-nization joined the ranks of the ever-expanding Canadian defence research community as its primary telecommunications research and experimentation facility.[25] Within the DRTE, atmospheric study and analysis was a central focus of effort, as the actual distribution of scientific human resources described below clearly demonstrates.

---

[21] D.J. Goodspeed, *A History of the Defence Research Board of Canada*, p.192.

[22] Ibid. 194.

[23] One must also consider the fact that the sound barrier was not broken until October 1947, and that very little was known about the effects of the upper atmosphere on machines and humans until fur-ther advanced research was undertaken throughout the 1950s.

[24] G.D. Watson, 'The Scientific Exploration of Space', *Canadian Aeronautical Journal.* 6:3 (March 1960), 87–88.

[25] D.J. Goodspeed, *A History of the Defence Research Board of Canada*, pp.195–197.

Fig. 1.4 Dr. Frank T. Davies served in naval intelligence during the Second World War, and was later Chief Superintendent of the DRTE from 1951–1969

The new DRTE consisted of two primary suborganizations. The recently renamed Radio Physics Laboratory (RPL) under the direction of Mr. J.C.W. Scott consisted of six subsections, including theoretical studies, atmospheric physics, radio prediction, and three specialized radio propagation units (high frequency, low frequency, and microwaves). Meanwhile, the Electronics Laboratory under the direction of J.W. Cox consisted of five sections, including transistor research, radio warfare, components research, navigation research, and radar research. In the mid-1950s, the DRTE later underwent a modest reorganization of both its radio and electronic research sections to include a third suborganization, the Communications Wing, under the direction of a recently-hired postgraduate, Dr. John Herbert Chapman.[26] This new group augmented existing research in radar and applied propagation, as well as focusing on new communications specific research and development.[27]

---

[26] John H. Chapman Papers [hereafter JHC Papers], Manuscript Group [MG] 31 J43, LAC. Vols. 2–3 contain his graduate and postgraduate work.

[27] Op. cit. Goodspeed, p.197.

The RPL staff concentrated on a number of projects, giving serious attention to the analysis of auroral disturbances such as aurora borealis that was particular to Canada's northern regions. Under the direction of Dr. Raymond Montabetti and Dr. W. Petrie, the RPL designed a set of prediction tables of the potential impact of ionospheric activity on radio wave propagation for use by the Royal Canadian Navy (RCN). In addition, the lab tested new communications systems, studied the effects of lightning on radio waves, studied the effects of birds on radar screens, and directed the research, development, and construction of the "McGill Fence" early warning line.[28] Finally, the RPL also devoted a portion of its resources towards the subject of radio astronomy.

Fig. 1.5 Dr. J.C.W. Scott, a former wartime Canadian air force squadron leader, became one of the world's leading experts in ionospheric studies during the 1950s

---

[28] Also known as the Mid-Canada Line, this early warning system employed McGill Fence technology developed at McGill University in Montreal. Construction began in 1954 and was completed in 1957 at a cost of $250 million. The line consisted of a chain of 98 radar stations, mostly unmanned, along Canada's 55th parallel. The Mid-Canada Line was closed in 1965.

Meanwhile, the Electronics Laboratory remained largely engaged in standardizing the communications requirements of the three services until 1951, when the initiation of Canada's defence Velvet Glove air-to-air guided missile program required this lab to shift most, if not all of its support, to this other program. When Velvet Glove was concluded in 1954, the Electronics Lab returned to supporting space research, just in time for the upcoming International Geophysical Year (IGY).

The International Geophysical Year

Upper atmospheric research and space science advanced rapidly during the 1950s, eventually leading both superpowers to launch manmade satellites into space in 1957–1958. While the Soviet Union and the United States clearly were the central actors in taking upper atmospheric research to the next level, Canada was similarly an important player in the military and scientific aspects of these events. Closely associated to American efforts to gain significant knowledge of the Earth's polar regions as well as push space scientific research out into orbit, Canada's own efforts as well as its cooperation with the United States were critical to the latter's security and successes at its earliest stages of space flight.

The idea of International Geophysical Year began, interestingly enough, during a dinner party held on April 5, 1950 in Washington D.C., in honor of the visit to the city by Professor Sydney Chapman, the renowned British geophysicist.[29] During the evening Chapman and his host, the noted American radio scientist Lloyd V. Berkner, proposed the initiation of a third International Polar Year.[30] The period chosen for the event was 1957–1958, as it followed the previous IPY of 1932 by exactly 25 years, and was also near a timeframe of known maximum solar activity.

Yet there were other motivations as well. In 1950, the United States government was about to release a science and foreign relations report authored by Berkner and others that very much emphasized the military and diplomatic importance of the world's polar regions. In addition to the obvious statement that "certain definite benefits which are highly essential to the security and welfare of the United States, both generally and with respect to the progress of science, stem from international cooperation and exchange with respect to scientific matters", the report's classified supplement also stressed the value and importance of increased American contact with foreign scientists to intelligence

---

[29] For general studies of the IGY see Sydney Chapman, *IGY: Year of Discovery* (Ann Arbor, MI: University of Michigan Press, 1959); Marcel Nicolet, 'Historical Aspects of the IGY', *Eos Transactions*, 64:19 (May 10, 1983); and H.E. Newell, *Beyond the Atmosphere: Early Years of Space Science*, SP-4211 (Washington: National Aeronautics and Space Administration, 1974).

[30] There remains some academic debate over who actually first had the idea for the third IPY, Chapman or Berkner. For views on the former man's claim see S. Chapman, *IGY: Year of Discovery*, while the latter's claim is supported in Allan A. Needell, *Science, Cold War and the American State: Lloyd V. Berkner and the Balance of Professional Ideals* (Washington: National Air and Space Museum, 2000).

Fig. 1.6 Canada's IGY logo

collection.[31] Again, both countries clearly identified science as a key to national security and other strategic interests in the early Cold War period.

The primary American agency overseeing its activities during the IGY was V-2 Upper Atmosphere Research Panel (V-2 UARP), a group of U.S. military and university interests responsible for that country's first sounding rocket program. From 1950 onwards, the V-2 UARP held a number of meetings with other American and international scientists to discuss a list of potential IGY activities. Canadian scientists were often invited to these early meetings, though their official participation on various committees was only formalized after a bilateral agreement was put in place to establish a large research rocket range in the Canadian north at Fort Churchill, Manitoba. In 1953, the V-2 UARP formed a Special Committee for the IGY (SCIGY) to work with the American National Research Council (ANRC) to oversee the development of the Fort Churchill research site. With the further assistance from the ANRC's Technical Panel on Rocketry, plans were executed to also have the United States Army set up an Aerobee rocket tower and a Nike-Cajun launcher at the Churchill station.[32]

---

[31] Lloyd V. Berkner, 'Science and Foreign Relations' (Washington D.C.: Department of State, 1950) as cited in A. Needell, *Science Cold War and the American State*, p.299.

[32] G.K. Megerian, Secretary, Minutes of the Rocket and Satellite Research Panel, reports 36–40, dated October 7, 1953 through Februar 3,y 1955, V-2 UARP, History Office Files, NASA.

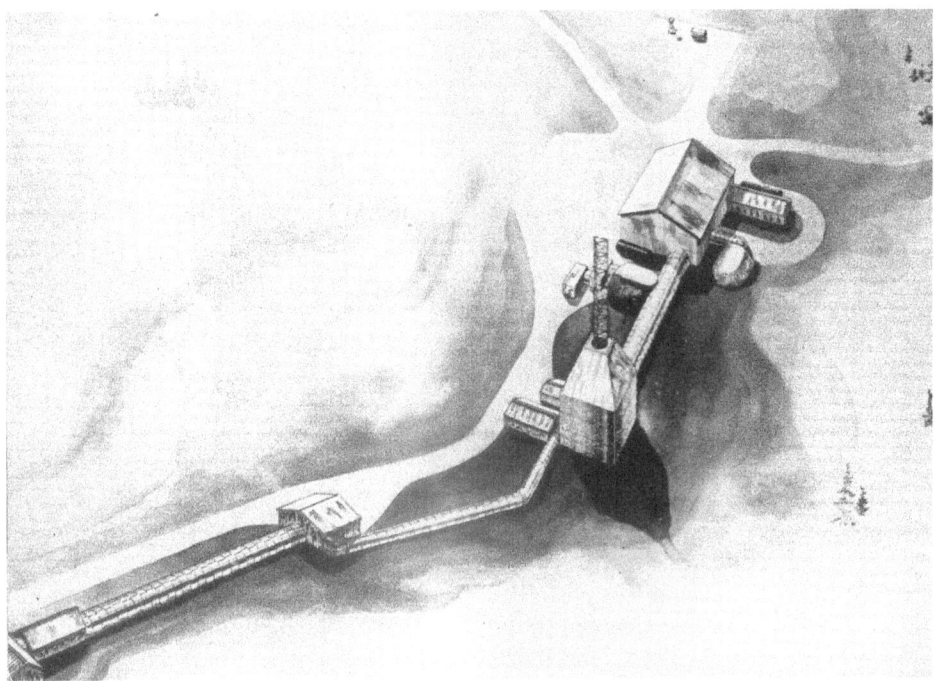

Fig. 1.7 Artist's illustration of the main rocket launching site at Fort Churchill c.1956–1958, with the Aerobee rocket launch tower clearly visible in the center

In Canada, Dr. Donald C. Rose, a prominent scientist with the NRC Physics Division and Chairman of the Canadian IGY organizing committee, directed activities with several projects dispersed among many research organizations and universities across the country.[33] Both the DRB and the NRC were directly engaged in a number of assignments, with the DRB concentrating its efforts at Fort Churchill, while the NRC supported those projects dispersed across the remainder of Canada. Interesting among the latter organization's efforts was the work of the Upper Atmosphere Research Section, a group headed by Dr. Peter Mackenzie Millman. An accomplished meteor astronomer, Millman had studied at the University of Toronto and Harvard prior to serving in the Royal Canadian Air Force during the Second World War. He had been transferred from the Dominion Observatory to the NRC in 1955, and took advantage of the IGY to revive the Council's meteor research program. He soon secured resources to build the Springhill Meteor Observatory south of the city of Ottawa, away from light pollution, from which first-rate meteor data could be collected and analyzed.

---

[33] Notice of Meeting—Coordinating Committee for the Canadian I.G.Y. Program dated October 28, 1958. File 9-2, Vol.9, MG30 J43, LAC.

Using visual observation and radar, Millman and his research teams collected a wide range of first-rate data on meteors re-entering the atmosphere at high speeds. They discovered that meteor trains were persistent not only visually but also maintained a 'visibility' to radar for some seconds after they entered the atmosphere. Millman assumed that this was caused by ionization created in the upper atmosphere as the object lost material during its descent. Combined with radar observations of the backscatter of echoes, Millman and his teams were able to ascertain the velocities of large numbers of meteorites using these techniques. It was an important discovery for Canadian scientists and engineers, as similar techniques were later applied to the detection of not just meteors entering the atmosphere, but also man-made objects re-entering the atmosphere from outer space.[34]

The primary DRB objectives for the International Geophysical Year included the use of a wide variety of sounding rockets to investigate spectroscopic and ionic characteristics of the upper atmosphere.[35] Two of these rockets were instrumented by personnel from Canadian Armaments Research and Development Establishment (CARDE) to measure infrared background levels of the upper atmosphere at altitudes up to about 250 km. Although perhaps a minor accomplishment compared to American and Soviet satellites being launched during the same period, it was the first time that Canadian instruments were successfully launched well past the Kármán Line and into outer space.[36]

The commencement of IGY planned activities at Fort Churchill completely altered the launch facility's organization and infrastructure. In early 1956, the Defence Research Northern Laboratory (DRNL) ended its seven-year partnership with two other Canadian research organizations at Fort Churchill, the Operational Research Group and the Biomedical Sciences Group, in order to prepare for its new assignment of geophysical research in support of space science and technology. As the two former organizations prepared for their departures, the DRNL initiated a survey of potential sites for the installation of equipment employed by Dr. Montalbetti's RPL scientists and the University of Saskatchewan for taking measurements of the ionosphere. Along with the recently-installed U.S. Army rocket launching sites, a total of four additional sites were needed for the outlying Canadian ionosphere measuring stations. Three of these were located south of Fort Churchill, while one other was placed to the north. Two of the sites were outfitted with special cameras for measuring the height of auroral displays employing the parallactic photography technique, while the other two were outfitted with an auroral recorder, a magnetometer, and an all-sky camera.[37]

---

[34] Richard Jarrell, *The Cold Light of Dawn: A History of Canadian Astronomy* (Toronto: University of Toronto Press, 1988), pp.156–157.

[35] JHC Papers. Speaking Notes—The International Geophysical Year Program in Canada. File 9–2, Vol.9, MG31 J43, LAC.

[36] G.D. Watson, 'The Scientific Exploration of Space', *Canadian Aeronautical Journal* (March 1960), 87–88.

[37] H.B. Lutz, 'Preparations for the IGY 1956–57', in A.M. Pennie (Ed.) *Defence Research Northern Laboratory, 1947–1965.* (Ottawa: Report No. DR179, DRB April 1966), p.84.

Fig. 1.8 Map. IGY main observation stations in Canada

The International Geophysical Year wrapped up at the end of 1958, just as the DRB formally entered negotiations with the United States National Aeronautical and Space Administration (NASA) with respect to a much more ambitious project. The IGY was extremely successful and had demonstrated the high degree of cooperation which had come to exist between the United States and Canada in the rapidly-developing fields of space science and rocketry. As a result of the IGY, the foundation was set for even greater Canadian–American space cooperation in the years to come.

At the Threshold of Space

Due to national security considerations and its commitment to the advancement of defence science and technologies, Canada achieved considerable success in the early development of its own upper atmospheric program in the first decade following the end of the Second World War. From rather conservative beginnings, the country took advantage of its new situation to first create the necessary means with which to adequately research and develop space-related technologies of interest and, second, employ its limited yet highly professional research and development resources to quickly gravitate towards the center of evolving Western space-related activity, unsurprisingly led by the United States. Further, there was nothing to suggest that Canada's agenda was accidental; instead, history has revealed that it was a rather well-orchestrated maneuver on Ottawa's part, and representative of what would eventually become the Canadian government's *modus operandi* for gaining access to future American space ventures.

The significance of Canada to the construction of American postwar security provided Ottawa with a considerable opportunity to advance its own rocketry and space science and technology agenda. From the very beginning of the Cold War the United States needed access to both Canadian territory and to Canadian science in order to develop its own technological response to the growing fear of the Soviets winning the space race. Canada, in turn, was prepared to cooperate with the U.S. to meet its own national security interests but, more importantly, the government's commitment to postwar scientific and technological innovation in defence gave the country the means to take full advantage of the evolving space race. By actively participating in both bilateral defence cooperation and international science and technology programs in the 1950s, Canada was able to demonstrate the maturity of its own ambitions and resources and ultimately leverage those capabilities along with its other defence contributions into negotiating its very own satellite project.

At a time when only the two great superpowers – America and the Soviet Union – had the technological means to break away from Earth's gravity and only these two nations had seriously contemplated launching satellites into space, here was Canada subtly inserting itself into a position where it would be the obvious best option for the continuation of the study of the upper atmosphere given it had access to certain additional resources. Fortunately these resources, such as launching capability, were things that the U.S. Department of Defense could easily provide to its ally. As a result, Canada became an early party to the exploration and exploitation of outer space, while for most other nations such notions remained the stuff of science fiction and fantasy.

As is true of other Canadian ventures in this period, the country effectively parlayed its own limited resources into achieving much larger objectives. What has not often been acknowledged, however, is the fact that Canada employed this tactic to ensure that it was well positioned at the end of the 1950s to play the leading third-party role as the world stood on the threshold of reaching outer space.

2

Forging a Space Nation: Policy and Program Development, 1957–1963

Canada entered the space age optimistic yet also uncertain about the role that rocketry and satellites would play in the country's future. National security and the threats of the Cold War dictated the immediate priorities of the space race, and for Canada the situation was at first no different. Projects such as those then under way at the Churchill Research Range, the Black Brant rocket project, the upcoming *Alouette* satellite project, and the Royal Canadian Air Force (RCAF) Space Development Plan were all born out of the Cold War space race between the United States and the Soviet Union. Apart from their obvious connection regarding defence science and technology applications, however, each of these projects evolved in relative isolation of each other, and were not considered equal or mutually inclusive parts of an overall national strategy for Canadian long-term rocketry and space development.

In fact, despite several proposals put forward by the Defence Research Board to organize and coordinate Canada's space activities more carefully, a general lack of knowledge amongst senior political leaders as to the potential long-term impact of space exploration and technology exploitation, and the distraction of more pressing domestic policy and government reorganization issues, resulted in little top-level direction for a formalized national space plan. Further complicated by competing agendas and political conflict between Canada's scientific, defence, and government communities during this period, Canada's rocketry and space program advanced not under a collective umbrella as many of its advocates had hoped, but rather in a piecemeal fashion that at times appeared disjointed, and which had serious ramifications for its future direction in the 1970s and beyond.

A New Space Agenda?

As Canada neared its own entry into the space age, many of those who guided the country through its first decade of post-war technological modernization retired from their posts. At the highest levels, Lester Pearson's Liberal Party lost the 1957 federal election to the Progressive Conservatives under John George Diefenbaker, and this would have a considerable impact on the formative years of Canadian rocketry and space activities. His party

© Springer International Publishing AG 2017
A.B. Godefroy, *The Canadian Space Program*, Springer Praxis Books,
DOI 10.1007/978-3-319-40105-8_2

came to power just as the space race began in earnest, and much of his country's early space development depended on the health of its relationship with the United States. Diefenbaker, however, came to power on an anti-American platform and seemed little interested in outer space or space cooperation with America unless he could draw good personal publicity from it.[1] Diefenbaker was observed to mishandle Canada's current space activities and future plans in his public speeches on many occasions, and was often accused by the scientific and engineering communities of deviating from facts or announcing new initiatives without substance or understanding of the details in order to simply make a splash with his audience. Further limited in his own knowledge of science, technology, and international relations while other members of his Cabinet had little or no knowledge of rocketry and space developments at all, Diefenbaker nevertheless assumed personal control of the critically important foreign affairs portfolio after coming to power rather than leaving it to the existing civil service deputy ministers, whom he greatly distrusted.

In 1959, Diefenbaker replaced himself in this portfolio with his Minister of Public Works, Howard Green. A veteran politician and confidant of Diefenbaker, but by no means adept at foreign affairs, Green was no better a choice to grapple with the issues of the space

Fig. 2.1 John G. Diefenbaker, seen here with U.S. President Dwight Eisenhower, served as Canada's Prime Minister from 1957 to 1963, critically shaping official space policy and programs during its early years

---

[1] File 12798-4-40, Vol.1, RG 25, LAC.

race and understood too little, if anything, about outer space beyond it possibly becoming a future battleground between the superpowers. To this end, he encouraged political initiatives that advocated for more international control of outer space without ever really understanding what or how exactly such a situation might influence Canada's own interests going forward.

Disappointments within the Canadian space community soon followed. Notwithstanding the considerable evolution within the scientific and technological communities, the political importance of space and its technological development in Canada remained unappreciated by Diefenbaker's government. For example, the Prime Minister rejected informal proposals from his senior ministers and advisors for the appointment of a science advisor within Cabinet (something similar to James Killian's recent appointment as special assistant for science and technology to the Eisenhower presidency in the United States), and remained opposed to enlarging the office of the Prime Minister with any additional bureaucratic advisors, scientific or otherwise.[2] He also seemed little interested in supporting technological development and engineering, and personally cancelled a number of high-profile national technology efforts early in his tenure, including the infamous Avro Arrow fighter interceptor as well as the national High Energy Project. The general consensus among later historians remains that Diefenbaker made science and technology in Canada a low national priority.[3]

Still, despite political inactivity, overall Canada was investing more in science and technology. By 1961, federal expenditures alone on science and technology development activity exceeded $220 million, almost seven times the amount the country was spending at the end of the Second World War. Nearly 18,000 Canadians worked in professional scientific and technology organizations, which themselves had also greatly multiplied as the country's population and economy grew. This assembly of new scientifically-oriented establishments, however, was represented within Cabinet by only two small committees, the Privy Council Committee on Scientific and Industrial Research and the Advisory Panel for Scientific Policy. Neither was very effective in being taken seriously by the senior political leadership, nor at advocating for space projects beyond those seen as contributing directly to national defence.

Keeping track of space developments within Canada's government at the end of the 1950s ultimately fell to Norman Robertson, a veteran diplomat then recently appointed back to Ottawa as Canada's Under-Secretary of State for External Affairs. Having previously served as Ambassador to the United States, Robertson had some experience in dealing with missile, rocketry, and space issues, but without any ministerial oversight within Cabinet he relied heavily on the chairmen of the NRC and the DRB for advice on this subject. Robertson's attention to every aspect of Canadian statecraft ensured that space activities would not be entirely ignored by the decision makers, but at the same time little more than passing attention could be expected given that it was not perceived to be a high priority with Diefenbaker's government.

---

[2] G.B. Doern. *Science and Politics in Canada.* (Montreal and London: McGill-Queen's University Press, 1972), p.144.

[3] Ibid.

Pearson's Liberal party returned to power in 1963, albeit with only a minority govern-ment. While certainly better prepared and equipped to handle nearly all aspects of Canada's international affairs – including that related to science, technology, and space programs – neither Pearson nor his Cabinet seemed particularly any more interested than their prede-cessors in developing a national space agenda or policy.[4] After 1963, the Pearson government turned away from many international issues dealing with science and technol-ogy, as internal government reorganization and professionalization of its civil service became a top priority and captured most of the leadership's attention.[5] As Canadian politi-cal scientist Bruce Doern later noted, "…Pearson's views of science policy tended to be characterized by a genuine, but superficial, belief that science had to be given greater structural recognition in the inner circles of decision making. Pearson's ultimate agree-ment to create a Science Secretariat and a Science Council seems to have been the product of internal advice…rather than any indigenous initiative developed by Pearson and the Liberals in their opposition days…". Even under new leadership, Canadian science and technology policy, and thus space policy, was still left to the subordinates of government departments to guide and develop.

The first appointed director of the newly-created Science Secretariat, Dr. Frank A. Forward, rarely met with the Prime Minister and had little influence in shaping the country's national science programs. In fact, from his initial appointment on April 30, 1964 to May 1965, he alone comprised the entire membership of the secretariat. Though he received three deputy directors, an executive secretary, and a small professional staff from June 1965 onwards, all of Forward's efforts during the next year were focused on special studies, legislative studies, and reviews of science policy in other countries rather than advising the Prime Minister on a national scientific strategy for Canada. "The attitude of Pearson, to both the place of science and the need for advisors", observed Bruce Doern, "seems to have been one of general sympathy and benevolent encouragement, without much of a disposition for the machinery itself."[6] Yet another organization, the Science Council of Canada, replaced the existing Science Secretariat on May 12, 1966, and Dr. Forward left his brief post as Canada's top scientific advisor having made very little differ-ence to the country's national rocketry and space agenda.

Pearson remained in power until his own retirement from politics in 1968, but during his tenure Canada's own space policy did not congeal as many had hoped it would. The absence of a clearly-defined mandate or ministerial advocate during both Diefenbaker's and Pearson's terms resulted in a disjointed approach to research and development, as the formation of policy and programs was left up to the discretion and competing

---

[4] Despite holding office at a remarkable period in the space race, there is not a single mention of space activities in Pearson's official memoirs. See Rt. Hon. L.B.Pearson, *Mike: The Memoirs of the Rt. Hon. Lester B. Pearson.* 3 vols. Toronto: University of Toronto Press, 1975.

[5] Canada returned to international space issues when it signed and ratified the 1967 Outer Space Treaty and the 1968 Rescue Agreement.

[6] Doern, G.B. *Science and Politics in Canada*, p.145. Doern's work on science policy and politics was published in 1972, soon after these events took place. He had considerable direct and confiden-tial access to many of those involved in the decision-making process, and his views should be con-sidered authoritative in the absence of other published sources.

agendas of those agencies actively engaged in developing space projects. Yet even within these agencies changes of the guard were taking place amongst those who had chosen the original course, and those who followed brought their own new spin on the direction and priority of Canada's future space efforts.

Both the NRC and the DRB also came under new leadership during this period. Dr. E.W.R. "Ned" Steacie, Director of the Division of Chemistry, succeeded Dr. C.J. Mackenzie who retired from his chairmanship of the NRC in 1952. C.J. Mackenzie was the wartime defence science guru with close friends in Cabinet and considerable influence with the government. Dr. Steacie was a very different character, generally distrustful of any government involvement in scientific affairs, much less diplomatic, and not afraid to accuse the politicians of trying to give directions in a field where Steacie felt they had little understanding or right to meddle. Under Steacie's brief appointment as chairman, the NRC often was at odds with the government on issues dealing with national research and development.

Fig. 2.2 Dr. Edgar William Richard 'Ned' Steacie served as President of the NRC from 1952 to 1962, and was a key figure in early Canadian space science policy development

Similarly, at the DRB, in 1955 Dr. Omond Solandt retired from his post as chairman and was succeeded by Dr. A. Hartley Zimmerman, who was then serving as vice-chairman of the board. Dr. Solandt had founded the DRB and shaped it during its first post-war decade, but Zimmerman was not an original member of the defence research establishment and only joined the DRB in 1951 as the Department of Defence Production representative to the board. Made vice-chairman of the DRB in 1955, he formally took over from Dr. Solandt the following year. Also, whereas Dr. Solandt was a trained scientist, Dr. Zimmerman was an engineer, and his post-war perceptions of science and technology were shaped by his return to civilian business after the war, not by national level programs in research and development. The result was a pragmatic but at times short-sighted chairman whose approach to running the DRB favored immediate and predictable returns from programs such as Alouette satellite program over supporting long-term and perhaps riskier research and development goals such as an indigenous launch capability.

Reaction to the Sputnik Spaceflight

The Canadian public was no less surprised than the American public when it read on the front pages of newspapers across the nation on the morning of October 5, 1957 that Russia had successfully launched the first manmade object into orbit. For readers of the Toronto daily *Globe and Mail* in particular, the front page was also filled with irony. Just below the news outlining the success of *Sputnik*'s spaceflight was a story and photo detailing the rollout of the first Canadian super jet fighter interceptor, the AVRO CF-105 *Arrow*. Across Canada's breakfast tables, readers read about the advent of one technological wonder as it foreshadowed the demise of another.

The reactions of Canadian and American leaders to the Soviet launching of *Sputnik* seemed incredibly different. Though senior American politicians and advisors were not necessarily surprised that a satellite had gone into orbit (recall that the United States had planned to launch a satellite as part of the IGY), they were considerably impressed by the technological magnitude of *Sputnik*. The United States was concerned that the success of *Sputnik* suggested that the Russians had developed an ominous capability to launch nuclear weapons over great distances, far enough to reach North America, or even put them into orbit. As a result of Moscow's achievement the United States brought a science advisory capability right into the White House itself so that President Eisenhower could have immediate consultation on matters related to science and defence.[7] In contrast, there is little to suggest that Canada's government was equally concerned about Soviet technological achievements. Neither Diefenbaker nor Pearson nor Pearson make mention of the event or

---

[7] Dr. J. Killian. *Sputnik, Scientists, and Eisenhower* (Cambridge: MIT Press, 1977), p.2. Dr. Killian, the first special assistant to the President for Science and Technology, later noted in his memoirs, "That a satellite had gone into orbit really did not surprise me...the real significance of the news for me lay in two words: 'Russian' and '184 pound'." By contrast the first American satellite, *Vanguard*, weighed in at roughly just over three and a half pounds.

its potential implications for Canada in their memoirs.[8] As well, Cabinet records reveal no significant decisions or statements regarding the Russian launch, and there is no serious mention of outer space in the House of Commons debates until the following year.[9] Diefenbaker did not appoint any additional specialized scientific counsel, although arguably he already had the Privy Council Committee on Scientific and Industrial Research and the Advisory Panel for Scientific Policy, as well as the senior leadership of both the NRC and DRB at his disposal. Similarly, the government did not indicate at first any plans for an organized response to the Russian event. The aloofness of Canada's government to such a monumental historical event is curious and lacks simple explanation.

Fig. 2.3 Designer's concept illustration of the AVRO Arrow CF-105 fighter

---

[8] For Pearson's account of the period, see Rt. Hon. Lester B. Pearson, *Mike: The Memoirs of the Rt. Hon. Lester B. Pearson, Vol.2 1948–1957.* (Toronto: University of Toronto Press, 1975); see also Rt. Hon. John G. Diefenbaker, *One Canada: Memoirs of the Right Honourable John G. Diefenbaker,* 3 vols. (Toronto: Macmillan of Canada, 1976).

[9] Canada. House of Commons Debates Index 1957–1958.

Still, not all parties within government showed complete disinterest. Within Canada's defence research community, for example, there was a real concern amongst the senior staff over the implications of recent Soviet missile and space achievements, because the scientists and engineers knew better what the Russians had truly accomplished. "The announcement of the first flight testing of the Russian ICBM on 26 August, followed on 4 October by the launching of the first man-made satellite," noted one DRB scientist in a secret internal report to fellow board members, "has not only pointed up the illusion of believing that the West has a well-established technical superiority, but in fact stresses the urgency of developing a thoroughly realistic approach to all of the complex problems of the next six, eight, or ten year period, in as short a time as possible."[10] The DRB staff was also concerned about keeping pace with Soviet resources, research, and development. "If the USSR maintains her output of scientists and technicians at the present rate, the race will be lost to the West in point of numbers", the same report noted, "Our hope must therefore lie in conservation of effort and concentration on high quality."[11]

Quality came at a price, however, and Canada's Defence Research Board was increasingly struggling to meet the large debts it constantly incurred in the pursuit of research and development of high technology. The government's ongoing cuts to Canada's defence spending heading into the 1960s in the face of increasing salaries, wages, construction, and equipment costs, was hard felt at the Defence Research Board. In the 1956–1957 fiscal year, approved salary increases alone cost the DRB $1.3 million in funds it had originally allocated towards research. The additional costs were not covered by the DND, of course, and instead were covered by deferring all new construction including a much-needed wind tunnel, reducing contracts with industry, and restricting the purchase of essential laboratory equipment required for proposed new programs including many space research projects. Overall, the situation was not overly promising at the time.

Nevertheless, the importance of missile, rocketry, and space research did not entirely escape senior-level planners at the DND. From a purely military perspective, the protection of Strategic Air Command (SAC) bases in North America by means of an integrated Canadian–U.S. air defence plan remained a priority and this would eventually include proposals for an anti-ICBM system. Similarly, strategic surveillance and reconnaissance was essential to early warning, intelligence analysis, and force protection, and this capability was rapidly transitioning towards secret space-based platforms. The DND needed to be able to defend against missile and space-based threats, if for no other reason than the fact that one plausible scenario for a Soviet attack on the United States indicated that it would likely be met somewhere over Canadian airspace. Yet before military planners could proceed with preparations and training, they needed clear policy guidance from Cabinet and the Privy Council on what Canada's missile and space priorities would be. Was Canada's DND expected to prepare for an imminent Soviet attack, or was it reasonable to assume that the United States would provide protection for Canada and Canadian assets and facilities essential to their own survival? Therefore, defining a space strategy was the first challenge to the evolution of Canada's defence and civilian space programs going into the 1960s.

---

[10] Memorandum (secret) 'Some Factors Affecting Defence Research Policy – A Report to Board Members, October 1957', dated October 23, 1957. File DRBS 173-1 (CDRB), RG 24, LAC.

[11] Ibid.,p. 4.

Fig. 2.4 A Royal Canadian Air Force CF-100 returns to Ascension Island after completing a mission to collect U.S. ballistic missile re-entry data. Canada was heavily involved in such projects throughout the 1950s and 1960s

Early National Agendas for Space

From the outset, Canada's own rocketry and space policy options were limited by politics and the size of its economy. Unwilling and unable to keep technological pace with the rapidly-expanding agendas of either Russia or the United States, Canadian political decision makers sought instead to build a modest yet relevant space program through the development of niche capability, leveraging cooperation with the Americans, and increasing Canada's own influence in international space cooperation and control by attempting to impose itself as the champion of all third-party space interests at international organizations such as the United Nations. Some of these efforts brought success while others did not, but all in some way influenced the early development of Canada's national space policy and agenda in its first decade.

Prime Minister Diefenbaker's Cabinet and the Privy Council first considered a national agenda for space in the summer of 1958, and among the leading figures in government examining the issue at the time was a bureaucrat named Douglas V. LePan, one of Norman Robertson's assistant undersecretaries. Realizing that the superpowers were preparing to dominate outer space both militarily and commercially, he was seriously concerned that Canada, lacking similar technological capabilities or resources, might soon find its own space interests and future plans, whatever they may be, restricted or even threatened. Worse, it appeared that those who already had major launch capabilities, and therefore guaranteed space access, would dictate both the law and rules of space exploitation, while those countries that did not would be forced to accept whatever the major space actors decided. Unsure of the current status or future potential capability of Canada's rocketry and space program, LePan sought out expert advice at both the National Research Council and the Defence Research Board whilst developing his own agenda for Canada's input at the UN.

Consulting administrators, scientists, and engineers from May through August 1958, LePan concluded that the best way for Canada to gain space influence was to promote and, if possible, codify an agreement that made space control an international responsibility.[12] This way, third parties such as Canada could secure guaranteed space access from launching nations, and possibly influence how outer space would be used by all nations going forward. The obvious venue at which to suggest such an option was the United Nations (UN), and LePan immediately set out to prepare his argument for review by the Secretary of State for External Affairs and, afterwards, the Prime Minister himself.

At first glance, LePan's plan looked straightforward. Hoping for the creation of a peaceful outer space environment where Canadian national interests were secure and the country could prosper, LePan suggested that, first, the international community declare space a complete sanctuary and, second, that Canada could play a lead role by offering to build an International Space Flight Development Station (ISFDS) where scientists and engineers from all nations could converge and share their research. Further, in his memorandum he suggested to the Canadian Secretary of State for foreign affairs that perhaps the existing Churchill Research Range, then still under combined Canadian–American joint military control, could be transformed into the new ISFDS.

Although it was probably a well-intentioned plan, LePan seemed horribly uninformed about evolving political trends in rocketry and space events when building his proposal. Though there was no data to prove it, he placed a great emphasis on the notion that the lack of Van Allen belt effects on Churchill Research Range due to its northerly location would make it a preferred launch facility for all future space flights. As well, LePan underestimated both the acrimonious attitude of the Soviet Union towards any space cooperation, as well as the amount of influence that Canada could possibly have in trying to force both superpowers – the Soviets and the United States – into relinquishing their obvious advantages in controlling access to space. He does not appear to have fully appreciated the larger political factors surrounding East–West Cold War relations either, preferring instead to trust without question his own sources. Whatever the reasoning, it was poor advice and a poor appreciation of the situation.

---

[12] Memorandum for Mr. Holmes from D.V. LePan, International Control of Outer Space, dated August 20, 1958. File 12798-4-40, Vol.1, RG 25, LAC.

LePan admitted in private correspondence to the Canadian Secretary of State for External Affairs at the time that he based a large part of his proposal upon informal conversations with Dr. John E. Keyston, Deputy Director of the Defence Research Board, and an anonymous paper prepared by concerned scientists from the DRB who worried about the potential weaponization of outer space.[13] Formal consultation was only planned for later once the Prime Minister approved the plan. Although he acknowledged that there could be some difficulties in convincing the government to transfer the Churchill Research Range away from American and Canadian military control, he felt that it should and could be done. What is odd is that this suggestion came almost at the same time as the Cabinet approved the renewal and expansion of the Canadian–American joint test facilities at Churchill out to 1962. As well, the DND and the RCAF were in the process of formalizing a military space agenda, largely at the approval of the Minister of National Defence through the Prime Minister. The obvious dichotomy between strategies under consideration in the Department of External Affairs and what the Cabinet was actually endorsing at the time is both interesting and demonstrative of the disjointed nature of Diefenbaker's government policy during a period of increasing East–West tensions.

At home, however, the issue remained on LePan's agenda. Cabinet requested that detailed studies on Canada's current and proposed future space program be completed, and Dr. Zimmerman submitted two papers outlining Canadian present and future activities in space to the Secretary of the Committee of the Privy Council on Scientific and Industrial Research in December 1958.[14] Both titled "Space Science and Space Technology – A Summary of Points Affecting Canada's Future Position", the first was an executive summary of all Canadian activities to date, while the second provided a more detailed yet still nontechnical analysis of potential space options for Canada going forward. Both papers recounted the natural advantage of Canada's geography in contributing towards the evolution of space science, and highlighted the obvious advantages that space assets could also provide to Canadian defence. To this effect the report advocated the development and promulgation of a national space policy, followed immediately by the creation of an official organization or agency to administer and control Canada's growing rocketry and space activities. Though the focus of the report was, interestingly, directed at expanding the military space capability of Canada, the paper marked the first request to the government to formulate an official space policy for the country. At the time, however, Cabinet was still evaluating all its options and subsequently opted to retain the DRB as the national level space advocate until at least the outcome of LePan's proposal on international space control then tabled at the United Nations.

---

[13] Ibid, p.1.

[14] Summary and paper "Space Science and Space Technology – A Summary of Points Affecting Canada's Future Position", dated December 17, 1958. DRBS 170-80/A16 (CDRB), File 4145-09-1, Vol.1. Box 112, Department of External Affairs [DEA], RG 25, LAC.

Fig. 2.5 A U.S. Nike-Hercules rocket lifts off from Fort Churchill c.1960 – a site that Canadian diplomats later proposed could be converted into a International Space Flight Development Station

Perhaps a bit overoptimistic about its diplomatic influence at the UN in the wake of its skillful resolution of the Suez Crisis, Canada fell into the bitter quagmire of international space politics in early 1959.[15] At the beginning of the 1960s, outer space still lacked any formal or legal definition and there were few internationally recognized rules that governed how it would be explored and occupied. Both superpowers realized early on the strategic importance of outer space, and neither were inclined to encourage legal boundaries in it, especially if it meant the possible forfeiture of their own unobstructed access to exploit space both militarily and commercially to their own benefit. While both superpowers publicly called for the peaceful use of outer space, neither was truly interested in making it a sanctuary. National security through space was simply too valuable to be left in the hands of the endlessly quibbling United Nations Security Council.

---

[15] This contest is examined in detail in Ilya V. Gaiduk, *Divided Together: The United States and the Soviet Union in the United Nations, 1945–1965.* (Washington D.C.: Woodrow Wilson Center Press, 2012).

Equally important, the nature of outer space exploration represented serious challenges to the concept of controlling it internationally. For example, space technology had a duality of purpose that made legal definition difficult. The United States *Redstone* and *Atlas* launchers were employed at the time both as ICBMs and as boosters for its manned space flight program. Similarly, the Russian *R-7* rocket also acted as both an ICBM and the launcher that placed Cosmonaut Yuri Gagarin into orbit. Early satellites had similar dual uses. If either nation was to agree to only using space for peaceful purposes, how then could anyone arguably build a launcher with the lift and capability of sending humans to the moon? Could not the same launcher lift a weapon of mass destruction into orbit? Who would adjudicate over a nation's space and technical programs? What international laws could apply? Just as with air law, the duality of the technology could not realistically deny that outer space would at some point be used for defence purposes, making Canada's call for international control of space far more altruistic than realistic.

However popular with Prime Minister Diefenbaker's Cabinet the international control of space might have been, Canada's UN delegation actually argued that LePan's proposal was simply unviable. While Arthur G. Campbell, then serving in Canada's UN Secretariat in New York, reported back to Ottawa that Canada could champion a third-party space fraternity "to give reality to the claim to equal rights in outer space and to gain a position of influence for negotiating international control of space to ensure that it is used for only peaceful and scientific purposes", he conceded that the generally unfavorable atmosphere within the UN on the development of any international cooperation in outer space would likely kill any such initiative quickly.[16] Back at the Department of External Affairs, Norman Robertson confirmed this in another memorandum prepared for the SSEA the same week, in which he detailed in some depth the status of present outer space activities as well as the announced future plans for many countries including that of the United States. There was nothing to suggest that either the United States or the Soviet Union would at that time suddenly reverse course and endorse international space control at the expense of its own national security or prestige.

Other options were therefore needed. In Ottawa, Robertson recommended to Cabinet that instead of pushing for international legislation Canada should first focus on the development of its own national agenda, which could then evolve further through international forums. He offered that the SSEA support an upcoming Defence Research Board proposal for the creation of a national space program in Canada, if for no other reason than to put Canada in a favorable position as junior partner to the United States and gain access to the technological and economic benefits that would surely arise from the anticipated large-scale American program to go to the moon. Regardless of whatever other objectives might be gleaned from the promotion of international space control, at that time the issue of Canadian benefit clearly stood out as the Department of External Affairs' primary aim.

---

[16] (Confidential) Draft memorandum to the Prime Minister on Outer Space – A Possible Canadian Initiative, dated March 16, 1959. File 12798-4-40, Vol.1, RG 25, LAC.

Norman Robertson's approach to emerging Canadian space strategy and policy was calculating if not shrewd. Championing international space control was seen as a way to "enhance Canada's prestige and status in the international community", while, "foster[ing] Canadian scientific progress and provide a focus of interest to maintain and attract scientific and technical manpower".[17] Surely there was political interest in the promotion of outer space as a sanctuary, but not if it jeopardized Canada's own agenda or its relationship with the rapidly-expanding United States space program.

The proposed way ahead for Canada's own space program was not to match the United States' missile, rocketry, and space efforts but rather cooperate with American plans writ large and see what could be gained from it. Far from encouraging some form of altruistic multilateralism, the Department of External Affairs instead sought forums where Canada might leverage its own capabilities to obtain an advantage. Robertson noted in a confidential memorandum to Cabinet commenting on proposed American plans for space that:

> "One point which appears to be relevant to the consideration of a Canadian programme [sic] is that, if Canadian industry is to have a reasonable chance at securing a share in the benefits of the expected United States [space] effort, it behooves us to ensure that government agencies are enabled to keep abreast of the course of developments so that Canada may claim status as a junior partner, at least in the area of space science, with an eye to the possibility of reaching ultimately a production-sharing agreement in the space technology area. Another point is that space exploration seems likely to prove one of the most prolific sources of stimuli to new ventures in many scientific disciplines and accordingly a suitable programme [sic] should serve to strengthen science generally in Canada."[18]

Ultimately, Ottawa chose a middle road. The Prime Minister consented to the pursuit of a Canadian initiative for the international control of space so long as Canada could strongly influence the process, and ordered both LePan and Campbell to proceed, albeit carefully.[19] Consequently, over the summer directives were issued to both the DRB and the NRC to initiate studies on the technical feasibility of a Canadian initiative to improve the outlook for international cooperation in space research and peaceful uses.[20] On June 5, the DRB nominated John E. Keyston to act as representative at the working group with Dr. William M. Cameron, the DRB's Director of Plans, as his alternate. A week later the NRC replied with its own nominations to the working group consisting of Dr. Donald C. Rose as well as the noted physicist Dr. John D. Babbitt. All parties were asked to examine the details of LePan's original proposal, improve upon it wherever they could, and return a final feasibility report to Cabinet no later than the end of August for further consideration by the government.

---

[17] Ibid., p.3.

[18] Proposal for Possible Canadian Initiative. (Confidential) memorandum for the minister dated March 12, 1959. File 12798-4-40, Vol.1. RG 25, LAC.

[19] (Confidential) Draft memorandum to the Prime Minister on Outer Space – A Possible Canadian Initiative, dated April 15, 1959 with corrections; and (confidential) memorandum from the Office of the Secretary of State for External Affairs signed by H.B. Robinson and dated May 1, 1959. File 12798-4-40, Vol.1, RG 25, LAC.

[20] (Confidential) letter sent to the President of the NRC and the Chairman of the DRB, International Cooperation in the Peaceful Uses of Outer Space, dated June 1, 1959. File. 12798-4-40, Vol.1, RG 24, LAC.

Ultimately the proposal was deemed to be a failure. Most, if not all, of the current or likely space-faring nations at that time had already developed their own indigenous launch facilities and industries, and since many western countries had already been invited to work out bilateral agreements with the United States on the pattern of the DRB-NASA project, it was doubtful how attractive the remotely-located Churchill Research Range would be to countries with a developed scientific and technological capability of their own.[21] Then there were the more pragmatic problems of staffing and financing. Both the DRB and NRC were already fully engaged at the time with a wide variety of projects, and could not realistically draw any support away from them to fuel new ventures without jeopardizing the existing Black Brant, topside sounder (Alouette), and related space science projects.

Additionally, the United States alone had already invested $14 million in the Churchill Research Range, and both American and Canadian military interests wanted to retain the range solely for rocket launchings and strategic defence missile testing. To suddenly open the range to the United Nations would mean either having to duplicate its military test facilities elsewhere or asking the United States to leave. Neither were considered very feasible options for the Canadian government.[22] Both Dr. Zimmerman and Dr. Steacie did agree, however, that the creation of a Canadian operating facility that attracted requests from international partners was still a possibility, but both also agreed that any such venture must remain entirely under Canadian, not United Nations or some other international form, of control.

If the 1960s were not going to result in a multilateral focused space strategy for Canada, then what road would it take? Various political attempts at championing a multilateral internationalist approach to space access and control, albeit selfishly with the intent of placing Canada in a more favorable position vis-á-vis. The American and Soviet domination of space, at the start of the space age had completely failed. That said, where Canada had succeeded was in forging a strong bilateral cooperative relationship with the United States in both rocketry and space flight. Therefore, while Canada tinkered with middle power politics elsewhere globally during the 1960s, in the exploitation of outer space it chose to remain firmly allied to that country from whom Ottawa could benefit the most.

The adoption of a strong, bilaterally focused space strategy in lieu of some manifestation of internationalism made sense. Senior decision makers within the DRB and the NRC did not foresee Canada undertaking significant space exploration programs on its own given the low level of political attention that such activities had generally received within government thus far. Canada would participate in the great exploration of space, surely, but by 1960 it already lacked any resources to make investments in large-scale endeavors similar to those witnessed in the United States or the Soviet Union. Complex space research, orbital research, space stations, and lunar and planetary exploration were all politically perceived at the time as well beyond the scope of Canadian financing or national necessity.[23] Indeed, where Canada did see itself focusing was on very specific projects that delivered clear benefits to the country while at the same time encouraging cooperation with its main ally.

---

[21] Ibid., p.3.

[22] Ibid, p.3.

[23] No author. Outer Space – Proposal For Possible Canadian Initiative 1959. File. 12798-4-40, Vol.1, RG 24, LAC.

The Black Brant and Other Early Rocketry Projects

Prior to the Second World War, Canadian interest in rocketry was largely limited to amateur groups. Notable among these was the Central Technical School rocket club in Toronto, formed in 1936 and led by a talented young German refugee named Kurt Stehling. Only 15 years old at the time, Stehling and about twenty other rocketeers constructed and flew small gunpowder-propelled rockets, assembled cardboard spaceship models, corresponded and shared ideas and concepts with other groups including the British Interplanetary Society and the American Rocket Society. In 1939, Stehling's club caught the attention of a local reporter, and he subsequently gave an interview on the CFRB radio station. The outbreak of the Second World War ended Stehling's rocketry ambitions for the time being, but after the war the industrious young man eventually went on to work alongside Werner von Braun and James Van Allen in the U.S. Space Program.

Despite the enthusiasm of amateur rocketeers like Stehling, the established academic scientific community in Canada, such as it was, were generally skeptical of the many claims made by these enthusiasts, and like most scientists of the day openly challenged the ambitious claims made by rocketry enthusiasts in various public forums. In the April 1932 issue of *Canadian Defence Quarterly*, for example, the noted pioneer Canadian chemical engineer Ernest A. Lesueur wrote, "We have in recent months been treated in the daily papers to thrilling accounts from certain, not precisely shrinking, enthusiasts as to what is to be expected from rockets and 'rocket planes'", but he then cautioned, "…the average man doubtless believes that it is only a question of time before transatlantic hops will be made by rocket. Heretofore, so far as I know, none of these prophecies has been put forward by an accredited engineer."[24] His article continued in painful detail with the many complexities surrounding velocity and gravity, summarizing, rather unclearly, that space flight with a rocket was essentially theoretically impossible.

Further to this, no evidence suggests that rocketry received official research support from the Canadian government prior to the Second World War. While the National Research Council's (NRC) Division of Physics and Engineering pursued experimentation with ballistics, official histories reveal investigations into rocketry or propellants were noticeably absent from its list of study fields during the 1930s.[25] There may have been many reasons for this, not the least of which was the more pressing concern of widespread economic depression. Historical writing on the development of science and technology in Canada during the 1930s similarly makes no mention of any organized research in rocketry, thus it is unlikely that there were any large-scale projects under way at the time.[26]

[24] Ernest A. LeSueur B.Sc. 'Rocketeers', *Canadian Defence Quarterly,* 9:3 (April 1932), 374. For a brief overview of LeSueur's professional career see Hugh J. Anderson, "Ernest A. LeSueur: Pioneer Canadian Chemical Engineer", *Journal of Chemical Education*, 72: 5 (May 1995), 390–393.

[25] M. Thistle, *The Inner Ring: The Early History of the National Research Council of Canada.* Toronto: University of Toronto Press, 1966, 345–349.

[26] For studies on science in Canada between 1880 and 1945 see Yves Gingras, *Physics and the Rise of Scientific Research in Canada.* Montreal-Kingston: McGill-Queen's University Press, 1991; Richard A. Jarrell and Yves Gingras (Eds), *Building Canadian Science: The Role of the National Research Council.* Toronto: Canadian Science and Technology Historical Association; and M. Thistle. *The Inner Ring.*

In contrast, government-sponsored guided missile and rocketry technology research and development sharply increased in Canada after the Second World War. Strategically situated between the two emerging postwar superpowers, the United States of America (USA) and the Union of Soviet Socialist Republics (USSR), Canada sought security through its support for science and technology including the employment of guided missile systems for strategic defence. In 1947, for example, the DND formed the Guided Missile Advisory Committee (GMAC), which undertook detailed studies of missile systems that might be employed from static defensive positions or as weapons for other military platforms. Around the same time, ballistics, rocketry, and propulsion technology research and development was initiated at the Canadian Armament Research and Development Establishment (CARDE) at Valcartier, Quebec. This important work later formed the basis of the new Canadian Rocket Propulsion Program (CRPP).[27]

The aim of the CRPP was to contribute to the existing Canadian program of applied defence research, as well as produce a group of indigenous experts in the field who could assist in guided weapons system studies with the armed services.[28] While the CRPP was perhaps the first focused effort to consolidate applied research, it was not the first official Canadian foray into rocket-assisted ballistics. During the Second World War, some graduate science students at Canadian universities were mentored in rocket propellant-related research, and the Canadian Army fielded a limited number of rocket assisted artillery systems towards the end of the war

In addition, the CRPP intended to provide a limited production facility for rocket propellants in Canada, so that early small-scale requirement for Canadian-produced short-range missiles could be economically met.[29] Once the infrastructure had been built and was functioning, the DRB could then potentially contribute to larger-scale projects, including the possible establishment of an indigenous rocket capability within Canadian industry.[30]

The CARDE rocketry and propulsion projects in the early 1950s focused on the development of solid rocket fuels for Canadian-designed short-range weapon systems. Specifically, defence scientists and engineers first concentrated their efforts on the design of a new semi-active radar homing air-to-air missile named Velvet Glove, which was being designed for use with the Canadian-designed CF-100 fighter interceptor. At roughly ten feet long and just under twelve inches in diameter, it was an ungainly product of the pre-miniaturization age and ultimately depended on a microwave radar proximity fuse to detonate its sixty-pound warhead. After considerable testing, full-scale production of the Velvet Glove began in 1953, and approximately 130 missiles were built before the termination of the project three years later. Originally designed to shoot down Soviet bombers

---

[27] This program was also referred to as the Canadian Rocket Development Program (CRDP) in some Canadian government reports. See John H. Chapman et al., *Upper Atmosphere and Space Programs in Canada.* (Ottawa: Science Secretariat Privy Council Office, February 1967), hereafter referred to simply as the Chapman Report.

[28] For discussion on wartime research see R.C. Fetherstonhaugh, *McGill University at War, 1914–1918 and 1939–1945.* Montreal, McGill University Press, 1947, pp.321, 336–337. Early post war missile studies are briefly covered in D.J. Goodspeed. *The Defence Research Board,* 127-133.

[29] R.F. Wilkinson, "Rocket Research in Canada", *Canadian Aeronautical Journal,* April 1959, 138.

[30] Confidential Résumé of Major DRB Activities up to 1962, Scientific Program Rocket Propellant Research and Development, 3. DRBS 173-2 pt.1. vol.7407 RG 24, Accession 1983-84/167, LAC.

at subsonic speeds out to 4500 meters, the missile became rapidly obsolete as newer plane designs transcended the sound barrier and were able to fire their ordnance from much greater distances than the Velvet Glove could reach. Still, defence projects such as Velvet Glove and others resulted in the creation of a solid core of knowledge and experience within CARDE that allowed the organization to pursue larger and more complex propulsion and ballistics projects in later years.[31]

In 1956, Dr. Adam Hartley Zimmerman was appointed as the new chairman of the Defence Research Board, succeeding Omond Solandt who had held the post since the end of the Second World War. Zimmerman initiated a revision of the existing CRPP to focus on more robust solid-state propellants that could be employed in larger rockets and guided missiles similar to those then in use by the United States.[32] The Canadian military was already engaged in joint arctic weather testing of various American-designed surface-to-air missile systems at Fort Churchill, and there was then the strong possibility that similar systems would be built in Canada for strategic defence. Dr. Zimmerman tasked the Aerophysics Wing and the Explosives Wing (later renamed the Propulsion Wing) at CARDE to initiate a revised program to improve the physical characteristics and performance of Canadian solid fuel rocket propellants then in use.[33] The research establishment was the logical choice to pioneer such work, as it had in particular the experience, facilities, personnel, and sustained government funding to undertake just such a program.[34]

Still, CARDE had its work cut out for itself. Although the establishment had developed and tested a number of short-range guided weapon systems after the war, the organization had yet to design solid propellants for use in anything larger than the Velvet Glove missile. Since the aim of the revised CRPP was to design propellants for large-scale vehicles which could launch heavier payloads to high altitude, scientists at CARDE chose to build a new 7.4-meter long Propulsion Test Vehicle (PTV) that was later named Black Brant.[35] It would become the first of a new generation of small payload launch vehicles and marked the official beginning of rocketry production in Canada.

---

[31] D.J. Goodspeed, *A History of the Defence Research Board of Canada*. Queen's Printer, Ottawa, 1958, pp.127–133.

[32] DRB List of Technical Fields, July 1959. Zimmerman replaced Dr. Omand Solandt, the DRB's first Director General who served from 1947 to 1955. File 73/778 Vol. 3, Appendix A, Acc. 1983-84/167, RG 24, LAC.

[33] Proceedings of the Second Meeting of the Associate Committee on Space Research, Ottawa, April 8, 1960. Annex C – Description of Black Brant I and II by R.P. Blake. File 12798-2-40 pt. 1, Vol.7841, RG 25, LAC.

[34] Secret memo to Chief (Sciences) from G.D. Watson reviewing progress of Defence Research Board activities dated June 24, 1958. DRBS 173-1, vol.7407, Acc. 1983-84/167, RG 24, LAC.

[35] I.R. Cameron, "Manufacture and Testing of Black Brant Engines", *Canadian Aeronautical Journal*, February 1961, 61. Dr. Cameron was then serving as superintendent of the Propulsion Wing at the CARDE.

Fig. 2.6 The first PTV/Black Brant I was transported to Sounding Rocket Complex #1 using vehicles left behind at Churchill Research Range by the U.S. Army Engineer's Arctic Test Detachment. Note the small Black Brant painted on the upper stage just below the nose cone

The conceptual design phase of the PTV was shaped by a number of factors. Ongoing advances in Soviet ballistic missiles, rockets, and high-speed aircraft during the 1950s heavily influenced the Canadian decision to pursue a solid- rather than liquid-based rocket propellant program. This was because the DRB was interested at first in developing a propulsion system which could be employed mainly in air defence weapons, and despite the weight constraints imposed by solid propellants, the incredibly short early warning of an impending enemy bomber or missile attack made an instant state of readiness essential. As well, the logistical difficulties and time constraints associated with fueling and emptying liquid fuel rockets at the time simply made them unsuitable as a quick reaction defence weapon.

Another important factor for the Canadian rocket design was stability. Because of its northerly location, high-performance composite rockets had to be serviceable under extreme weather conditions such as those encountered around Fort Churchill. Solid propellants were less volatile than liquid fuels, making them much more suitable for use in extreme temperatures. In addition, given that Canada's main bases were often situated in remote areas and difficult to access or supply easily or regularly, solid fuels were also considered easier to logistically maintain locally for longer periods of time. All of these factors came to shape the decision to pursue a solid booster, despite the limitations that it also imposed on the size and scope of the rocket design project.

The first thing the CARDE engineers needed to do was to begin testing all of the solid rocket fuel variants then in use by similar launchers. The team acquired the details of the British 0.44-meter diameter Skylark from Bristol Aerojet Company of England, which served as the basis for the Canadian PTV design.[36] As with the Skylark, the Canadian PTV model was kept relatively simple in order to facilitate the manufacture and reliability of more advanced models as the CARDE teams gained experience in rocketry development. Consisting of three basic elements, the first PTV personified simplicity with only a motor, the propellant, and the casing. Bristol Aircraft (Western) Ltd., situated in Winnipeg, supplied the rocket casings while Canadair Ltd., of Montreal, supplied the nose cone and the tail fin stabilizers. The parts were shipped to Fort Churchill, where they were assembled locally at the CARDE facilities under the direction of Dr. Ian R. Cameron, who then served as Superintendent of the Propulsion Wing.

Although all of the sounding rockets in the Black Brant series were designed to be capable of carrying a payload, the purpose of the first two launchers built in the series, PTV/Black Brant I and Black Brant IIA, was simply to prove the design. Both rockets employed a highly reliable Canadian-designed motor within a solid composite propellant casing. Known as the 15KS25000 motor, the machine generated 15 s of 25,000 pounds thrust, making it capable of boosting up to 108 kg of payload to an altitude of approximately 97 km. Although the original design relied on thick layers of polyurethane-mica to contain the hot propellant gases, later versions of the rocket incorporated in-situ molded asbestos-phenolic mats instead.[37] Overall, the rocket design proved very reliable in testing, malfunctioning only once during its first twenty-two static firings.[38]

Equally successful was the research team's solid fuel design. The CARDE appropriately named their creation CARDEPLEX, and this solid propellant was used in all subsequent Black Brant designs throughout the project. Based on a solid crystalline oxidizer and an organic polymerizable binder rather than the usual nitrocellulose and nitroglycerine components used in other solid rockets at the time, CARDEPLEX was specifically designed to meet casing requirements that standard nitro-based propellants in use at the time simply could not achieve. The manufacture of CARDEPLEX solid fuel was a two-step process. First, ammonium perchlorate oxidizer was ground, sifted, and then mixed with a smaller amount of another fuel based on carbon, hydrogen, oxygen, and nitrogen, to create the crystalline oxidizer. Then, after blending and curing, the oxidizer was again blended with a polyurethane binder before being poured into the PTV engine casing. The casing itself was internally coated with a heat barrier restrictor and mica filled polyurethane bonding agent that applied easily to both the steel wall and to the propellant.

The CARDEPLEX propellant formulation was finalized in October 1958, and by December Dr. Cameron's teams had completed construction of the first PTVs. Testing of the new vehicle began soon after, with the first static firing of the PTV taking place at the CARDE facility in February 1959.[39] The results of the test, while not perfect, were very

---

[36] CARDE. Technical Note [TN] 1421/61 *General Information*. (Unclassified) dated September 1961.

[37] CARDE. TN 1525/63, *Summary of Performance of the 15KS25000 Rocket Engine Used in the First Sixteen Black Brant Vehicles*; and TN 1528/63, *The 15KS25000 Black Brant Engine Ground Operations and Handling Instructions,* dated 1963.

[38] Ibid. CARDE. TN 1525/63.

[39] CARDE. Technical Manual [TM] 343/60. *CARDE Black Brant I Vehicle Trials.* (Valcartier: CARDE, 1960).

promising and gave the team confidence that it had made good design decisions. The design team engineers made corrections to the PTV design, and more tests were conducted throughout the spring and summer, including testing at different temperatures and environmental conditions. With a little effort, it was anticipated that a real launch could be attempted at the Churchill Research Range sometime during the autumn of that year.

The Churchill Research Range proved to be an ideal site for launching Black Brant rockets. Based at Fort Churchill, Manitoba, the northern region military base already served as the home of the Defence Research Northern Laboratory (DRNL), a research establishment tasked with testing military capabilities in extreme conditions. The location provided a huge natural safe zone of impact that was necessary for the conduct of the American and Canadian missile tests, and thus was more than suitable for the operation of Black Brant rockets.[40] There was also the great scientific benefit of Churchill lying near the middle of the zone of maximum auroral activity, and this was even further augmented by Churchill's proximity to the north magnetic pole. This gave both the civilian and the defence scientists and engineers an ideal location from which to conduct ionosphere-related studies that were crucial to the development of both rocketry and space flight.

As previously described, Canada and the United States concluded an agreement to build and maintain a research and test facility at the CRR in 1955 to test fire under extreme cold weather conditions the American NIKE AJAX missile system.[41] This test range was subsequently employed during the International Geophysical Year the following year, and remained under American control until the conclusion of related IGY projects in December 1958. During the IGY period itself, the United States bore the entire cost of the installation (USD$7.3 million), and provided all 86 sounding rockets needed for international scientific experiments. Two of these rockets were designated specifically for Canadian use, and were flown in November 1958 mounted with nose cones instrumented by scientists at the CARDE.[42]

In 1959, the range and facilities were scheduled to revert to Canadian ownership less any equipment the United States chose to remove; however, the United States Department of Defence Research and Development Office (DRDO) expressed an interest to the Canadian government in reopening the site for further joint Canada-U.S. missile and rocketry testing. A new agreement was subsequently completed between the two governments, after which the U.S. Army tasked the staff at the White Sands Missile Range to put the northern facility back into operation. The U.S. Army constructed several new buildings and modernized the launch facilities which would later prove more than suitable for all Canadian rocketry needs. Having previously cooperated with the Americans

---

[40] Government of Canada. Chapman Report, 22.

[41] LAC. RG 24, Vol.25, File No. 1200 pt.2 VII. Secret Memo HQS 6001-Guided Missiles TD 8160 (CGS), Surface-to-Surface Missiles and Fort Churchill. Drafted by LGen. S.F. Clark, Chief of the General Staff, Ottawa, dated September 29, 1959. See also Pennie, A. M. *Defence Research Board: Defence Research Northern Laboratory, 1947-1965.* (Ottawa: Report No. DR179, DRB April 1966).

[42] ATI LAC. RG25, Vol.1 File No.12798-4-40. Outer Space – Proposal for Possible Canadian Initiative. (Confidential) memorandum 'International Implications of a Canadian Space Research Program', prepared for the Under-Secretary of State for External Affairs, dated August 27, 1959, p.2.

on the use of the CRR, there was little work now required to allow Canada to test and fire its own rockets from the same launch pads then employed by the U.S. Army, and later, the United States Air Force (USAF).[43] The bilateral cooperation that had started between American and Canadian militaries at Churchill before the space age was now set to last several more years.

A new and inexpensive inclined launch rail platform for the Black Brant I was later designed and installed on one of the existing launch pads at the CRR in the latter months of 1957. Though somewhat makeshift and temporary in measure, the launch rail sufficed for early design and test integration of the rocket and launch platform. The concept consisted of a simple elevating boom that, when horizontal, could have a PTV underslung from three guide rails. The boom and PTV were then simply elevated to any desired launch angle between 70° and 82°, and the forward braces were bolted to ground anchors. Once everything was stabilized and secured the PTV would then be ready for launch.[44]

The first permanent inclined launch rail platform was installed at the CRR in the fall of 1958, approximately 150 m south of the American NIKE-CAJUN assembly building that had been previously built there during the IGY. This new platform provided an almost due east trajectory for Black Brant rocket firings, taking them out over Hudson Bay, which was necessary since there was always a requirement for maximum safety during launches. The PTV was a non-recoverable launcher; therefore, should any problems occur during a firing it was far less likely for falling debris to cause any serious material damage or civilian casualties.

The first group of four PTV/Black Brant I rockets was scheduled for launch from the Churchill Research Range in September–October 1959. Since the rockets were not designed to carry actual payloads but simply test the quality and characteristics of the CARDEPLEX propellant, the first two vehicles were ballasted heavily to ensure that a large stability margin was maintained at Mach 6, while the second pair of rockets was lightly ballasted to what was considered the minimum telemetry weight.[45] Telemetry was measured using a standard 30 by 30 PDM-FM (pulse duration modulated – frequency modulated carrier) radio system with 28 active channels. Accelerometers and skin thermistors were added to supply additional data on the vehicle performance and on the effects of flying at high Mach speeds through the lower atmosphere. Finally, a radar beacon was installed in the nose cone of the PTV to aid tracking of the vehicle throughout its trajectory, to evaluate drag, and to verify its high altitude performance.[46]

The first two flights of the Black Brant I (vehicle No. CC601 and CC602) in 1959 provided the CARDE engineers with the benchmark from which to proceed with the rest of the program. While it was noted that the telemetry from rocket CC601 was not considered of value due to the high roll rate of the vehicle during its flight, rocket CC602 provided much valuable data to crews back on the ground. The two experiments flown aboard CC602, a sodium photometer and an infrared photometer, both performed well beyond

---

[43] DND. *Operation Probe High, 20 July 1963*. Churchill: Fort Churchill pamphlet, 1963, 16–17.

[44] I.R. Cameron, "Manufacture and Testing of Black Brant Engines", p.66.

[45] Ibid., p.66.

[46] Ibid., p.65.

initial expectations greatly satisfying research crews on the ground.[47] With these flights successfully completed, Canada's rocketeers moved onto the next task.

The next rocket design was essentially a mature version of the Black Brant I. Engineers at the CARDE had originally designated this evolved rocket the 'Snow Goose', but the moniker was soon dropped from official references and reports in favor of the simpler name of Black Brant II. The new rocket was based on the main requirement to attain from a near vertical launch a minimum altitude of 228,600 m with a max payload of 68 kg. The design philosophy behind the new rocket essentially was that the structural weight was to be a minimum consistent with high efficiency structural design. As a result, the Black Brant II was also slightly longer than its predecessor at approximately 8.43 m from tail to top, and the length of the nose cone was also extended out to 2.18 m. While the Black Brant IIA model continued to employ the 15KS25000 motor originally designed for the Propulsion Test Vehicle, the Black Brant IIB variant was designed to use a newly-designed longer burning 23KS20000 propulsion unit.

As rocket designs evolved and testing proved successful, more sustained operations and support for Canadian rocket launches were also beginning to take shape. With the entry of the Black Brant II into regular service, the DRB, the CARDE, and the various other agencies sponsoring each flight collaborated on the development of formal procedures for the check-out, ground handling, and launching of each rocket in an effort to ensure satisfactory firings from the pad every time. The CARDE engineers and the Canadian military organized dedicated casings and transport vehicles for rocket parts, and put requests into Army Headquarters for specifically-designed vehicles required to support any expansion of the existing program. Finally, launch platforms designed for use with the PTV/Black Brant I were modified to increase the maximum launch angle form 70° to 85°, while another Aerobee rocket launcher was also modified to support Black Brant II series rockets. Though none of the launchers had yet been augmented to support year-round firings, these small increases greatly expanded the potential range of operations for Canada's new "rocketeers".

After the completion of ground, engine, and instrumentation testing in early 1961, Black Brant IIA test flights began and, overall, generally proved successful. The rocket design continued to perform consistently well in flight, and was subsequently utilized in no less than fifty-five experimental launches between 1961 and 1966. The main employer of the Black Brant IIA rocket was the Canadian university researcher community, usually sponsored by the NRC, though a small number were also put to use by the United States Air Force Cambridge Research Laboratory. The Black Brant IIA rocket was the first workhorse of the Black Brant fleet and only finally retired from service in the mid-1970s, though its motor was still used later in both Black Brant IV and VA rocket configurations.[48]

---

[47] NRC-DRB Permanent Joint Committee on Space Research – Proceedings of the First Meeting of the Associate Committee on Space Research [ACSR], Ottawa, 2 October 1959, p.2. File. 12798: 2-40 pt.1 DEA, Vol.7841, RG 25, LAC.

[48] A.W. Fia, "Canadian Sounding Rockets: Their History and Future Prospects", *Canadian Aeronautics and Space Institute Journal*, 20:8, October, 1974.

Fig. 2.7 DRB engineers and technicians preparing a Black Brant II for launch at Fort Churchill Research Range, c.1960

As the Black Brant IIB rocket was designed to meet a specific technical requirement to achieve higher altitudes using the new 23KS20000 motor, its production was terminated after only four tests flights in 1963. Once the engineer's objective had been achieved, instead of continuing production of the IIB model, the lessons learned from this interim launcher were incorporated into an improved IIA model as well as the forthcoming Black Brant III rocket.[49]

With the Black Brant IIA and IIB projects well under way, the Defence Research Board next turned its attention to planning the next generation of sounding rockets. The initial success of the first two rockets in the series had demonstrated that Canada clearly had the potential for building a more sophisticated family of sounding rocket for both civilian and defence scientific research. The United States Department of Defense and NASA had also expressed an early interest in Canada's Black Brant program. While the CARDE had taken the lead in the initial project, it simply did not have the resources or authority to undertake a large-scale

---

[49] Ibid. See also F. Jackson, "Development of the 23KS20000 Motor for the Black Brant IIB Vehicle", *Canadian Aeronautics and Space Journal*, Vol.11:12, December 1965, 377–383.

production of Black Brant launchers. Dr. Cameron therefore made a recommendation to the DRB that perhaps one of the Canadian civilian companies involved in the project could be made the prime contractor for manufacturing additional rockets.

The Department of Defence Production agreed with this recommendation and subsequently conducted extensive market surveys in 1957 and 1958 to establish which companies would be best suited to take over the main portion of Black Brant rocket production. A major concern amongst decision makers remained – would Canadian industry be able to provide adequate technical support for an all-Canadian launch program? It was soon determined that Bristol Aircraft (Western) Ltd. (BAL), one of the CARDE's original rocket suppliers, was capable of meeting the basic requirements of the Black Brant program. Further extensive market surveys conducted by both Bristol and the government in 1959 ultimately led to a solid proposal for a Canadian rocket industry start-up in November. A year later, the government formally awarded a contract to Bristol for three new versions of the Black Brant, to be developed as a joint DRB-BAL project over the next few years. A few signatures later, Canada's rocket industry was officially born.

Fig. 2.8 A Bristol Aerospace Limited Information Manual for the advanced generation of Black Brant rockets

Like the CARDE, due to its technical expertise Bristol Aircraft was an obvious choice to undertake a pioneering role in Canada's emerging rocket program. The company, headed at the time by Stanley Haggett, possessed both the technological base and the manufacturing capability to produce the rockets. The company's engineers and scientists, then under the management of Murray Auld, possessed all the specialist techniques and skills needed in working with high tensile and heat and corrosion resistant metals like those employed in launcher systems.[50] Finally, Bristol's manufacturing plant was also located reasonably close to the Churchill Research Range where most Black Brant rockets would be launched.

A young talented engineer named Albert Fia was chosen to head the Black Brant rocket technical team at Bristol. Mr. Fia held a degree in Electronic and Electrical Engineering, and had extensive experience in the development of missile systems for the Canadian Army. He was also a member of both the Manitoba and Ontario Associations of Professional Engineers, and was considered by his peers at Bristol to be an excellent team leader who could successfully direct the complex challenge that now lay before BAL. Interestingly, Albert Fia joined Bristol in 1958 just as the launcher contracts were tendered to industry, and he soon found himself assigned to head the newly-created Special Projects Group that would design and built the next generation of Black Brant rockets.[51]

The three new versions of the Black Brant to be designed by Albert Fia's Special Projects Group covered a range of altitudes and payload weights based on the requirements of potential clients as well as the environment to which the research instruments would be subjected. Both the DDP's and Bristol's own market surveys had identified that research scientists and engineers desired a minimum-cost, highly reliable rocket able to carry payloads of 5kg to 135kg to heights of eighty through to a thousand nautical miles with little dispersion in technical performance.[52] Equally important, designers had to keep in mind that the environmental conditions of northern Canada limited the acceleration loads on the instruments carried in rockets to forty Gs, and the temperature to no higher than 125°F. Such demands were optimistic, and from the very beginning compromises were necessary in order to achieve any success within these design limitations.

The first alteration to the existing design specifications was a reduction of the maximum desired altitude from one thousand nautical miles to just six hundred nautical miles. Next, Fia's design team simplified the development process and the time needed to complete a new plan by incorporating a number of existing CARDE components into the three new designs. It was decided that, for the time being, Bristol could and would continue to employ the Black Brant I's 15KS25000 motor and the DRB's CARDEPLEX propellant rather than attempt to create entirely new components and fuels. Finally, Bristol overcame the remaining design limitations by developing and manufacturing new steels with increased tensile strengths and improved ablative coatings that provided rigidity-to-weight

---

[50] Anon, "Black Brant: Canadian Bristol Aerojet's Family of Sounding Rockets", *FLIGHT International*, 7 January 1965, 14.

[51] Anon, *Bristol Aerospace Limited: 50 Years of Technology, 1930–1980, Volume Two – The Second Quarter Century*. Winnipeg: Bristol Aerospace, 1985, p.66.

[52] Ibid., p.15.

ratios very close to the desired rocket specifications. All in all, the Special Projects Group achieved remarkable engineering design success given the difficulties in meeting the original parameters set out by the clients.[53]

The first new rocket in the series, the Black Brant III, was designed to be a scaled-down version of the Black Brant II rocket using a 25cm diameter vehicle in place of its predecessor's 43cm casing and the CARDE's newly-designed 9KS11000 rocket motor. Approximately 5.8m long, the Black Brant III was capable of carrying a 18kg payload to a height of 178km.[54] Testing of the new rocket motor began in late 1961, with fifty-three static firings carried out at the CARDE, followed by another twenty structural and aerodynamic tests at Bristol.[55] By May 1962, everything on the Black Brant III had been tested, re-tested, and tested once more leaving only the final exam for the rocket – proving that it could actually fly.

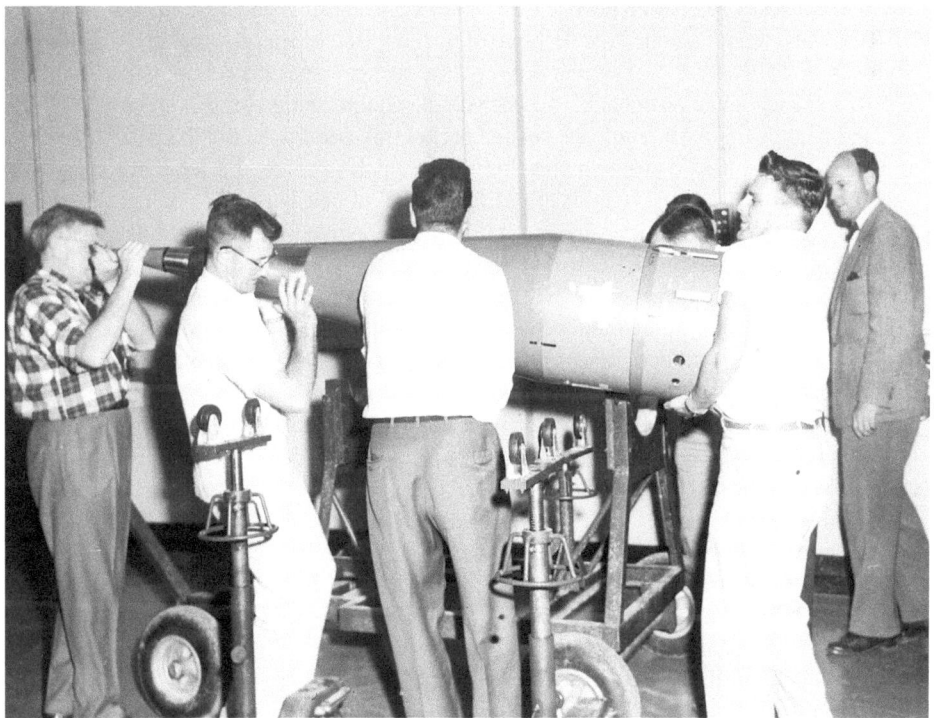

Fig. 2.9 NRC and DRTE scientists, engineers, and technicians prepare to load a scientific instrument nose cone for a Black Brant rocket. From left to right: Dr. D.C. Rose, Bud Budzinski, Dr. Ian McDiarmid, Don Awry, and Dr. Peter Forsyth

---

[53] Ibid., p.15.

[54] Peter Alway, *Rockets of the World: Third Edition*. Ann Arbor: Saturn Press, 1999, pp.343–346.

[55] A.W. Fia, "Canadian Sounding Rockets: Their History and Future Prospects", *Canadian Aeronautics and Space Institute Journal*, 20: 8, October 1974, 398–399.

The original schedule to launch the first Black Brant III from the Churchill Research Range had to be scrapped after a terrible fire devastated a good portion of the rocket test facility in February 1961. Desperately, the Bristol team sought an alternate launch site to keep the program on track, and fortunately they were able to secure a launch pad at the American Wallops Island rocket range just off the Virginia coast. The entire CARDE/BAL rocket team and four Black Brant III rockets were soon transported down to the American test facility aboard a Royal Canadian Air Force C-130 Hercules transport plane. Ralph Bullock, an electronics engineer serving with Bristol at the time, was on the flight and remembered that on arrival at Virginia the pilot of the C-130 executed a pre-landing show of aerobatics that he considered hardly appropriate for a transport plane let alone its precious cargo. Fortunately, however, all eventually arrived safely on the tarmac and the Bristol team spent the next few weeks conducting the final assembly and checkout of the four Black Brant III rockets.[56]

On June 15, 1962, after nearly four years of designing and testing, the combined CARDE/BAL team launched the first two of the new Black Brant III rockets. Carrying net payloads of 42.1kg and 43.1kg respectively, the rockets performed impressively but less well than the team had estimated. The first vehicle attained a maximum velocity of 1700m/sec, and a range of 98km, while the second rocket traveled faster at 1705m/sec, to a range of 158km.[57] On-board instrumentation on both vehicles recorded a large but short-lived pitching disturbance at six seconds after lift-off that cut nearly twenty percent off both the rocket's peak altitudes. The Canadian rocket team realized that this problem had to be rectified in order to stabilize the rocket and allow it to fly properly.

Despite the rocket team making deliberate adjustments to the remaining rockets, the next two vehicles suffered similar problems during their flights. A third vehicle, launched on June 19, 1962, had been fitted with additional fin cuffs to reinforce the stabilizers, but the rocket's lighter payload was wrenched so severely by a pitching disturbance seven seconds after lift-off that all telemetry with the vehicle was soon lost. Ground tracking films later revealed that the Black Brant rocket had righted itself after the disturbance and continued flying on to an estimated altitude of 144.5km, but the problem of instability remained unsolved.[58]

The launch of the fourth and last Black Brant III rocket brought to Wallops Island took place a week later on June 28, 1962. For this flight, the Canadian launch team attempted to keep the vehicle properly positioned by intentionally spinning the rocket at three revolutions per second whilst it was in flight. It was hoped that this gyroscopic spin would produce sufficient stabilization to keep the last Black Brant III from suffering the same fate of its predecessors. As well, the rocket was fitted with a heavier nozzle in order to give it additional stability. The tail stabilizer fins were canted accordingly and the rocket fired while the Canadians held their breath and their American hosts looked on with some reservation.

---

[56] Anon. *Bristol Aerospace Limited: 50 Years of Technology*, p.68.

[57] Anon. "Black Brant: Canadian Bristol Aerojet's Family of Sounding Rockets", p.17.

[58] Peter Alway, *Rockets of the World*, p.343; see also Black Brant III Firings Table in Anon., "Black Brant: Canadian Bristol Aerojet's Family of Sounding Rockets", p. 17.

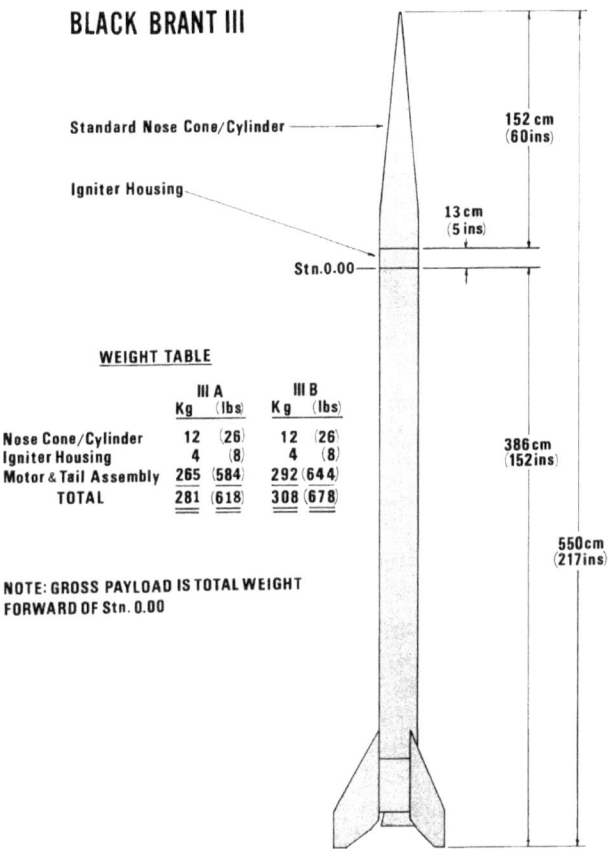

BLACK BRANT III

Standard Nose Cone/Cylinder

Igniter Housing

Stn.0.00

152 cm
(60ins)

13 cm
(5 ins)

386 cm
(152 ins)

550 cm
(217 ins)

WEIGHT TABLE

	III A Kg	(lbs)	III B Kg	(lbs)
Nose Cone/Cylinder	12	(26)	12	(26)
Igniter Housing	4	(8)	4	(8)
Motor & Tail Assembly	265	(584)	292	(644)
TOTAL	281	(618)	308	(678)

NOTE: GROSS PAYLOAD IS TOTAL WEIGHT
FORWARD OF Stn. 0.00

Fig. 2.10 Black Brant III technical diagram

At first the flight went well and things looked promising. The Canadian team let out a sigh of relief as the final Black Brant III passed the point of previous instability without incident and continued to head skyward. As the rocket continued to climb, however, it began a graceful if not desirable helical motion during which, as one observer later described, the rocket used up a great deal of sky before recovering itself and heading off in a completely new direction.[59] The rocket reached a mere 27.6km altitude before plummeting back down to Earth. Downhearted but not defeated, the Canadian team decided to call it a day. The initial tests of Canada's newest rocket design were done.

The CARDE/BAL launch team returned to Canada in July to begin painstakingly to pore over the massive amounts of telemetry and tracking film recorded during the four launch tests. There was much work to be done by the Bristol Special Projects Group. The largest problem – and the one that needed to be dealt with first – was the stabilization

---

[59] Anon, *Bristol Aerospace Limited: 50 Years of Technology*, 68; and Peter Alway. *Rockets of the World*, p.343.

of the rocket, and until this issue was rectified production of the Black Brant III could not proceed. Among the corrections, the engineers replaced the original fin stabilizer assembly on the rocket with three stronger, lighter, aluminum single wedge fins employing a more pliable plastic insulation known as Avcoat. As well, the fiberglass wrap originally used on the outside of the motor and nose cone to keep it cool during flight was replaced with an internal payload-insulating blanket. Design changes were also made to the motor case liner to give the rocket more thrust and a lighter nozzle. It was hoped that all of these changes would prove successful in improving the rocket's flight.

Two of the newly-redesigned Black Brant III rockets were brought back to Wallops Island, Virginia, where they were successfully launched on December 13, 1962. Both of the Black Brant III rockets achieved near perfect flights and returned full telemetry right until splashdown. The Canadian launch team congratulated one another and returned home, this time much more confident that Black Brant III was now ready to be manufactured for commercial use. Team frustrations returned in July 1963, however, when in preparation for competing for a United States Navy contract a Black Brant III fired from Point Mugu Test Range lost control just shortly after take-off once again due to lateral disturbances. The telemetry and nose assembly was also lost, and although ground-based tracking films showed that the rocket recovered and continued its flight, no other data were returned to the Canadian launch team.

The failed test in July also gave little reassurance to the launch team going into the American sounding rocket competition scheduled for that winter. However, their next launch on November 7, 1963 was a near perfect flight. Sporting a spin-balanced nose assembly the Black Brant III outperformed the team's expectations, but the U.S. Navy was left unconvinced of its long-term reliability and an American order for Canadian sounding rockets never materialized. Demoralized yet not defeated, Albert Fia's Special Projects Group carried on with a final test launch the following spring as the data returned would prove valuable to other work. Fired from a newly-reopened Churchill Research Range, the last of the initial Black Brant III rockets was sent skyward on April 21, 1964. The rocket performed well overall, but still left the Bristol engineers with questions about their design concepts. Answers would have to be sought in the company's next test phase.

Unlike the Black Brant III rocket, the Black Brant IVA was conceived as a two-stage rocket incorporating the Black Brant VA as a first stage and the Black Brant III for its second stage. The new rocket was expected to carry up to 18kg of payload to altitudes of 856km, much higher than the Black Brant III or its successor, the Black Brant V. The Black Brant IV was also the first two-stage launcher design attempt by Albert Fia's Special Projects Group, and he tasked a trusted and experienced colleague, an engineer named Harry Sevier, to lead the effort. Sevier assembled a ten-man team to tackle the Black Brant IV project. Utilizing Black Brant VA and IIIA motors, the goal was to marry up these two stages to produce a light yet robust launch vehicle. While the two separate stages were structurally resilient, however, the staging joint between the two was another matter. The design and redesign of this critical part of the Black Brant IV rocket required several attempts before getting it just right.

The dramatically increased burnout altitude needed of the second stage of the Black Brant IV rocket – approximately 35km – called for a longer and larger diameter exit cone for the nozzle. Also, instead of fins the upper stage now required an odd-looking titanium

Fig. 2.11 DRB engineers inspect a BBIV on its launch rail c.1963

conical device to keep the second stage rocket directionally stable in its flight.[60] To facilitate separation of the two stages, instead of physically connecting them, the top portion rested on the bottom portion using a sliding fit. The design team then fit a drag ring on the first or 'booster' stage, so that when it stopped firing its higher drag would cause the first stage to decelerate more rapidly than the upper stage and thus simply pull away from the remainder of the rocket. The goal was to achieve a smooth separation in flight and the design appeared valid, at least on design board.

While the first Black Brant IVs were being readied for testing, preparations were also being made to collect as much telemetry from the flights as possible. Though the Churchill Research Range, now fully recovered from the February 1961 fire, was capable of providing all the necessary diagnostics for the upcoming Black Brant IV launches, the high altitudes expected from the rocket merited the use of backup trajectory recorders then located at the tracking facility at Prince Albert, Saskatchewan. Preparations were completed and the rockets were moved to the launchers in May 1964 for their final checkout. Anxious and nervous, the Bristol team held their breath as tests began and the Black Brant IVA rocket vehicle No.01 left the launch pad on June 24, 1964.

---

[60] *Bristol Aerospace Limited,* p.70.

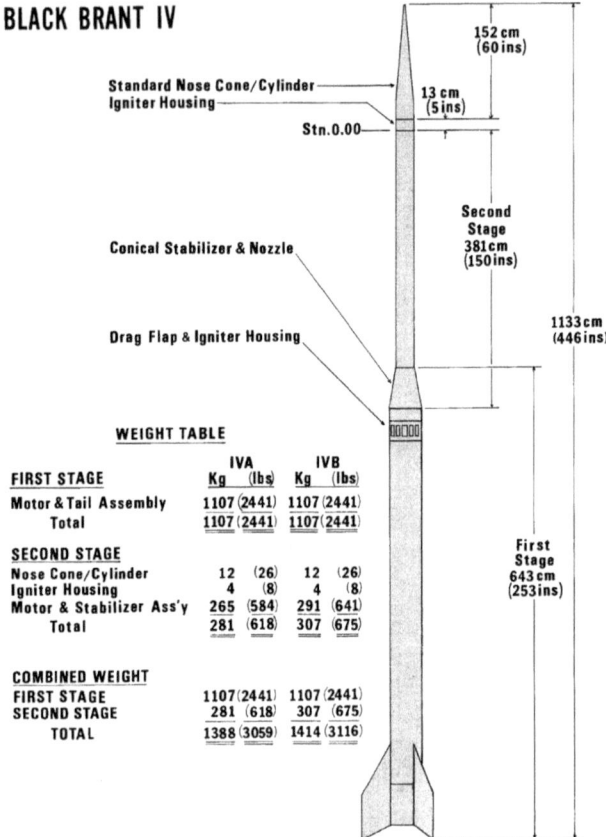

Fig. 2.12 Black Brant IV diagram

Despite their best efforts to prepare for any and all design contingencies, the Bristol team experienced another launch failure. The Black Brant IVA shot into the sky without difficulty and after a short while it was reported that the first stage appeared to be firing as it was designed to. When the time came for the second stage to separate at thirteen seconds into the flight, however, disaster suddenly struck. The ground team saw the exhaust trail wobble and then heard a loud explosion. Flight telemetry was suddenly lost. Still the sustainer engine on the second stage ignited and carried on, but by this point the Black Brant IVA was so far off trajectory that it barely reached an apogee of 470 km versus the anticipated 734km.[61]

---

[61] Ibid., p.70.

The next attempt to successfully launch a Black Brant IV rocket was made on July 2. This rocket suffered a similar fate, but imbedded rocket telemetry later revealed that the rocket had simply staged too early, and that that the internal pressure recorded almost twice the outside ambient pressure immediately prior to separation of the stages.[62] The Bristol Aerospace Limited official history adds further explanation, noting that the initial Black Brant IV flights had failed due to:

> '...inadequate inter-stage pressure venting, which in the absence of any structural joint between the stages, had prematurely pumped the rockets apart. As soon as the separation began, trapped inter-stage air escaped and the thrusting booster immediately drove up into the sustainer nozzle, producing violent oscillations – and collapse of the nose.'[63]

Fig. 2.13 DRB technicians recover the remains of a Black Brant IV rocket somewhere northeast of the Churchill Research Range, July 1964

---

[62] Ibid., p.70.

[63] Ibid., pp.70–71.

As a result of the initial failures a small number of changes were made to improve the system, namely the installation of a proper inter-stage venting process, a solid explosive bolt to physically connect the two stages, and flush-mounted booster drag flaps that would deploy at the same time the inter-stage bolt was cut. In essence, the redesign of the rocket allowed the Black Brant IVA to separate only when permitted, but to be able to do so very quickly when the command was executed.[64]

The Black Brant IVA design and launch team returned to the Churchill Research Range in January 1965 with two completely revamped rockets and the objective of determining whether or not the design upgrades were sufficient to make the rocket a success. After months of effort Bristol finally reaped the rewards of dedication and hard work. Both the BBIVA rocket vehicles No. 3 and No. 4 flew textbook flights, and two follow-on flights, also faultless, brought an agreeable conclusion to the Black Brant IVA project. An augmented version of the rocket, the Black Brant IVB, later went onto become a very successful commercial sounding rocket used by customers around the world.[65]

With each configuration, the Bristol scientists and engineers altered design parameters and introduced new motors and equipment with every intention of subjecting the rocket to rigorous testing that would push its aerodynamic limits. While success was hoped for with each flight, the engineers accepted it as unlikely. Yet, with each test failure came valuable data and lessons learned that, when applied, ultimately resulted in the overall successful completion of the Black Brant IV program.

The final phase of the DND-sponsored rocket project was the design and flight of the Black Brant V launch vehicle. Both the Black Brant VA and VB model differed little in their external appearance, both at 43 cm in diameter, 7.3 m long, and each tailed with three stabilizing fins. With the intent of lightening the structure to increase the overall range and altitude of the new rocket, the Black Brant VA model was designed utilizing the mechanical interchangeability of the motor cases used for the 15KS25000 and the 26KS20000 engines. This flexibility in the design allowed for a lighter casing using a motor that could deliver almost twice the rocket performance. The former engine was also subsequently employed on the Black Brant VB model.[66]

Still, the new motor design caused some initial problems, with a failure during the first static test revealing inadequacies in the design of the liner, which was required to insulate the highly stressed motor tube from the extreme temperature of the burning propellant. It took several re-designs of the engine and an additional 12 static firings before the DRB and Bristol engineers felt confident that rocket and its follow-on VB model were both ready to fly.[67]

---

[64] Ibid., p.71.

[65] Some flights took place from Northwest Territories, Peru, Brazil, Spain, Kauai Hawaii, and Greenland.

[66] Ibid., p.128.

[67] A.W. Fia, "Canadian Sounding Rockets: Their History and Future Prospects", *Canadian Aeronautics and Space Institute Journal*,.20:8, October 1974, 128; and Anon., *Bristol Aerospace Limited*, p.71.

Other ways to lighten the structure of the rocket were explored. The large and heavy magnesium tail fins normally fitted to Black Brant IIs and IIIs were replaced on the VA model with smaller and lighter units consisting of a thinner aerofoil section that produced much less supersonic drag. The tail fin itself was constructed not out of the usual solid sheet of metal, but instead was made of an aluminum honeycomb composite with bonded aluminum sheet skins. The whole construction was then coated with the plastic insulate Avcoat – a sheet only 60,000's of an inch thick, but capable of reducing 1000 °F external temperature to a surface temperature of only 300 °F. Many of the scientists and engineers were somewhat skeptical of the feasibility of the new design, but from the very outset the new fins worked well and often beyond expectations.[68]

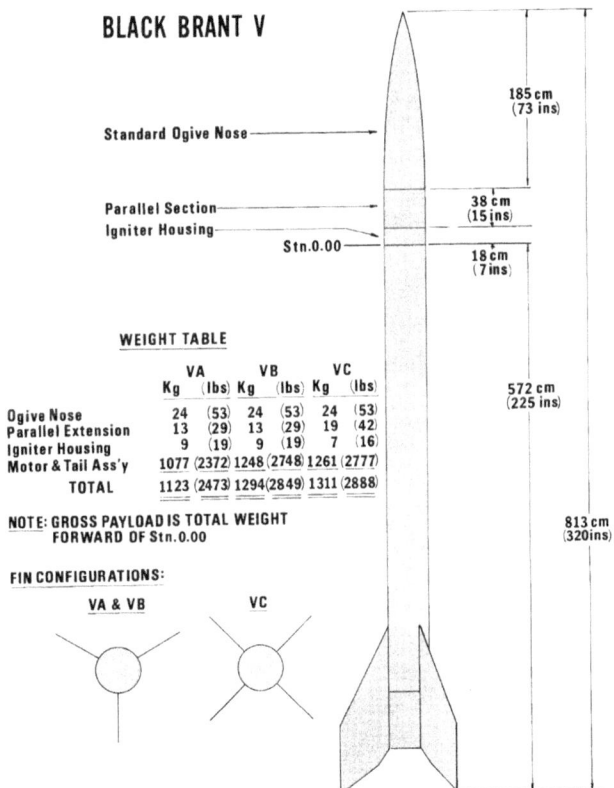

Fig. 2.14 Black Brant V diagram

---

[68] *Bristol Aerospace Limited*, p.71. See also Chapman Report, pp.61–62.

Built concurrently with the Black Brant IV, the first low-performance Black Brant VA was launched nearly two months before its sibling rocket on April 16, 1964. The rocket performed admirably and paved the way for the first Black Brant VB rocket tests. The first Black Brant VB launched from the Churchill Research Range on June 12, 1965, carrying 140kg of instruments to an altitude of approximately 378km. It was a flawless flight, easily demonstrating that the DRB-Bristol engineering team had honed their skills to the point where few design mistakes, if any, were made. The Black Brant VB rocket continued to perform without error in subsequent testing and was quickly adapted to carry scientific payloads even before the testing was finished. On August 16, 1966, a Black Brant VB rocket laden with a research payload from the Max Planck Institute of West Germany was lofted to 391km above the Earth where it released a cloud of barium. The resulting artificial aurora was visible as far away as Winnipeg, and resulted in a boon of data for the experiment's scientific researchers. The experiment was repeated three days later with another Black Brant VB rocket, which in turn produced another textbook flight and returned much valuable data.[69]

The remaining Black Brant VB rocket test flights were equally successful, bringing a satisfying conclusion to nearly ten years of challenging rocket design and testing. It was a bittersweet end to what many Black Brant engineers called "a labour of love".[70] When the last of the Black Brant V rockets had flown, both the DRB and Bristol could look back with a sense of pride in what was achieved. Essentially, in a little over a decade, scientists, soldiers, and engineers had given their country its very own access to high altitudes and outer space.

The Associate Committee on Space Research

The initial success of the CRPP, and the subsequent successful flights of the Black Brant I and II series rockets, only whetted the appetites of both the defence and civilian scientific and engineering communities for an expanded Canadian space research program. Yet in the absence of an overarching national-level body to oversee and coordinate all aspects of Canada's space research and development, the agencies that sought to employ Black Brant rockets were, at the most, still only loosely organized to coordinate amongst themselves what was sure to be an expensive and complicated task. Realizing that a more official government-led forum was required for the longer-term planning and execution of Canada's emerging rocket and space program, Dr. Donald C. Rose from the National Research Council called together a group of pioneering space scientists, engineers, and advocates from across Canada to participate in an official space program committee.

---

[69] Peter Alway, *Rockets of the World – Third Edition,* Saturn Press 1999, pp.349–350.

[70] Email interview with Dr. Lorne George Mason, Black Brant rocket engineer (1963–1965), December 2000.

Fig. 2.15 Dr. Donald C. Rose, a veteran defence scientist, served as the first chairman of Canada's Associate Committee on Space Research

This newly-formed government advisory body was named the Associate Committee on Space Research (ACSR), and it met officially for the first time in Ottawa on October 2, 1959.[71] Dr. Rose acted as chairman of the twenty-member group, which included senior academics from a dozen universities as well as scientists and engineers from the NRC, the DRB and its subordinate organization, the CARDE, as well as the Department of Transport (Table 2.1). After calling the group to order, Dr. Rose introduced Dr. E.W.R. Steacie, then serving as President of the NRC, who in turn welcomed everyone to the meeting before briefly outlining the proposed objectives of the group assembled before him. Dr. Steacie announced to those assembled, "This committee will provide a mechanism for participation in rocket firings by the universities as well as government departments. It will also serve as the Canadian national committee for the International Council of Scientific Unions (ICSU) Special Committee on Space Research (COSPAR) and will ensure that Canadian representatives at the United Nations are kept informed of the views of Canadian scientists regarding activities in space."[72] While UN-related space activities were deemed important, however, the subsequent record of the early years of the ACSR clearly demonstrates that the issue of rocketry for science and defence was at the center of the group's attention and efforts.

---

[71] Unofficially, an ad hoc meeting was held at the NRC on April 7, 1959, which resulted in a written proposal for the formation of the ACSR from Dr. Rose to Dr. E.W.C. Steacie on May 13, 1959. See space research folder Vol.7841 File. 12798: 2-40 pt.1 Exhibit "Q", RG 25, LAC.

[72] Ibid., p.2.

Table 2.1 Member of the Associate Committee on Space Research, 1959–1963

Chairman
Dr. D.C. Rose, Division of Pure Physics, National Research Council, Ottawa

Secretary
Mr. B.D. Leddy, Division of Administration and Awards, National Research Council, Ottawa

Members
Dr. J. Auer, Medical Research Council, Ottawa
Dr. J.H. Chapman, Radio Physics Laboratory, Defence Research Board, Ottawa
Dr. R.F. Chinnick, Defence Research Board, PO Box 1427, Quebec
Mr. J.W. Cox, Directorate of Physical Research, Defence Research Board, Ottawa
Dr. P.A. Forsyth, Department of Physics, University of Western Ontario, London
Prof. C. Fremont, Department of Physics, Laval University, Quebec
Dr. G.M. Griffiths, Department of Physics, University of British Columbia, Vancouver
Dr. A. Kavadas, Department of Physics, Dalhousie University, Halifax
Dr. D.P. McIntyre, Air Services, Meteorological Division, Department of Transport, Toronto
Dr. D.W.R. McKinley, Radio and Electrical Engineering Division, National Research Council, Ottawa
Mr. G.S. Murray, United Nations Division, Department of External Affairs, Ottawa
Dr. R.W. Nicholls, Department of Physics, University of Western Ontario, London
Dr. G.N. Patterson, Institute of Aerophysics, University of Toronto, Toronto
Dr. H.I. Schiff, Department of Chemistry, McGill University, Montreal
Mr. M.M. Thomson, Dominion Observatory, Department of Mines and Technical Surveys, Ottawa
Mr. F.R. Thurston, National Aeronautical Establishment, National Research Council, Ottawa
Mr. H.J. Williamson, Telecommunications Branch, Department of Transport, Ottawa
Dr. B.G. Wilson, Department of Physics, University of Alberta, Calgary

Source: Andrew B. Godefroy, *Defence & Discovery: Canada's Military Space Program, 1945–74*

Though such a focus may at first have appeared to be naive, when placed within the context of the period it made perfect sense. In 1959, space exploration was still in its infancy, and even the two main space race adversaries, the U.S. and the USSR, were still mastering the art of making escaping Earth's gravity a routine affair. Without the successful development of rockets, nothing – alive or otherwise – was getting into outer space, into orbit or off on a lunar journey. As well, defence planners and engineers knew that rockets designed to reach the upper atmosphere or farther were simply ballistic missiles without their warheads yet attached. As such, the rocket had huge potential not only as a platform for exploration and science, but also for weapons and defence. Thus, with a continuously acrimonious relationship evolving between the two super powers, the Canadian government saw investment in rocketry development not just being scientifically informed, but militarily prudent as well.

Once an assessment was made of the current status of Canada's rocket-based research program, the ACSR then concentrated on the development of plans to coordinate the nation's future activities. While the National Research Council and the universities were primarily interested in expanding their overall research in space science, Canada's defence sector was keenly interested in upper atmospheric research, atmospheric seeding experiments, atmospheric effect on the re-entry of objects, and larger-scale rocket studies and testing.[73] Together, it was feasible that both sectors could work together cooperatively; it was now just a matter of determining exactly how.

There were also many questions that remained unanswered. It was uncertain whether the NRC or the DRB should become the primary agency responsible for the overall design and construction of rockets and payload experiments, and the details of how nose cones with their delicate cargos would be transported to the remote firing sites in Churchill for checkout and launch had yet to be determined. The DRB and the Department of Defence Production (DDP) had initiated a series of studies on establishing some form of indigenous rocket supplier in Canada, perhaps in cooperation with industry, though it would be a year or more still before any detailed report on the subject would be ready. There were also growing concerns over launch facilities and range control issues, and whether or not there were enough human resources available to successfully engage in sustained Canadian launching activities. All of these concerns were raised during the first meeting of the ACSR, immediately providing the members with a number of issues requiring their best efforts and attention. Lastly, details of the first meeting and the nature of issues discussed by the group were treated as secret, with the Chairman reminding all present that no one should communicate anything to the press or public until it was deemed appropriate to do so.[74] The order, not intended with severity, was rather a subtle reflection of the great sensitivity with which any issues related to rocketry and space exploration were treated during the dawn of the space age. With that final comment from Dr. Rose, the first meeting of the Associate Committee on Space Research was adjourned.

Though the creation of the ACSR certainly focused the wide range of diverse interest in Canada's rocketry program through a single committee and empowered it with some political legitimacy, the executive committee of the ACSR was still faced with the difficult task of validating its plan to expand the current rocket program through government. Since the NRC and the universities had no way of initiating sustainable production, launch, and control facilities on their own, another central agency would have to provide these essential capabilities. The Canadian military already had some of these services in place, but hopes that the defence community alone would drive the expansion in Canadian rocketry development were quickly extinguished. At an executive meeting of the ACSR held on December 10, 1959, Dr. Keyston, then serving as Vice-Chairman of the Defence Research Board, explained why.

There was perhaps some hope that Canada's military would develop strategic missile systems similar to those then being developed in the United States. Canada was in the process of acquiring and equipping the nuclear-capable Honest John Surface-to-Surface Missile for its land forces stationed in Europe; therefore, it might have seemed plausible to the Canadian rocketry community that the RCAF would follow suit with its own launch system. Cabinet previously stressed to DND, however, that the delivery systems currently within the

---

[73] Ibid., p.5. Mr. Chinnick also described efforts by the CARDE to design and test a single stage rocket capable of reaching an altitude of 250km with a payload of 68kg, designated the Snow Goose and later, the Black Brant IIA.

[74] Ibid., p.3.

arsenals of Canada's Armed Forces were deemed more than adequate at the time for its limited stock of high-yield nuclear ordnance, and as such the government had no intention of spending large sums to develop larger multistage rockets for military use of any kind.[75]

It was an important if not somewhat confusing government decision. While the government encouraged some development of rocketry and missile systems, at the same time it was setting limits on the level of capability it expected to achieve and sustain. With these restrictions the DRB then had no official mandate to maintain expert knowledge in sophisticated rocketry and launch systems to advise the military staff, but was instead directed to employ the small group of ballistics personnel it was currently employing at the CARDE to act as its rocketry subject matter experts when needed.[76] Those personnel who had worked on the previous Velvet Glove project and were currently engaged with Black Brant program were expected to suffice if another program was started. If further subject matter expertise beyond this was required, the plan was to have the RCAF recall any of its personnel currently on exchange with space and missile systems divisions in the United States.

Nor could mass production of rocketry be justified for defence research alone. Since the CARDE only needed a small number of research rockets every year for its own programmed activities, there was no requirement to build any additional production or control facilities that larger rockets demanded. Neither was Canada yet in the business of conducting multiple satellite launches, the other main impetus for creating a sustainable indigenous launching operation. Unfortunately for the ACSR and other rocket program advocates, any further justification for an expanded rocket program would have to come from Canadian civilian and commercial, not military, needs.[77]

Another suggestion was tabled at the committee meeting. If the ACSR could provide Dr. Keyston with a submission stating the national need for a Canadian rocket, and hence the production and control facilities that go with it, he would be in a better position to have the DRB request that additional funding and resources be made available over and above those already committed to the Churchill Research Range. Further, Keyston was interested in knowing when a two-stage or larger rocket might be necessary for Canada, as this would influence the level of resources required. The issue of opting for the use of American rockets as an alternative was also discussed, but this option immediately raised questions among the group about availability, maintenance, fitting into their handling facilities,

---

[75] Proceedings of the First Meeting of the Executive Committee of the Associate Committee on Space Research, December 10, 1959, p.1, file 12798-2-40 pt.1, vol. 7841, RG 25, LAC; Outer Space – Proposal For Possible Canadian Initiative. (Confidential) memorandum 'International Implications of a Canadian Space Research Program', prepared for the Under-Secretary of State for External Affairs, dated August 27, 1959, p.2, File No.12798-4-40, Vol.1, RG 25, LAC. Evidence suggests the decision was taken for economic reasons, not as a result of disarmament ideology.

[76] Throughout this entire period DND maintained a small cadre of military and civilian personnel outside of Canada on official exchange to American missile-related programs and project offices.

[77] Ibid., p.1.

Canadian know-how, and others.[78] In the end, it was decided that the best way to proceed for the present was with a program based on existing Canadian capabilities and resources and then build from there.[79]

Subsequent meetings of the ACSR during the year 1960 produced a tentative budget and agenda for the first series of regular Canadian scientific rocket flights. The number of Black Brant rockets needed for initial university space science experiments was determined as was the substantiation for the development of a number of multistage versions of the rocket for larger scientific payloads. If this activity was sustained or expanded over the next couple of years, it was hoped that larger multistage launchers such as the American Scout rocket might then become a more viable option for Canadian use. The CARDE was again identified as the best government agency to lead Canada's rocketry development; however, it was acknowledged that commercial contractors would very likely also be required to support various production and integration stages of the program.[80] As well, it was determined that arrangements for the transfer of launching facilities at Churchill from the United States back to the Canadian military be initiated as soon as conveniently possible, with an expected handover date arranged for some time in 1962 or 1963.[81]

First Satellite: The Topside Sounder Project

The idea for Canada's first space satellite evolved within the space science community of the Defence Research Board and select Canadian universities during the early 1950s, but came to fruition after the U.S. National Academy of Sciences made a call for international scientific proposals to be flown on a satellite. Originally tasked by his government to coordinate both internal and international efforts in future ionospheric research, Dr. Lloyd Berkner issued an open invitation in early 1958 on behalf of the NAS to the Western scientific community to submit proposals to them for possible satellite-borne experiments that could explore the characteristics of the 'topside' of the ionosphere. A number of countries returned submissions for topside sounding experiments, including a rather clever offer from the Defence Research Telecommunications Establishment (DRTE) then located in Ottawa.

---

[78] Ibid., p.2. Interestingly, however, was the fact that the United States owned and operated the Churchill Research Range during the period when the ACSR intended to execute its own rocket research program there. As such Canada was essentially depending on American facilities and assistance already.

[79] It was also agreed that American Nike-Cajun rockets, which were previously employed by Canadian scientists at Churchill, would continue to be used for some future experiments.

[80] Proceedings of the Second Meeting of the Associate Committee on Space Research, April 8, 1960, p.7. File 12798-2-40 pt.1, vol. 7841, RG 25, LAC.

[81] First a fire, then politics, delayed the transition between governments until January1, 1966. The range continued to be funded and operated jointly by the United States and Canada until the late 1960s.

Fig. 2.16 DRTE main buildings at Shirley's Bay west of Ottawa c.1961

Within the DRTE, Dr. Eldon Warren, Dr. Colin Hines, Dr. John H. Chapman, and others had seriously considered the idea of building and launching a Canadian satellite for some time, as had Dr. Peter Forsyth and fellow science colleagues at the University of Saskatchewan. When the DRB began officially canvassing its own research establishments for potential interest, however, at first satellite experiments were not an easy sale to any of its defence scientists. Dr. Hines of DND's Radio Physics Laboratory (RPL) later recalled, "[DRTE Chief Superintendent Dr. James C.W.] Scott called me in and asked if I wanted RPL to do satellite-borne studies. I replied that in due course I did…He told me that there was a proposal for the DRTE to build a baseball sized chunk of equipment for topside sounding to be carried aboard someone else's satellite…I simply replied that perhaps it would be a good thing for us to do in another year or so."[82] Dr. Hines politely spurned the offer in order to properly address other priorities and existing commitments within the RPL, so Dr. Scott instead turned the project over right away to the Electronics Laboratory (EL) for consideration.

Dr. Eldon Warren, however, then currently involved in 'bottom side' sounding of the ionosphere, immediately warmed to the idea. He further developed the concept along with Dr. John Chapman and other talented scientists and engineers in the EL and organized a proposal and briefing for the DRB's senior managers to review. The initial design was

---

[82] University of Western Ontario Space Workshop. Supplementary Materials: Alouette Satellite, Comments from Colin Hines in adding details to a letter on the origins of the Alouette concept from Dr. P. Forsyth to Dr. J. Scott, September 22, 1981.

considered both original and very technically sound, and subsequently the Electronics Laboratory team got their proposal approved for submission to the NAS.

In their proposal, the engineers at the DRTE reasoned that their submission would meet with greater approval if the satellite demonstrated advanced engineering capabilities. Thus, instead of submitting a single frequency investigation experiment like other international organizations, the Canadians proposed to build a second-generation satellite that could employ a topside sounder capable of sweeping through a wide range of frequencies. When all of the interested agencies met at the American Space Science Board's Working Group on Satellite Ionosphere Measurements in Boulder, Colorado, in September 1958, the DRTE concept indeed demonstrated advanced engineering capabilities that none of the other invited parties had considered in their own designs.[83]

A special independent meeting to specifically consider the topside sounder experiment in detail was called by Dr. Henry G. Booker of Cornell University in October 1958, attracting the attention of at least seven interested groups including the Canadian team from DRTE.[84] Again their proposal met with a favorable response from the committee and the Canadian experiment was eventually selected as the preferred option. Unfortunately, however, just a short time after this happened the NAS cancelled its participation in these experiments and the offer to the DRTE "farm team" was officially rescinded.[85]

Undeterred by this setback, however, the DRTE engineers approached other interested parties in the United States in late 1958, this time their target being the United States Department of Defense. Former DRTE scientist and Communications Research Center historian Mr. LeRoy Helms wrote a popularized account of their visit to the Pentagon in 1958 as follows:

"It is said that they [Dr. Scott and Dr. Warren] were greeted at DoD by a big Texan USAF Colonel who listened attentively, Cowboy boots on his desk and smoking a large cigar, as these boys from the far north explained their proposal. When they had finished, he put his feet down, snuffed out his cigar and said, "Sure we'll launch for you. But there's a new agency just starting up – called NASA – who are supposed to do international space research projects. Probably won't amount to anything but you'd better go see them first. But if they aren't interested y'all [sic] come right back and we'll look after you."[86]

---

[83] Canada Department of Communications, *Alouette 1: Canada's First Venture into Space*. Ottawa: Information Services Booklet, June 1974, p. 6; and J.E. Jackson, R. Knecht, and S. Russell, "First Results in the NASA Topside Sounder Program", in NASA. *Publications of the Goddard Space Flight Center, 1959–1962, Volume II: Space Technology*. Washington: NASA HQ, 1963, p.41.

[84] Evidence infers that the meeting was held at Cornell University at Ithaca, New York, in October 1958. Details available from transcript of presentation made by Colin A. Franklin at the IEEE International Milestone in Engineering Ceremony, Shirley Bay, Ottawa, May 13, 1993.

[85] Transcript of presentation made by Colin A. Franklin at the IEEE International Milestone in Engineering Ceremony, Shirley Bay, Ottawa, May 13, 1993; see also DOC. *Alouette 1: Canada's First Venture into Space*, p.6.

[86] Leroy Nelms, "DRTE and Canada's Leap into Space: The Early Canadian Satellite Program", p.2.

Finally, in early 1959 Dr. Scott and Dr. Warren approached a newly-created United States civilian space authority named the National Aeronautics and Space Administration (NASA). Again the DRTE engineers impressed their potential sponsor – the fact that they already had a carefully planned proposal in hand from their previous attempts certainly helped – but NASA still received the proposal with a degree of skepticism. The American space agency experts were concerned about both the satellite power-supply and antenna design. The Canadian proposal needed to continually generate power throughout its orbit, and called for the deployment of four robust antennas measuring between 23 and 45 m in length yet weighing no more than 4.53kg. The Canadian engineers even later admitted, after the project had succeeded, that they initially estimated that the satellite would operate for no more than a few hours.[87]

Nevertheless, the Canadian scientific proposal was an excellent fit with the emerging mandate of NASA, and after further negotiation the two organizations agreed to cooperate on the launch of the Canadian topside sounder satellite experiment. The two countries made a joint announcement of the arrangement on April 20, 1959, with an official exchange of letters between the DRB and NASA following later in the year on August 25, November 18, and December 6, 1959.[88]

With the advent of orbital capability, Canadian defence scientists once again had the opportunity to expand their investigation of the upper atmosphere. The DND through the DRB in turn supported its ongoing endeavors as the Canadian government saw it as a positive way of acquiring advanced space technology that was then considered critical to future national defence. From the mid-1950s onwards, knowledge of the Earth's ionosphere played an ever-increasing role in the design of modern ballistic missile defence systems, now largely based on wireless communications. As well, the development of advanced weapons systems and defences was increasingly dependent on technologies similar to those developed for space systems including solid-state electronics, miniaturization, and computers. When considering the military value of the data derived from satellites at the time, namely imagery and scientific data, there was little difficulty for the defence science community in convincing their military and political leaders of the need for continued research and financial support.[89]

---

[87] J.E. Jackson et al., "First Results in the NASA Topside Sounder Program", in NASA, *Publications of the Goddard Space Flight Center, 1959–1962, Volume II: Space Technology,* pp.41–42. C.A. Franklin and others have suggested that NASA was originally highly skeptical of the DRTE proposal and internally assessed that the satellite would not likely function for more than a few hours. Others have suggested the move was a stalling tactic to ensure that a similar American project, known as S-48, would be launched first. Contemporary American- and Canadian-related archival evidence does not substantiate any of these claims beyond the anecdotal.

[88] Copies of these agreements may be found in J.H. Chapman, P.A. Forsyth, P.A. Lapp, and G.N. Patterson, *Upper Atmosphere and Space Programs in Canada: Special Study No.1.* Ottawa: Science Secretariat, February 1967.

[89] The idea that these innovations might be diffused from the military out into the civilian economy was another factor that drew sustained support. The government saw space technology not just for defence but also as an industrial opportunity. The challenge remained, however, on how best to exploit that opportunity.

Project S-27: Initial Planning

Concurrently with the making of formal political arrangements with Canada, NASA issued a request to the U.S. Central Radio Propagation Laboratory (CRPL) of the National Bureau of Standards to examine the DRTE (and other) proposals for their scientific merit and engineering feasibility. In considering the Canadian proposal, the CRPL suggested that it might be rather ambitious for a first attempt at examining the ionosphere. Instead, it offered that a fixed-frequency system should be launched as a first-generation experiment, while the DRTE swept-frequency system proposal was developed separately as a second-generation satellite. All parties agreed to this approach and planned for the first Canadian satellite launch sometime in the 1963 timeframe.[90] For the time being, NASA officially designated the Canadian DRTE experiment as Project S-27. Christening the satellite as *Alouette 1* came later.

In January 1960, NASA received another proposal jointly from the CRPL and the Airborne Instruments Laboratory (AIL) for a fixed-frequency experiment. The AIL was to design, construct, and test the satellite payload while the CRPL was to provide scientific supervision and analyze the resulting data. Work began on this experiment, designated S-48, on May 9, 1960. This project would now supposedly precede the DRTE satellite, but being similar in objectives and technique, it was still to complement the Canadian follow-on effort. The main difference between the two satellites was in the instrumentation. The American-designed S-48 experiment emphasized the study of cross-sections through the ionosphere employing both Canadian and American telemetry stations along the 75°W meridian. It also had a low resolution and a fast profile acquisition rate, employing six fixed frequencies providing a downward pulse transmission and echo reception in the 3–9 megacycle (Mc) range. By contrast, the Canadian S-27 experiment emphasized the investigation of polar, arctic, and auroral effects that produced the very complex ionospheric conditions existing over Canada. For this purpose, telemetry stations were established at specific points across the country. As well, instead of employing a low resolution and fast profile acquisition rate, the Canadian satellite intended to work in the exact opposite manner.[91]

The range and scope of the Canadian experiment soon attracted interest from the United Kingdom, which in turn expressed a desire to also participate in the topside sounder program. In return for access to data, the United Kingdom offered the use of three of its own ground telemetry stations – one at the Falkland Islands in the South Atlantic, one at Singapore, and a third station at Winkfield, England, for the collection and distribution of satellite data. Both NASA and the DRB accepted this offer and the United Kingdom officially joined the topside sounder program in 1960. The experiment in turn came under the overall management of the Goddard Space Flight Center, which coordinated the efforts of the various countries and organizations involved.[92]

---

[90] J.E. Jackson et al., "First Results in the NASA Topside Sounder Program", pp.41–42.

[91] Ibid., pp.42–44.

[92] L. Wallace, *Dreams, Hopes, Realities: NASA's Goddard Space Center, The First Forty Years*. Washington: NASA History Office SP-4312, 1999.

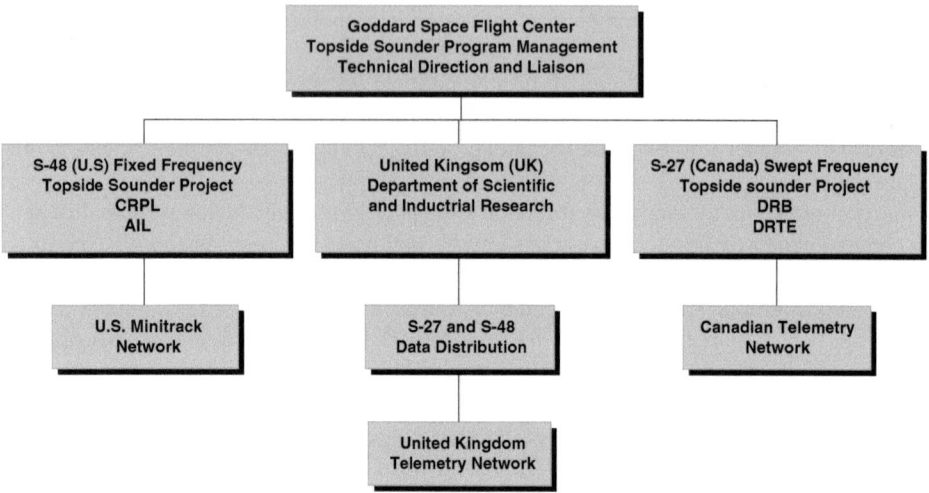

Fig. 2.17 Project management and oversight for the S-27 project, 1960–61

Design, Construction, and Testing

In the late 1950s, space systems design, construction, and testing remained a largely unproven process, drawing mainly from the existing experience of the aerospace industry. Management and control of technology research and development continued to evolve, and scientists and engineers were figuring out a viable process for coordinating large-scale technology development. At the same time, their leaders were looking for organizations that could innovate, learn, adapt, and sustain adaptation in order to achieve long-term scientific and technological goals.[93] It was within this context that the DRTE was expected to produce Canada's first full-scale satellite.

Understandably, the approach to designing Canada's first experimental satellite was at first conservative. The original concept called for a spacecraft size along the same lines as the first American satellites, roughly no bigger than a grapefruit or basketball. As the S-27 Project evolved, however, the size and complexity of the satellite also grew. The DRTE Electronic Lab satellite team also originally wanted only a single role for the satellite – to measure the state of the ionosphere directly below the satellite as it orbited the Earth. Yet as the project progressed, the designers and engineers added an additional three experiments to the satellite: a sounder receiver to measure cosmic noise, a frequency receiver for "listening" to radio noise in the range of 1–10 kilohertz, and an experiment to measure primary cosmic ray particles such as electrons, protons, and alpha particles, outside of the

---

[93] S. Johnson, *The Secret of Apollo: Systems Management in American and European Space Programs.* Baltimore: The Johns Hopkins University Press, 2002, pp.2–4.

Fig. 2.18   The Alouette 1 satellite undergoing vacuum chamber testing at the DRTE labs, c.1960

denser portions of the Earth's atmosphere.[94] In the end, *Alouette 1* resembled an oblate spheroid measuring approximately 107cm in diameter with a height of approximately 86.5cm. The total weight of the final design was just over 145kg.[95]

As with most satellites designed during this period, the basic shape of *Alouette 1* resembled a clamshell. The satellite design itself consisted of four main components: the structure, the spacecraft electronics, the antenna, and the four experiment payloads. The internal backbone consisted of a pair of thrust tubes, one above and one below, with a pair of circular structure disks between these which served as the mounting areas for the electronic components of the experiments described above. Between the two structure disks was also housed the four erectable antenna units. Surrounding the center of the structure was the solar cell shell that consisted of a pair of spinnings upon which were mounted support channels to carry the flat solar cell panels. Inside the spinnings were diaphragm rings that

---

[94] Technical specifications for *Alouette* may be found in DRTE Annual Reports 1962 through 1967; see also Department of Communications. *Alouette 1: Canada's First Venture into Space*. Ottawa: Information Services Booklet, June 1974. Numerous technical papers were also published by DRTE staff in various scientific journals, and lists of these are available within the DRTE Annual Reports.

[95] DRTE, *Alouette Satellite 1962 Beta Alpha One*. Shirley Bay: DRB, October 1962.

served the two-fold purpose of stiffening the spinning and providing the attachment between the spinning and the structure disks.[96]

The structure of the satellite, which in turn held all the other components together, posed the largest design challenge. It had to be able to withstand the violent vibrations of launch and the vacuum and radiation of space yet still function to collect and return its data. Several building materials were considered, including sophisticated materials such as micarta, polyurethane epoxy, Teflon, aluminized Mylar, unbonded glass fiber paper, etc., and even some not-so-sophisticated materials such as commercially-available brown wrapping paper. To cope with the vacuum conditions, the DRTE engineers had to avoid materials with high partial pressures that sublimed easily. As a result, aluminum was chosen as the primary material for the body of the satellite, held together with steel and stainless steel fasteners.[97] The structure of the satellite was completed by an cap at either end in order to prevent sunlight from striking through to the interior of the satellite and causing overheating of the electronic components inside.[98]

Temperature, or more importantly the control of temperature, affected every aspect of the design of the spacecraft. Everything from shape, to material used, to launch times, was considered to ensure that the satellite and its precious payload experiments remained within acceptable tolerances during launch and orbit. Up to that time many spacecraft had died quickly in the harsh environment of outer space as one side of the satellite literally cooked while the other half froze to death. The engineers at DRTE were most concerned about their satellite surviving long enough to return useful data to scientists on the ground.[99]

It was estimated that the most critical period for the satellite's life would be during and shortly after launch. Given the size and weight of the satellite as well as the intended orbit, the launch vehicle chosen for *Alouette* was an American two-stage Thor-Agena B booster. Already a workhorse in the American surveillance satellite program by 1960, the upgraded B variant was designed to push larger payloads off the launch pad into orbit and was a preferred choice for the *Alouette 1* experiment. The rocket's first stage – Thor – would propel the Canadian satellite for approximately 165 seconds into the upper atmosphere. Once the Agena B upper stage had successfully separated from the Thor lower stage and resumed its own burn, the Agena would shed its payload shroud leaving *Alouette* firmly attached yet totally exposed to the vacuum of space for two and a half minutes. During this time the satellite would not be spinning, therefore solar heat could not be equalized around the entire shell of the spacecraft. As a result, it was planned to schedule the launch for a time when the satellite would be in the Earth's shadow during its ascent, but even then there remained the problem of aerodynamic heating as *Alouette* reached towards outer space.[100]

As described above, the outer shell of the structure was designed so that the power plant was partly located on the outer surface. Employing 6480 small solar cells arranged in

[96] J. Mar and H.R. Warren, 'Structural and Thermal Design of the Topside Sounder Satellite', *Canadian Aeronautics and Space Journal*, September 1962, 163.

[97] Department of Communications, *Alouette 1: Canada's First Venture into Space*. Ottawa: Information Services Booklet, June 1974, pp.13–15.

[98] J. Mar and H.R. Warren, 'Structural and Thermal Design of the Topside Sounder Satellite', 163–164.

[99] Ibid., 166–167.

[100] Department of Communications, *Alouette 1: Canada's First Venture into Space*. Ottawa: Information Services Booklet, June 1974, 12–14.

groups of forty-five that almost completely covered the outside of its structure, *Alouette*'s outer skin was designed to charge the batteries located inside the satellite. In order to provide adequate charging currents regardless of the satellite's orientation with respect to the Sun, it had to be able to consistently expose the same number of solar cells at all times.

Fig. 2.19 Dr. John H. Chapman poses with the Alouette satellite at DRTE labs c.1961. A brilliant scientist and satellite pioneer, he would lead Canada's space program development until his early death in 1979

The solar cells were then further covered with paper-thin chips of glass using a special non-reflective and spectrally selective coating that passed ultra-violet light but not infrared light which could cause heating of the spacecraft. In a sense, the solar cells and coating acted as thermal insulators but still let the much-needed light through,[101] and also protected the solar cells from potential micrometeorites and harmful radiation.[102] This constraint influenced the overall semi-spherical design of the satellite, and the power

---

[101] The glass covers, attached to the solar cells with an epoxy-based adhesive, together with the satellite's spin were designed to keep the cells within a temperature range of –20 degrees Celsius to +50 degrees Celsius. In operation, the spacecraft temperature at times rose as high as +75 degrees Celsius. See also DOC, *Alouette 1: Canada's First Venture into Space*, p.12.

[102] J. Mar, 'Meteoroid Impact on the Topside Sounder Satellite', *Canadian Aeronautics and Space Journal*, November, 1962, 237–240.

requirements of the payload in turn dictated the overall surface area size of the spacecraft. Lastly, the semi-spherical shape also contributed to the temperature control of the spacecraft. The intention for *Alouette 1* to have a slight spin in orbit, roughly two revolutions per minute, not only reduced spacecraft oscillation but also helped avoid the danger of the satellite getting "hot spots" from overexposure to direct daylight.[103]

The design of the interior electronic components highlighted some of the technological challenges still faced in the early 1960s when dealing with advanced systems. Although extensive miniaturization of the electronic content was not considered absolutely necessary for *Alouette 1* it was desirable as much as possible, if for no other reason to lessen the weight of the craft and/or potentially make more room for the onboard experiments. To consider the employment of vacuum tube technology meant possibly providing more power but doing so at the expense of reliability, higher power consumption, greater weight, and ultimately the need for more space in an already small satellite. Instead, the *Alouette 1* team sought out the most modern electronics available to them at the time, building the spacecraft using then ultra-modern solid state transistors which provided less power but were more reliable and provided greater semi-conducting efficiency.

The antennae posed a particular challenge for the engineers at first but in the end resulted in a novel and uniquely Canadian solution. The *Alouette 1* design called for four erectable sounding antennae; two crossed dipoles a hundred and fifty feet from tip to tip, and two crossed dipoles measuring seventy-five feet from end to end. All four antennae were designed to extend in a traverse plane at the center of the satellite exactly 90° to one another, and had to be housed within the satellite in such a manner that the entire payload would fit into its confined payload bus fairing atop the rocket. At first inspection, there was no way the antennae would fit inside the satellite, and they were too fragile to fold and pack alongside the main structure. Further, a multijointed object meant that any one of these could fail, resulting in a partial deployment of the antenna or even no deployment at all. All antennae had to deploy perfectly for the satellite to complete its assigned task.

The solution to the problem was derived from a tool first conceived and developed by Mr. George J. Klein, an engineer with the National Research Council since the Second World War. A taped length of spring steel, previously heat treated and opened flat, was wound on a drum and placed within an antenna assembly with a guide sleeve and an electro-static shield. Altogether, the antenna assembly unit was no more than approximately a foot in length and fit comfortably into the satellite structure housing. Once in orbit, the antenna deployed by pulling the spring steel of its storage drum by means of a simple drive belt, and once guided through the antenna sleeve the metal tape sprang back into its natural tubular shape with about 180° of overlap. Even at a hundred and fifty feet, the antenna proved extremely robust with considerable bending strength.[104] In fact, the concept, later known as the Storable Tubular Extendable Member (STEM) system, worked so well that this method of antenna deployment was employed on nearly all subsequent Canadian and American satellites and spacecraft throughout the next two decades.[105]

---

[103] DOC, *Alouette 1: Canada's First Venture into Space*, 14–20.

[104] DRTE, *Annual Report 1962*. See also J. Mar and H.R. Warren, "Structural and Thermal Design of the Topside Sounder Satellite", 164.

[105] Grease and oil were used only in the sealed ball bearings of the antenna extension mechanisms; they had only to operate successfully once to deploy the antennas.

Fig. 2.20 DRTE employee Mrs. V. MacDowell holds a partially deployed STEM antenna unit c.1962

Testing this system on the ground, however, was likewise a difficult proposition. The philosophy employed by NASA at that time was that an identical prototype of the flight version of the unit had to survive a program equaling a hundred and fifty percent of the highest expected design loads. As well, there were several vibration tests that had to be passed where the dynamic balancing of the satellite was proven. Also with *Alouette* and subsequent Canadian-built satellites employing STEM technology, the antennae had to be tested, which in turn proved a real challenge in Earth's gravity. An instrument known as a full extension rig was employed to extend the antennae along a series of hung carriers, while engineers measured rate of extension, drive motor current, and voltage.[106]

So much of the *Alouette* mission depended on the successful deployment and operation of the antenna. To further ensure their success, the design was flight tested in June 1961, when a pair of the STEM were mounted in the nose cone of a U.S.-built Javelin rocket and launched to satellite altitude. Though some improvements were indicated as a result of the test, the experiment was overall very successful and considered proven for a final deployment of the STEM antenna on *Alouette 1* itself.

---

[106] J. Mar, J. and H.R. Warren, 'Structural and Thermal Design of the Topside Sounder Satellite', 165.

The Alouette 1 Satellite Launch

Canada's first satellite project remained on schedule and the completed spacecraft was moved to Vandenberg Air Force Base, California, in the late summer of 1962 to undergo final checkout and transfer to the launch pad. *Alouette 1* was ultimately scheduled for a night flight; the planned launch window of 2330hrs to 0130hrs on September 28/29, 1962 was finally chosen, as it allowed for the mission scientists to get as many soundings as possible on the first few orbits after lift-off before *Alouette 1* was fully exposed to the Sun. There was still some concern that the spacecraft might suffer a catastrophic malfunction or failure when it heated up for the first time, so the intent was to ensure that at least some data were returned as soon as possible. This flight plan also allowed mission scientists to ensure that the onboard passive temperature control system was actually working, as it should, by slowly exposing the satellite to progressive amounts of sunlight with each passing orbit.

Fig. 2.21 Dr. John H. Chapman stands before the Thor-Agena B rocket that carried Alouette 1 into orbit, September 1962

After years of planning, design, construction, and testing, *Alouette 1* was launched into outer space from California just before midnight on September 28, 1962. Dr. John Chapman, head of the DRTE *Alouette* team, was there to see her into orbit and later was quoted as saying, "I had my fingers crossed, my legs crossed, and everything else crossed.

At that time, there was still a fifty percent chance of failure in launchings."[107] Dr. Chapman and his team, however, had little to worry about. The launch of the Thor-Agena B took place without difficulty and lit up the evening sky as the American booster lofted its *Alouette* payload into orbit. Accompanying *Alouette 1* into orbit was another smaller payload known as TAVE – the Thor-Agena Vibration Experiment. This small technology satellite returned valuable data that assisted engineers in continuing their improvements to the design of the rocket and develop further versions for other use.

As the rocket rose into the sky all aspects of the flight were closely monitored from the ground. Arrangements were made prior to launch by NASA through the United States Navy to have a tracking ship pre-positioned in the Indian Ocean to monitor *Alouette*'s antenna extension once it reached orbit, but at the critical moment there were equipment and operator troubles aboard the designated ship. As a result, the Canadian launch team were kept in suspense for some time after launch until a later report arrived from a ground tracking station in Johannesburg, South Africa, verifying that the *Alouette 1* antennae had fully extended as the design had intended.

The satellite that some NASA scientists had felt would last but a few hours went on to surpass all expectations. *Alouette 1* enjoyed a near textbook deployment and continued operating well beyond when experts thought it might fail. Within weeks of the launch, scientists on the ground were flooded with detailed data on ionospheric structure collected by *Alouette*. Having "optimistically" planned for a three-month mission (*Alouette* was designed for a nominal lifespan of one year), the scientists collected and processed as much data as possible, but it soon became apparent that the satellite was functioning well and would continue to deliver data for some time. When *Alouette* passed its first birthday in space senior administrators at both NASA and the DRTE were impressed. When the satellite celebrated its fifth anniversary in orbit, both groups were simply amazed. Even then, no one would have presumed that the satellite would continue to function for another five years.[108]

The success of *Alouette* was certainly a great achievement, one that the Canadian defence and scientific community could be very proud of. Compliments came from most international and national scientific organizations, but particularly from the skeptics at NASA and the NAS who earlier had supposedly questioned some aspects of the mission. The admiration for Canada's *Alouette 1* was stated very succinctly in the 1963 publication series from NASA's Goddard Space Flight Center. One author wrote:

"The success of the NASA program of topside ionosphere studies is evidenced by the considerable amount of knowledge obtained from *Explorer VIII*, *Ariel 1*, and high altitude rocket soundings. Perhaps the most spectacular of these accomplishments to date has been the Canadian swept-frequency topside sounder, *Alouette*,

---

[107] T.R. Hartz and I. Paghis, *Spacebound.* Ottawa: Department of Communications – Minister of Supply and Services Canada, 1982, pp.60-61.

[108] Alouette was actually decommissioned after its tenth year in orbit and shut down by ground command. Arguably, the satellite would have otherwise continued to function for possibly many more years even. It was a remarkable technological achievement.

which will probably yield more data about the upper ionosphere…than all the other programs combined."[109]

The experimental satellite project also demonstrated what Canadian defence scientists could achieve when given a clear mandate and the resources with which to carry out the task. The design and construction of *Alouette*, however, should not be perceived as simplistic or something that was easily repeated. The initial concept in place when Dr. Scott was Chief Superintendent called for the design and construction of a very small payload. But as Dr. Hines noted many years later in an historical interview, as the size of the project grew so did the political and financial headaches:

> "I don't recall when the baseball expanded into a basket ball, as it did in mid preparations, or then to the ultimate size of *Alouette* itself, requiring all the time more and more resources. [Dr.] Frank [T.] Davies replaced [Dr.] Scott as Chief Superintendent, DRTE, part way through, and from time to time bitched about this Albatross Scott had hung around his neck. The program ultimately took over EL [Electronics Lab], all of the finances and manpower that could be pulled together inside DRTE, and ultimately required massive subsidy from the DRB itself."[110]

Undoubtedly, the *Alouette 1* satellite came at a considerable cost – approximately $3 million – and it drained resources, scientists, and engineers from all other projects at the DRTE; those not involved with *Alouette* were understandably somewhat resentful of the impact that the high-profile satellite project had on other quieter, less spectacular, fundamental defence science research activities ongoing at that time. Some even suggested later that the success of the *Alouette* satellite ultimately contributed directly to the demise of the DRTE in the late 1960s when all satellite expertise was transferred out of DND over to the newly-created Department of Communications.[111]

Nevertheless, the data returned from the first *Alouette* satellite about the ionosphere, aside from proving to be extremely valuable to future defence telecommunications research, also revealed that there was much more still to learn about the upper atmosphere than at first realized. In turn, this meant that there was a requirement for additional experiments involving other ionospheric parameters, "which could only be satisfied by subsequent satellites."[112] Canada's scientists and engineers were keen to build and launch those satellites.

---

[109] L.J. Blumle et al., "The National Aeronautics and Space Administration Topside Sounder Program". *Publications of the Goddard Space Flight Center, 1963, Vol.II: Space Technology.* Washington: U.S. Government Printing Office, 1963, p.13.

[110] *UWO Space Workshop.* Supplementary Materials: Alouette Satellite, Comments from Colin Hines in adding details to a letter on the origins of the Alouette concept from Dr. P. Forsyth to Dr. J. Scott, dated September 22, 1981. Accessed at http://quark.physics.uwo.ca/~drm/history/space/space_history.html.

[111] Leroy Nelms, "DRTE and Canada's Leap into Space: The Early Canadian Satellite Program", p.23.

[112] T.R. Hartz and I. Paghis, *Spacebound*, p.64.

Fig. 2.22 Satellite telemetry receiving station personnel pose for a staged photo as they process data from the Alouette 1 satellite, December 1962

Conclusion

While no official Canadian space policy was put in place during the early 1960s, neither was any policy that might compromise its existing missile, rocketry, and space cooperation efforts both at home and with its American partners. Instead, the DRB and NRC signed a number of cooperative agreements on space technology development that traded Canadian niche capabilities for more general access to the larger American space program.[113] These agreements were critical as both of Canada's current efforts at that period, the DRB space science program (*Black Brant* and *Alouette*) and the RCAF Space Defence Program, relied heavily on American support for assured access to outer space.

---

[113] For details of Canada-U.S. space cooperation agreements between 1945 and 1974, see A.B. Godefroy, 'From Alliance to Dependence: Canadian-American Cooperation Through Space, 1945–1999' (Kingston: MA Thesis Royal Military College of Canada, 1999), and by same author, *Allies in Orbit: Canadian-American Defence Cooperation Through Space* (Ottawa: Department of National Defence, Directorate of Space Development, 2000). See also J.H. Chapman et al., *Upper Atmosphere and Space Programs in Canada: Special Study No.1*. Ottawa: Science Secretariat, February 1967.

Apart from those agreements already connected to the Churchill Research Range, several other accords were exchanged between the DRB and NASA concerning the evolving *Alouette-ISIS* program between 1959 and 1964. Three of these were Letters of Agreement (LOA) directed at the *Alouette* program in 1959, formalizing Canada's first official cooperation effort in the launching a satellite and exchanging data obtained from a satellite. A Memorandum of Understanding (MOU) was similarly put in effect in May 1963 between the DRB and NASA concerning the follow on *ISIS* satellite series, a joint program in ionospheric research by means of satellite, which was followed up in May 1964 with a further exchange of notes. The *ISIS* program (detailed below) would continue until 1971.

Canada also signed a number of accords related to satellite tracking. In 1960, an agreement was completed concerning the placement of an American satellite tracking station at St. John's, Newfoundland, with further amendments made to the agreement in 1962 to update the equipment and convert the existing facility into a minitrack station (essentially a modernized version of the original tracking facility). From this agreement Canada gained access to the scientific data obtained from the tracking station and very likely also received certain space intelligence from the American collection point.

General agreements and exchanges of notes with respect to communications satellites were made between the two countries in 1963 and 1964, the latter agreement aiming to secure arrangements for Canada to cooperate in the eventual establishment of a global commercial communications satellite system (GCCSS).[114] At the time, no country in the world had yet established a domestic communications satellite capability, with the United States only having just passed its own Satellite Communications Act in Congress in 1962. Canadian participation in this effort led to its further involvement with the recently-created Communications Satellite Corporation in the United States, and later with the International Telecommunications Union (ITU) and the Interim Communications Satellite Committee. The latter organization was responsible for the establishment of the space segment of the GCCSS generally known as INTELSAT.[115]

Lastly, it is equally important to note the fact that Canada's cooperative space endeavours during this period were very nearly exclusive. With the exception of some aspects of its satellite communications evolution, between 1957 and 1967 Canada did not sign any agreements regarding rocketry or space development with any other country, and had even retreated on more than one occasion from entering into serious negotiations with France and other Western European countries on space cooperation at this time. Though it was interested in the exploration of various cooperative options with Western Europe, Canada simply would not enter into any official agreements that might compromise its already established bilateral arrangements with the United States. A proposal for launcher and satellite cooperation with France and West Germany in 1967, for example, was declined as it appeared to compete for services already provided by the United States. Subsequent proposals from these two countries for agreements were also stalled as Canadian officials

---

[114] For technical and legal details of these agreements see Chapman Report, pp.145–200.
[115] Ibid, 2.3 Satellite Communications, pp.14–15.

suggested it would be inappropriate to begin signing space cooperation accords with West Germany when none yet existed for cooperation with closer allies such as Britain.[116] Britain had already received some special consideration in this respect, however, but then cooperation existed only in the exchange or sharing of information and scientific intelligence rather than actual projects or programs, such as had been done previously with the telemetry received from *Alouette*. In the end, Canadian-American space cooperation surpassed all other efforts.[117]

This held true even at the international level. After the failed attempt at instituting international control of space through the UN, Canada maintained only two space-related memberships through this organization. One was with the Committee on Space Research (COSPAR), whose Canadian membership was held by the NRC's Associate Committee on Space Research, and the second was the UN General Assembly's Committee on the Peaceful Uses of Outer Space, whose Canadian membership was organized through the Department of External Affairs.[118] At the time, both committees were more politically esthetic than functionary, as space was still a Cold War battleground and was not yet ready for international laws or control. Essentially impotent, neither committee had any serious influence on the development of Canada's space agenda until the end of the decade.[119]

---

[116] Committees and Boards – Canadian Military Space Group pt.1. Proposed Agreement with France Regarding Cooperation in Space Science dated November 30, 1967. Acc 83-84/232 Vol.46 File 1150-110/M16 pt.1, RG24, LAC; and 'Proposed International Space Science Agreement – France,' dated December 5, 1967, Vol.46 File 1150-110/M16 pt.3, RG 24, LAC. Part of the problem was also due to the fact that the DRB, Canada's de facto 'space agency' was itself an arm of the country's Department of National Defence. While this did not present serious problems when dealing with the United States with whom Canada already had a close defence and security relationship, the DRB's involvement in defence matters would very likely send the wrong political signal when dealing with other European nations.

[117] It was suggested that the appearance of Canada's space program depending too heavily on the United States might hinder its ability to form relationships with other European countries, but there is no evidence of action being taken to distance Ottawa from Washington in order to appease European partners or improve options for space cooperation at this time. It simply was not in Canada's national interest to do so.

[118] For a detailed history of COSPAR see Gordon Shepherd and Agnes Kruchio, *Canada's Fifty Years in Space: The COSPAR Anniversary*. Burlington: Apogee Books, 2008.

[119] Canada's signing and ratification of the 1967 Outer Space Treaty is often identified as the origin of Canada's non-weaponization of space philosophy, but the fact that Canada joined such space committees and endorsed various resolutions is often given much more weight than it deserves. Given the nature and scope of the Canadian space program at this time, both the 1967 Outer Space Treaty and the 1968 Rescue Agreement made little difference to Canada's overall agenda, and were treated with no more grandeur of purpose than any other diplomatic activity at that time. For general notes on Canada and space law see A. Beesley et al., "Canada's Contribution to Outer Space Law and Arms Control in Outer Space", *Space Strategy: Three Dimensions*. Toronto: Canadian Institute of Strategic Studies, 1987, pp.94–110.

3

Challenge and Commitment: Canada's Space Program in Transition, 1964–1974

Having successfully broken the bonds of the Earth in late 1962, Canada entered a critical stage in its emerging space program just as the country itself also entered a period of widespread government transformation. Approximately halfway through Prime Minister's Diefenbaker's tenure in office, a Royal Commission on Government Organization was appointed to examine the whole of government services with a view towards identifying and eliminating duplication of effort and uneconomic operations and practices, and to recommend improvements in decentralization as well as more efficient management practices. Chaired by John G. Glassco, the Glassco Commission as it soon became known, included a number of recommendations concerning the future of Canadian science and technology policy, management, research and development. Events such as the Cuban Missile Crisis and the subsequent federal general election of 1963, however, overtook much of the commission's work and its initial recommendations concerning changes to Canada's space program were not acted upon until some years later.

In the meantime, the frustrations grew. The continued absence of clear senior-level political direction or motives for its space program meant that Canada's early space advocates were largely left to their own devices in promulgating national interests, research, and development during most of the 1960s. Canada–U.S. bilateral space agreements continued to be signed on a project-by-project basis, with no real attempt to weave each agreement into a larger Canadian agenda or vision for outer space. One might argue that this piecemeal approach *was* the agenda, for Cabinet appeared content with letting lower-level defence organizations like the DRB and the Royal Canadian Air Force lead in their own way, or was otherwise too preoccupied with other matters of state to devote more energy to defining a clearly Canadian role in outer space. Not even direct advice from senior science and engineering advisors could greatly change the nature of evolution in Canada's space program at this time. Evidently, the relationship between science, engineering, and government mirrored that which then existed between government and defence – largely ignored except when necessary – and even then only when bureaucracy needed it to satisfy a political purpose rather than a military or scientific goal.

© Springer International Publishing AG 2017
A.B. Godefroy, *The Canadian Space Program*, Springer Praxis Books,
DOI 10.1007/978-3-319-40105-8_3

   Though repeated attempts were made by the DRB, the NRC, and the RCAF to encourage Cabinet to approve a national space policy and create an official Canadian Space Agency, each time the government responded that it preferred to proceed with the status quo. In July 1961, Dr. Donald C. Rose submitted a proposal to government to create a dedicated space research organization in Canada, but his argument bore no fruit. Further efforts to solidify an official space policy after the successful launch of Canada's first satellite, *Alouette*, in September 1962, a time when enthusiasm for space exploration was high within government, also failed to gain increased Cabinet attention once the initial enthusiasm had waned. On December 3, 1962 the DRB issued a persuasive position paper arguing for the creation of a Canadian Space Agency (CSA) and an official space policy to direct future efforts. The study cited the myriad space activities that were already under way in Canada and the obvious advantage of coordinating government and industry efforts as reasons for creating the CSA, but despite laying out the argument so clearly the government still stalled in taking any further decision on the issue. Cabinet just was not interested.

   The slower pace of Canadian aerospace technology development ultimately resulted in some of Canada's best and brightest moving south of the border to work in the much larger American space program. Some of these talented men and women went on to occupy very senior posts in NASA and elsewhere. For example, Bryan Erb, Stanley Galezowski, Bruce Aikenhead, Owen Maynard, David Ewart, and Dwight Owen Coons, all of whom at one time or another were involved in the AVRO Arrow fighter jet program, left Canada when the program was cancelled and went to work for NASA. Of this group, Maynard played a very important role in American space exploration and went on to senior tenures in both the Mercury and the Apollo program. In 1963, he was appointed head of the Lunar Excursion Module (LEM) engineering office in the Apollo Program Office. In 1966, while serving as the Apollo Program Chief of Mission Operations Division, he devised the 'A to G' sequence of Apollo test flights that would ultimately lead to the first lunar landing.[1]

   The motivation behind the latest request from the DRB to form an official civilian-led space agency was largely the result of changing circumstances. First, the DRB had a simple desire to see a Canadian Space Agency born. Though it enjoyed its position as the primary point of contact for Canada's scientific and civilian oriented space projects and cooperation, it itself was not a civilian establishment like NASA. This fact became an especially noticeable problem as the DRB ventured to expand its cooperative ventures with countries other than the United States. "Due to DRB involvement in defence matters", one memorandum to government cited, "its appearance to other nations when participating in such political arrangements...will have a different connotation than would the participation of a non-military research organization."[2] The DRB was acting like a national civilian space agency, but it was not and should not have been. Such activities split management priorities and invited a degree of risk, and it also took the DRB away from its primary mandate of providing for Canadian defence research and development.

---

[1] The story of these Canadian engineers and others is explored in detail in Chris Gainor, *Arrows to the Moon: AVRO's Engineers and the Space Race*. Burlington: Apogee Books, 2001.

[2] Committees and Boards – Canadian Military Space Group pt.3. (Canadian Eyes Only) Memorandum from F.S.B. Thompson, Director Requirements communications to SA/CTS - Proposed International Space Science Agreement – France, dated December 5, 1967. Acc 83-84/232 Vol.46 File 1150-110/M16 pt.1, RG 24, LAC.

In addition, the RCAF plan to formalize its own military space program, and the competition for scarce material and human resources it would create, certainly pushed the DRB to attempt to stake an early claim on the future national space agenda. Again, playing the role of civilian space agency encouraged it to see itself in direct competition with an armed service that instead it should have been directly supporting. Here one can also see the link between national defence and national welfare obscured and the relationship between science, engineering, defence, and government confused. Why did the DRB want to compete against a defence space agenda when it was itself a defence research organization? Obviously, somewhere within that defence science community the Faustian bargain had crumbled and DRB staff started seeing themselves no longer as a purely defence-oriented research establishment. Even if only a small portion of the board actually felt that way, it would explain why the DRB senior management perceived the RCAF military agenda as anathema to its own civilian-oriented priorities. The advent of Canadian government reorganization in the early 1960s could have rectified the matter, but in the end it only served to further complicate the situation between the DRB and the department it served.

Still, the ongoing Royal Commission on Government Organization studies also identified the serious fragmentation of Canada's space program at the time. Though seemingly less interested in considering Canada's military space plan, which may have been addressed under the Glassco Commission's reports dealing with the Department of National Defence, the committee was very concerned with amalgamating Canada's civilian and commercial space interests under a new single government authority. In January 1963, for example, the Glassco Commission called particular attention to the inefficiency and duplication of effort within Canada's non-military space research agenda, and sought recommendations for consolidating all civilian efforts into a single agency. The advocates for a single space agency were pleased. The commission and a subcommittee drew up guidelines for the analysis and a group of specialists was formed to investigate and assess four main issues. These included:

a. All upper atmosphere research conducted by the Defence Research Board which is not of direct significance to defence, whether or not it involves the use of rockets;
b. All upper atmosphere and satellite research now conducted in other divisions of the National Research Council, for example, that on cosmic rays;
c. The research now conducted in the Radio and Electrical Engineering Division on meteors and satellites and the operation of minitrack instrumentation; and,
d. Research on satellite communications on behalf of the Department of Transport, which should not be encouraged to set up its own facilities.[3]

Interestingly, the commission also provided a final recommendation, despite the fact that research had not yet begun on the matter, that "all non-military space and telecommunications research could be transferred to the Radio and Electrical Division of the National Research Council."[4] Such situated outcomes in advance of the completion of studies were common throughout Canada's review of national science and technology policy during this period, and indicative of wider opposition and public accusations that most of the science policy review was prejudiced to arrive at a predetermined solution for

---

[3] Canada. The Royal Commission on Government Reorganization, Vol.4, p.274.
[4] Ibid., p.274.

the future benefit of certain parties and agendas, and not necessarily for the advantage of future national research and development writ large.[5]

Amazingly, not even the Glassco recommendations could spark serious government interest in better situating the country's national space policy. The atrophy in this field of endeavor continued for another three and a half years, during which time both the DRB and the RCAF constantly clashed over who had priority in determining the direction and focus of Canadian rocketry and space programs. The DRB was clearly making headway on its own schedule, with the *Alouette 1* satellite continuing to operate for years, well after its expected lifecycle of several days, and the preparations for the launch of the *Alouette 2* satellite were well under way. In contrast, the RCAF Space Defence Program (SDP) remained largely in its conceptual stage during this time, while its proponents dodged constant challenges from the DRB, all while facing the additional difficulty of convincing Cabinet-level decision makers of the program's long-term political and military value.

The National Space Study

In early 1965, the ongoing debate over the future direction and shape of Canada's space program finally reached a watershed. The country was at an impasse in the evolution of its space program and faced the possibility of not being able to realize any further achievements beyond those currently under way. Though Canadian rocketry and outer space activities had demonstrated some degree of success thus far, the country's piecemeal approach to the maturation of its own program and the splintered management and control of its projects between several different departments and interests caused serious political strife, wasted valuable technical and human resources, withered funding, and retarded long-term national science and technological development.

No longer could the various proponents of space exploration and exploitation depend on the traditional Canadian relationship of individual scientists, engineers, administrators, or organizations dealing directly with those in government who controlled the funding. That era had passed. The advent of space exploration and the ongoing international space race between the superpowers to land a man on the moon by the end of the decade clearly demonstrated the enormous impact that science and engineering could have on wider political issues, government programs, and social values. It also brought public attention to the ever-increasing cost of 'big' science, and invited serious questions at home about how Canadian science and technology was governed, managed, financed, and directed.

In mid-1966, the Science Secretariat at last turned its attention towards Canada's space program and policy during its final days in existence. In May 1966, just as the Science Council of Canada Act Act introduced to the House of Commons was receiving assent, the Secretariat commissioned a comprehensive study based on those parameters first identified back in the 1963 Glassco Report. One would have expected both the RCAF and the DRB to receive equal attention in a national-level assessment but, as detailed in the Glassco Report, the recommendations specifically sought to address only non-military issues. Though it made perfect sense to evaluate a national space policy in primarily civilian

---

[5] G. Bruce Doern, *Science and Politics in Canada*, pp.211–224.

terms – after all, Canada was governed by civilian authority – this decision to focus just on civilian applications was a determining factor in ensuring priority and success for the DRB's civilian agenda over the RCAF Space Defence Program.

In May 1966, the Canadian government tasked a group of three distinguished scientists – Dr. Peter A. Forsyth (University of Western Ontario), Dr. Phil A. Lapp (de Havilland Aircraft of Canada Ltd.) and Dr. Gordon N. Patterson (University of Toronto) – to the Science Secretariat study group under the leadership of one of Canada's leading space engineers, Dr. John H. Chapman.[6] Already a well-known and respected figure in Canada's fledgling space community, Dr. Chapman was a skilled radio scientist and liaison officer then working in the Defence Research Telecommunications Establishment in Ottawa. He had overseen most of Canada's bilateral space technology cooperation with the United States since 1958, and was largely responsible for the overall success of Canada's *Alouette* and *ISIS* satellite projects.

The study group – also referred to as Ad Hoc Task Force and/or the Chapman Task Force – collected data by soliciting requests from various stakeholders in Canadian rocketry and space activities as well as through numerous public hearings and interviews. Additionally, written submissions arrived from universities, industry, government departments and agencies, technical associations, and various professionals associated with Canadian space projects. The study group presided over hearings at Halifax, Quebec, Montreal, Ottawa, Toronto, London, Winnipeg, Saskatoon, Calgary, Edmonton, and Vancouver between June 30 and October 31,1966. In total, the group received 112 briefings and presentations from a broad range of interested groups during these hearings.[7] These exchanges generated thousands of pages of ideas, options, technical notes, and status reports on the current state and possible future direction for Canada's space program. In terms of looking ahead to Canada's future space activities specifically, three points in particular repeatedly surfaced throughout the study group's inquisitions:

1. There was a need for a centralized organization for space activities in Canada;
2. There was a need for Canadian satellites for domestic telecommunications by 1970 or 1971 at the latest;
3. There was a need for an indigenous Canadian satellite launching capability.

The first and third items were old arguments made many times before. The Canadian space research community had advocated for the creation of its own space agency since nearly the very beginning of Canada's involvement in the space age, and the indigenous launch capability issue first appeared on the agenda of the Associate Committee for Space Research (ACSR) at its first meeting back in 1959. The second bullet, however, 'the immediate need for telecommunications satellites', was a new recommendation item and almost certainly included as much to suit the agenda of those who were conducting the program review as those who were being interviewed. Both Dr. John Chapman and Dr. Peter Forsyth had a long relationship with the government experimental communications satellite program and were keen to continue political support for research and development in

---

[6] Science Council of Canada, *A Space Program for Canada.* Report No.1 (Ottawa: Queen's Printer, July 1967), p.2.

[7] Ibid., p.3.

this field. In fairness, however, it was a logical priority for the committee to include, given Canada's expertise and rather limited resources, and it was a feasible goal around which to base the beginning of an official policy, program, and agency. Canada had achieved considerable success in satellite communications technology development throughout the 1960s and, when compared to the rest of its fledgling program, this domain was also the most advanced in terms of being a mature program. There was little doubt that telecommunications would receive considerable focus in the final report.

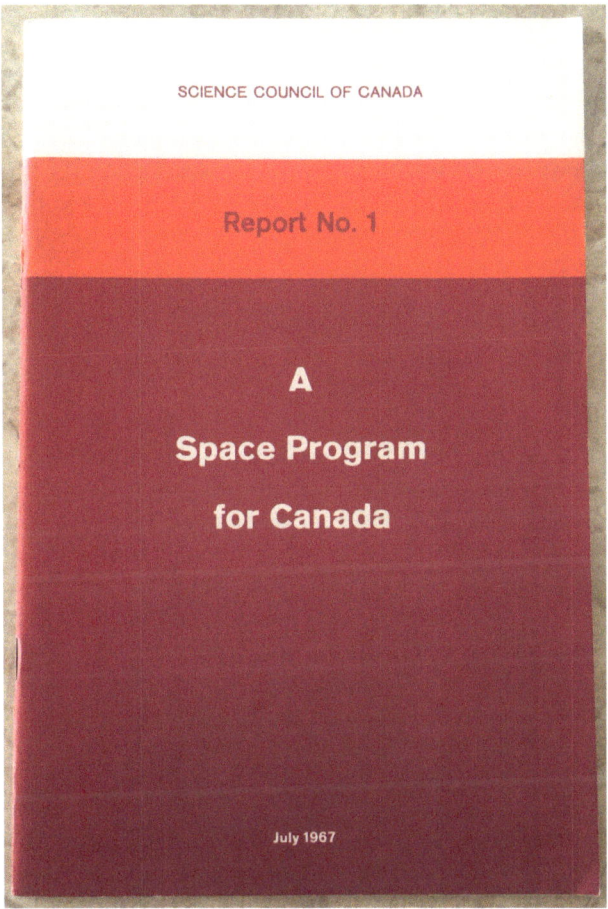

SCIENCE COUNCIL OF CANADA

Report No. 1

A

Space Program

for Canada

July 1967

Fig. 3.1 Science Council of Canada Report No.1, *A Space Program for Canada*, otherwise known as the Chapman Report, 1967

Preliminary copies of the special study, commonly referred to within government circles as the Chapman Report, were distributed to members of the Science Council of Canada (SCC) in advance of official meeting planned for January 16, 1967, to discuss the group's findings. At 258 pages, it was a thick and detailed report, divided into three separate sections. The first part described government, university, and industrial space projects created and planned for between 1961 and 1971. Every aspect of Canada's space program was included in this portion, from detailed descriptions of scientific experiments and equipment to budgetary summaries and human resource allocations. Similarly, the second part of the study presented detailed copies of the texts of space agreements, memorandums of understanding, and other international arrangements affecting or concerning Canada and in effect as of 31 October 1966. The last part, as an appendix to the first section, presented a detailed and comprehensive review of Dr. Gerald Bull's High Altitude Research Program then under way at McGill University.

The Chapman Report was well received by the Science Council of Canada at its official tabling in January 1967 and was officially published by the government as *Upper Atmosphere and Space Programs in Canada*, Special Study No.1 of the Science Secretariat in February. It also became a topic of debate and decision at subsequent Science Council of Canada meetings in March, May, and June.[8] This was finally followed in July 1967 by another official publication, Report No.1 of the Science Council of Canada titled *A Space Program for Canada*. It was from this last document, a more concise version of the Chapman Report delivered in February, that Cabinet largely based its plan for future space activities in Canada.

Second Series: The ISIS Program

Following the technical and scientific success of the *Alouette 1* satellite, an agreement was reached between the United States and Canada on January 10, 1963 to embark on a joint program of launching an additional four satellites devoted to ionospheric research. Known as the International Satellites for Ionospheric Studies (*ISIS*), these four satellites were subsequently designated *Alouette 2* (also known as *ISIS-X*), *ISIS-A*, *ISIS-B*, and *ISIS-C*. The plan called for the design and construction of these satellites in Canada, to be later launched by NASA at intervals during a period known as 'the half cycle' of sunspot activity from between the years 1964 and 1969.[9]

This new project continued to consume almost the DRTE's entire human and material resources, but the agreement to go ahead with this project depended on the stipulation that Canadian industry be employed to its fullest extent in the process. This clause was inserted in order to stimulate Canada's somewhat limited technical and industrial base during the 1960s, so that by the end of the decade a skilled industry for the design and construction of spacecraft would potentially exist in the country. It was a difficult objective to meet based on the construction of only four satellites, but proposing such national objectives was necessary in order to convince the government to commit itself both politically and financially to the task and provide for a special parliamentary vote to secure initial funding for the program.[10]

Planning and working groups for the new satellite project gathered momentum in early 1963. An overall plan for the four projected satellites was developed, with the satellite working group generally advocating for higher inclination orbits than *Alouette 1* in order

---

[8] Science Council of Canada, *A Space Program for Canada*, p.2.

[9] Chapman Report, p.11.

[10] Ibid., p.11.

to explore a greater range of altitudes. As well, it was planned to include direct measurement experiments in *ISIS* satellites so that mission scientists could collect data on the immediate environment surrounding the spacecraft. More detailed measurements were also desired of the charged particle influx into the ionosphere, and of its relation to the auroral processes and to the production of ionization.[11]

Work on the first *ISIS-X/Alouette2* satellite began in March 1963. Briefings were held to acquaint industry with the program after which DND and the Department of Defence Production selected companies to lead the project. The RCA Victor Company, Ltd. of Montreal was subsequently chosen as the prime contractor with de Havilland Aircraft of Canada Ltd. in Toronto as the associate prime contractor. In order to prepare these two companies for satellite manufacturing, their first assignment was focused on training personnel and gaining experience through the construction of *Alouette 2*. Industry personnel were requested to report to the DRTE at Shirley Bay near Ottawa for indoctrination and to set up the workshops where the *Alouette* and *ISIS* satellites were to be built. After spending the summer establishing these new workshops, industry commenced building *Alouette 2* in September 1963.[12]

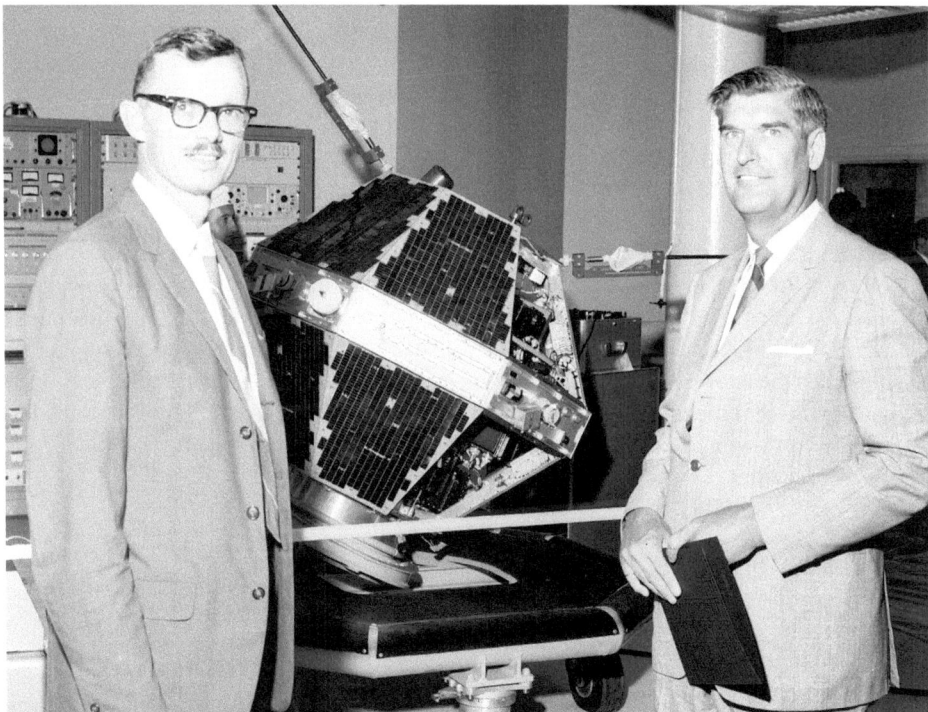

Fig. 3.2 Dr. John H. Chapman (left), and Dr. Robert Uffen, then serving as chairman of the Defence Research Board, stand before the first of the ISIS project spacecraft

---

[11] T.R. Hartz, andI. Paghis, *Spacebound*, p.65.

[12] Chapman Report, p.12.

The design and construction of the *Alouette 2* satellite was very similar to that of its predecessor. The machine used as its core the *Alouette 1* backup satellite that had been built concurrently alongside the prime spacecraft, to be used if that one was lost. The DRTE engineers modified the backup satellite to include a new probe experiment, and the frequency range of the sounder was increased to better suit the chosen elliptical orbit planned for the *ISIS* series spacecraft.[13]

The other three satellites in this series would officially carry the *ISIS* name. An ionosonde would serve as the principal experiment on the first two satellites in the series, with additional experiments being canvassed for from scientific agencies in both Canada and the United States. The initial plan was to have as many of the experiments on both satellites as possible, so that mutually supportive data of greater scientific value might be collected. Meanwhile, primary and secondary payloads for the third *ISIS* satellite remained unconfirmed for the moment, as it would depend to some extent on how things progressed with the first two.[14]

As with its predecessor, the construction of *Alouette 2* remained on schedule and the completed satellite was ready for launch in the summer of 1965. The final product carried no less than five experiments, consisting of the four same experiments as seen aboard *Alouette 1* plus a new one designed to measure the sheath or 'halo' of positive ions occurring about the spacecraft's antennae as it passed along the ionosphere. Room for the fifth experiment was made possible by the redesign and improvement of the nickel cadmium batteries employed in the original satellite, so that fewer were needed to provide the same amount of power. Removing the extras therefore provided space for the new instrumentation.[15]

Canada's second experimental satellite was successfully launched from the Vandenberg space launch complex 2E at 04:48GMT on November 29, 1965. Riding along with *Alouette 2* this time was an American scientific satellite, *Explorer XXXI*, another payload also designed to conduct investigations of the upper atmosphere. *Alouette 2* reached orbit in a textbook fashion and went into operation almost immediately. Following the launch, Canada's Minister of National Defence, Paul Hellyer, summed up the entire DRTE's feeling when he told public affairs personnel that *Alouette 2* was "an achievement in which all Canadians can be justly proud".[16] Indeed, the DRTE satellite team was more than pleased. It marked their second successful satellite project in just three years.

---

[13] The mission profile planned for an orbit apogee of 3000km. T.R. Hartz and I. Paghis. *Spacebound*, p.65.

[14] Ibid., 65-66.

[15] Flt. Lt. T.G. Coughlin, 'Alouette II – Canada's Second Satellite Probes the Mysteries of Outer Space', *Sentinel*, 2:1 (January-February 1966), 8-13.

[16] Ibid., 8.

Fig. 3.3 Alouette 2 lifts off from Vandenberg Launch Complex 2E in the early morning hours of November 29, 1965

The initial *Alouette* series flights had brought a group of professional and talented Canadian defence scientists, engineers, and technicians into the national limelight and underscored what might be possible within the wider context of Canadian aerospace industry. The DRTE Chief Superintendent, Frank Davies, very likely no longer felt *Alouette* to be such an albatross about his neck, and others such as Eldon Warren, John Chapman, Colin Franklin, M.A. MacLean, John Mar, and their colleagues were gaining both reputation and prominence within the space exploration community as a result of their efforts. Some of these men, perhaps most notably Dr. Chapman, would continue on to much higher posts in Canadian scientific administration and government. Others would stay with the DRTE through the years of transition into the civilianized Communications Research Center, while others still left when the DRTE dissolved and moved on to industry or other scientific pursuits. For the moment, however, they were undoubtedly the center of attention, much to delight of some and even at the disapproval of others.

Those organizations that disapproved of the increased attention and focus on the *Alouette* project included senior leaders in the Royal Canadian Air Force (RCAF). Desiring a greater role for space technology in Canadian defence, and in particular under the command of the RCAF, they met the success of *Alouette* with some reservation.[17] The experimental satellite program had completely consumed the DRTE and, to a larger extent, nearly all space technology development within the Defence Research Board. Though *Alouette's* mission did support defence research in some programs, obviously in defence communications but also in studies related to ballistic missile defence and to ballistic missile reentry into the atmosphere, it was clear to the RCAF and other DND planners that a DRB agenda was forming that advocated for the pursuit of pure scientific research at the expense of more defence-oriented science, technology, or application.

Not surprisingly, the DRB responded to the concerns voiced about their future direction in a predictable manner. Though the organization empathized with the DND's military requirements they fought hard to maintain their own agenda of scientific satellite research and development. With national political support behind them, the DRB no doubt felt that there was very little that the RCAF –or even the DND as a whole –could really do to alter its desired course. Interestingly, as the decade advanced it was federal Cabinet, not the DND, that forced the demise of the DRB agenda and, later, the whole DRB itself.

Ending the *ISIS* Program

Work on the remaining *ISIS* series satellites began in March 1964.[18] By this time the Topside Sounder Working Group (TSWG) had resolved its terms of reference and was well into planning and coordinating the overall mission profile for *ISIS*. Dr. J.E. Jackson of NASA steered the group until 1974 when Canada took over, assigning Dr. T.R. Hartz

---

[17] This is discussed in greater detail in Chapters 4 and 5.

[18] (Confidential) Record of Cabinet Decision – International Satellites for Ionospheric Studies, dated April 28, 1964. The Order in Council was passed accordingly (P.C. 1964-608). Privy Council Office, RG 2, LAC.

from the Communications Research Center to succeed him.[19] The task of the TSWG was to coordinate all operations planning and scientific aspects of the mission, facilitate meetings between various parties associated with the project as a whole, and de-conflict any scientific and engineering issues as they arose. The TSWG, although not specifically labeled as such, was in essence acting as the systems engineering and project management oversight for the whole *Alouette-ISIS* program. To a large degree, it was upon these members that the success of the entire project depended.

Fig. 3.4 A technician working on the *ISIS*-A (subsequently renamed *ISIS-1*) satellite as it undergoes integration testing, August 1968

Unlike the *Alouette* satellite series, the *ISIS* project posed a much greater technological challenge to the staff and resources of the DRB. The design of *ISIS* was more complex than *Alouette* and the engineers and technicians needed more time to successfully build, integrate and test its components. *ISIS-1* carried eleven different experiments aboard, including a fixed frequency sounder to accompany its swept frequency sounder and mixed mode sounder. Also in this satellite was a Very Low Frequency (VLF) receiver and transmitter test; a cosmic noise receiver; an energetic particle experiment; two Langmuir

---

[19] T.R. Hartz and I. Paghis, *Spacebound*, p.71.

probes that measured electronic temperature and density; two ion mass spectrometers; an ion probe to measure ion temperature and density; a soft particle spectrometer; and a beacon set at 137MHz.[20]

For this new project, the DRTE was also now expected to assume responsibility for its own satellite operations. That meant constructing new facilities to house the control center necessary for *ISIS* satellite operations, as well as training a number of specialized technicians and administrators to carry out the work needed to 'fly' the satellite. The DRTE satellite operations staff was expected to complete spacecraft checkout procedures after launch; inject the *ISIS* satellite into the desired orbit for the mission; monitor the health and status of the satellite in orbit; and change the oriented and spin-axis of the satellite to accommodate each of the onboard experiments as required. Mission controllers were also to be responsible for maintaining the accurate determination of the satellite position in orbit, and predict its position several weeks in advance as needed.[21]

There were other tasks for the satellite operations team as well. These included coordinating and de-conflicting all transmissions between the satellites and the ground stations and collecting and processing scientific data. Other organizations also assisted in this latter task as the volume of data became unwieldy for DRTE to manage alone. NASA's Goddard Space Flight Center in Washington assumed a data processing role for the *ISIS* missions, helping the DRTE to keep some control over the twenty-five kilometers of scientific data-laden magnetic tape received daily from the satellite at the height of its operation.[22]

Though technologically rewarding, the *ISIS* missions still proved a daunting task for the Canadian defence scientists and engineers. The satellites took longer to design and build, and consumed far more resources and greater funding than their predecessors. The adaptation of ground-based techniques to a spacecraft where weight, space, and power consumption was at a premium demanded ambitious and pioneering talent that was soon stretched thin both at the DRTE and the Canadian civilian industries contracted to support the missions. As well, the sheer magnitude of controlling and operating satellites on a daily basis far exceeded the existing resources available to the DRTE, and the DRB had to canvass both the DRB headquarters and the Canadian Forces Headquarters on a regular basis for additional resources and funding just to sustain ongoing operations.[23]

The difference in scale and effort between satellite projects, combined with other factors then affecting Canada's space program writ large, certainly had an impact on the timeliness of the immediate project. The first satellite in the *ISIS* series, *ISIS-1*, was originally scheduled to head into orbit in either 1967 or 1968, but schedule delays forced its launch back to January 30, 1969.[24] Taking off from the Vandenberg space launch complex

---

[20] C.D. Florida, *The ISIS Satellites*. DRTE Technical Note No.619. Ottawa: DOC-CRC, April 1969, p.11.

[21] T.R. Hartz and I. Paghis, *Spacebound*, p.73.

[22] Ibid., p.74.

[23] *UWO Space Workshop*. Supplementary Materials: Alouette Satellite, Comments from Colin Hines in adding details to a letter on the origins of the Alouette concept from Dr. P. Forsyth to Dr. J. Scott, dated September 22, 1981. Accessed at http://quark.physics.uwo.ca/~drm/history/space/space_history.html. Colin Hines made reference to Dr. John Chapman as the man who "carried the can at DRB/HQ in getting the extra funding".

[24] Also referred to in contemporary literature as *ISIS-A*.

Fig. 3.5 The *ISIS-2* satellite sitting atop its Delta rocket awaiting the installation of the payload bus shrouds, April 1971

2E in California at 06:46 GMT, the third Canadian satellite into space was lifted effort-lessly into the morning sky upon its *Thor-Delta E1* booster and entered flawlessly into its assigned orbit of 88.4° with an apogee of 3522km and a perigee of 574km. Early feedback from the satellite to ground control was promising and the spacecraft subsequently per-formed nominally and acquired good data throughout its planned lifecycle.[25]

The next satellite in the series, *ISIS-2*, suffered similar schedule delays and was not launched until March 31, 1971. While it too performed as expected, if not better, by the time it reached orbit the focus of Canada's space program on the ground was rapidly shift-ing away from further space science projects in favor of concentrating on the development and deployment of commercial space applications. As a result of these decisions the third planned satellite in the series, *ISIS-C* (*ISIS-3*), originally scheduled for launch sometime in 1974, was cancelled by the government in the summer of 1969 in favor of pursuing other more desirable space projects and missions.

Attempting a Satellite Launch Program

The ISIS satellite series was not the only program to feel the direct effects of the Chapman Report. As the testing of Black Brant V rocket drew to a close in 1966, discussions began amongst the DRB administration on what the next phase of Canada's rocketry program might consist of. Would the DRB and the NRC continue to support the development of larger research rockets, or would a commercial agency such as Bristol take the lead in manufacturing them? Though it remained uncertain exactly who or what agency would direct and oversee future launch operations in Canada, few within the rocketry community at the time seriously doubted that there would not be a next phase of some sort. After the success of Canada's sounding rocket program and the launch of its first two satellites, it seemed only natural that the development of an all-Canadian small satellite launch vehicle would come next.[26]

Proposals for future Canadian launch vehicles were plentiful. Those who were advo-cates for larger rockets tabled a number of alternatives ranging from enlarged Black Brant models to mimicking American solid- and liquid-fuelled multistaged launchers then already in use. Though many ideas for a launch vehicle were eventually presented, of par-ticular interest to Canadian space scientists and engineers in the mid-to-late 1960s were three concepts. The first two consisted of traditional rocket designs, while the third exam-ined the possibility of placing small satellites in orbit using a large bore gun.

While the concept of gun-launched rocketry was not in and of itself new, Canada made the first serious investigation of this method of launch during the 1960s. Under the direc-tion of a brilliant CARDE engineer named Dr. Gerald Bull, a series of gun-launched rock-ets and probes were experimented with as part of McGill University's High Altitude Research Project (HARP). Using large bore guns ranging from approximately 12 to 44cm, the HARP team successfully fired a number of fin-stabilized discarding sabot solid

---

[25] C.D. Florida, *The ISIS Satellites*, p.15.

[26] I.A. Stewart, "The Churchill Research Range and Canada's Satellite Program", *Canadian Aeronautics and Space Journal*, Vol.15, October 1969, 311.

propellant rockets, named Martlet, to altitudes between 96 and 112km. On November 19, 1966 the U.S. Army's Ballistics Research Lab, using a HARP gun designed by Dr. Bull, fired an 84kg Martlet rocket to an altitude of 179km. It was, and remains, a world record for any fired projectile.[27]

Despite having enjoyed considerable success in early testing, the system still had several disadvantages. First, the payload had to be slender enough to fit into the gun-barrel tube, severely restricting the types of probes and satellites that could be launched this way. Second, the Martlet rocket left the tube discarding its sabot at a super high velocity, at times accelerating to 10,0000 times the force of gravity. This automatically ruled out any option of attempting manned space flight from a cannon, and also eliminated the possibility of carrying sensitive payload instrumentation into orbit.

The development of the Martlet rocket under the aegis of the HARP proceeded to the point where subsystems were tested and fired under high force of gravity conditions, but funding support from both the government and McGill University was later terminated in favor of pursuing existing Black Brant rocketry and related systems. Although discontinued, it was and remains the most impressive concentrated technological effort to place an object into orbit using a cannon.[28]

Of the other two concepts focusing on more traditional rocketry, the first was built on existing Black Brant technology while the second involved the acquisition of the American Scout launch vehicle and then tailoring it to specific Canadian needs. Like the HARP, both ideas went through various stages of conceptual design and development with the combined support of limited government funding, the DRB, and to a lesser degree, university research.

In 1967, engineers at the High Speed Aerodynamics Section of the National Aeronautical Establishment of the NRC published a study on the potential employment of Black Brant VB rocket motors as booster building blocks for a larger satellite launcher.[29] The concept built on existing technology within the Black Brant program and established different configurations to construct three and four-stage launch vehicles. One configuration in particular, known tentatively as the 221X, was considered by its advocates to be more efficient than even the American two-stage Scout rocket, being able to lift 45.35kg and deploy

[27] B. Sterling, "Think of the Prestige", at www-istp.gsfc.nasa.gov/stargaze/Sgbull.htm. The United States military was very interested in Dr. Bull's work and jointly funded his research for several years. Dr. Charles Murphy, a scientist with the U.S. Army's Ballistics Research Lab, was a life-long friend of Gerald Bull, supporter of his work, and coordinator for U.S. military assistance to HARP.

[28] The fascinating story of Dr. Gerald Bull (1928–1990) and his quest to build a super cannon continues well after the end of his association with CARDE and HARP, but unfortunately lies outside the scope of this study. Briefly, he continued to produce large caliber long-distance cannons for various countries including South Africa and Iraq, of which his association with both put his life at risk. On March 22,1990, Gerald Bull was mysteriously assassinated outside his apartment in Brussels, Belgium. His killer was never found For more on Dr. Bull see J. Adams, *Bull's Eye: The Assassination and Life of Super Gun Inventor Gerald Bull.* New York: Times Books, 1992.

[29] See L.H. Ohman, 'A Satellite Launch Study Employing Black Brant VB Sounding Rocket Motors as Booster Building Blocks'. *Canadian Aeronautics and Space Institute Journal,*.13 November 1967, 427.

Fig. 3.6 Dr. Gerald Bull, a brilliant and controversial defence scientist who headed Canada's HARP, was later assassinated outside his apartment in Brussels, Belgium in March 1990, likely for reasons related to his work. His killers were never identified

it into 185km orbit. A flexible configuration, the 221X was designed so that the rocket could even exchange its last solid propellant stage for a (yet to be developed) liquid-oxygen-hydrogen last stage, increasing its lift to about 154kg.[30]

---

[30] Ibid., 427–429.

The following year, another proposal for the employment of the American-built Scout rocket was published.[31] Arguing instead that proposals for an all-Canadian system suffered, like all new technology programs, from a high risk of failure, unknown costs, and long development schedules, the Scout rocket lobby offered that based on the successful precedence of employing American launchers and facilities – for both Black Brant rockets as well as the Alouette-ISIS satellite series – pursuing an already-tested American launcher would prove less expensive and more timely. The Chapman Report itself quoted another report stating:

> "It should be emphasized here that the development of a Canadian launcher of the Scout class is not an overly ambitious undertaking for a country which already is producing and launching multistage sounding rockets of the Black Brant type. The progression from sounding rockets to satellite launchers is fundamentally one of providing the necessary guidance and control to incline the flight path horizontally and insert a payload into orbit. The basic elements already exist of rocket motor technology, staging design, and launch and tracking complex at Churchill.[32]"

The Scout rocket was also attractive for other reasons. Unlike most American launchers then in use, the Scout was specifically designed as a non-military rocket and was obviously unsuitable as a weapons carrier.[33] This limitation on the rocket's use facilitated its export to other countries, namely Canada. Similarly, the same limitation was likely to make it a more appealing choice for the government, whose foreign and defence policies at the time were moving the country away from the ballistic missile business altogether. Finally, the argument was further supported by the fact that recently-released strategic level direction on the future of Canada's space program did not mention any notion of pursuing a costly independent launch system.[34] As such, it appeared to the authors of the study that the Scout rocket was the preferred if not only real option.

The four-stage solid fuel Scout rocket stood roughly 23m tall and weighed 21,750kg fully loaded. A flight-proven design in flight, the Scout could successfully launch, depending on the azimuth, anywhere from 210–270kg of payload into a 185km orbit. The rocket also had the option of employing an upper stage designed for more difficult geotransfer orbits, allowing payloads to be placed in much higher geosynchronous orbits. Most important, perhaps, the Scout was still small enough that it could be launched from places like the Churchill Research Range without serious range operations, tracking, or safety concerns, allowing Canada to potentially place its own orbiting platforms into non-synchronous polar orbits as needed or desired.

---

[31] I.A. Stewart, 'The Churchill Research Range and Canada's Satellite Program', pp.311–313.

[32] Chapman Report, p.106.

[33] Ibid., p.105.

[34] The Chapman Report provided an assessment of *Black Brant* activity but did not reveal any plans for creating an all-Canadian launch system.

Fig. 3.7 The American-built *Scout* rocket was considered an attractive option for Canada due to its solid fuel configuration and its multistage payload capabilities

The Scout rocket proposal was revealed in tandem with another presentation to build and launch a Canadian-designed remote sensing satellite called CERES (Canadian Earth Resources Survey).[35] As a large payload – too large for Black Brant rockets – obviously

---

[35] See Appendix F for more details on this subject.

the second proposal could not fly without the first, and the two ideas worked hard together to seek joint approval from the DRB and government. Consequently, the CERES satellite project never received government approval, and with all other Canadian satellites then being lofted by American-based launchers, there was some doubt by the late 1960s as to what exactly the Canadian government needed the Scout rocket for.[36]

Both technology concepts held great merit, yet in the end neither was successful in gaining the necessary political or financial support for full implementation. For those space advocates who in the late 1960s held a vision of the future where Canada had its own indigenous launch capability, the painful fact was that nearly a decade earlier Cabinet had already concluded that it was neither affordable nor very practical to embark on such an endeavor. National pride aside, the Canadian government did not need such a launch system at the time and therefore could not justify spending the taxpayer's purse on such a project just to keep its distant third place in the ongoing space race. Though unquestionably there were many advocates within the DRB, the NRC, the CARDE, and even industry at the time promoting the creation of an indigenous launch capability in Canada, the government had decided early on in the space age against developing any serious capability in this domain.[37] There was already limited political support within Canada for the continued development of solid propellant rockets by the late 1960s, and no political support seemed to exist for initiating new projects involving liquid fuelled or large-scale multistage launchers. In a memorandum prepared for the Department of External Affairs a decade earlier in 1959, DRB Chairman Dr. A. Hartley Zimmerman noted, "…since Canadian geography possesses no unique advantage in this area, there is no plan to undertake satellite launching." Further along in the report under a section titled 'Facilities Required to Support Anticipated Canadian Program' Dr. Zimmerman added:

> "There is no intention to move toward Canadian manufacture or launching of very large multi-stage rockets to support any satellite program in which Canada would participate. Motors presently in use for this purpose by the United States are derived from the ICBM production effort of that nation. When it is considered that only because of the availability of the development and manufacturing skills of this large defence industry has the United States obtained its place in the space research field, the practicability of Canadian manufacture of such rockets for space research alone is out of the question."[38]

---

[36] I.A. Stewart, 'The Churchill Research Range and Canada's Satellite Program'. P.313; and Steven J. Isakowitz, *International Reference Guide to Space Launch Systems – Second Edition*. Washington DC: AIAA Publications, 1991. Further degrading *Scout*'s applicability was the fact that communications satellites, which were essentially the only type of satellite Canada planned to launch between 1969 and -1979, were too large for *Scout* and required much greater and more stable insertion orbits.

[37] See discussion in minutes of Associate Committee on Space Research, 1959–1963.

[38] Outer Space – Proposal for Possible Canadian Initiative. (Confidential) memorandum 'International Implications of a Canadian Space Research Program', prepared for the Under-Secretary of State for External Affairs, dated August 27, 1959, p.5. Vol.1 File No.12798-4-40, RG 25, LAC.

Instead, the DRB, and subsequently the NRC, retained a near-term focus on the development of its current strengths in the rocketry field – namely design, test, instrumentation, telemetry, and electronics – without concerning itself with any further mid- or long-term rocketry development planning unless the situation changed to merit a reevaluation of the current decision. The 1967 publication of the first comprehensive government review of Canadian space activities – later commonly known as the Chapman Report – only reinforced this position, when it stated, "We do not consider that Canada should attempt at this time to provide satellite launch facilities to meet all program needs", but as a result it, "will be necessary to purchase launches for communications satellites for at least the next decade."[39] Though not a very ambitious or independent way forward for the country given what was occurring in other spacefaring countries, it was nonetheless in line with other government decisions on strategic interests and served Canada's needs at the time.

Decision: Launch on Demand

With the successful completion of the final *Black Brant V* launch operation on April 1, 1967, the joint DRB-Bristol rocketry development program officially ended. Further versions of the *Black Brant* rocket were designed and built, but not under the aegis of further DRB or CARDE oversight, guidance, and control. Similarly, McGill University's High Altitude Research Project was concluded in June 1967, and with its ending, the era of defence-led rocketry design and development in Canada was over. The non-defence led rocketry initiatives that were left faced an uncertain future, as scientific research funding slowed in the following decade.

The Churchill Research Range, where Canada's official rocketry program was born, was officially transferred back from American to Canadian control on January 1, 1966, and temporarily placed under the management of the NRC and its Space Research Facilities Branch (SRFB) until a final decision on its future management could be made. In the meantime, the range continued to be funded jointly by both the NRC and NASA, with the costs shared equitably between the two countries as both made use of the facilities. As well, a Joint Range Policy Committee (JRPC) was formed to oversee all decisions related to firing specific rockets and payloads at the site. Consisting of eight members in total, each country nominated their own four representatives, one of whom was also designated a co-chairman. In 1966–1967, Canada's quartet of members consisted of three staff from the NRC (including a co-chairman) and one member from the Department of Public Works. The American representation consisted of two members from NASA (including their co-chairman) and two members from the United States Air Force. Finally, for the time being the range itself continued to be operated by the American conglomerate, Pan-American (PANAM) World Airways, with approximately 210 employees located at the site. The NRC also had eighteen personnel stationed at the CRR to support Canadian administration and flight operations.[40]

---

[39] Chapman Report, pp.102–104.

[40] Ibid., p.22–23.

Fig. 3.8 Bristol Aerospace aggressively marketed its rockets commercially after the initial test program completed and the government backed away from creating a national spaceport

Overall, the *Black Brant* rocketry program was remarkably successful and performed well beyond its original expectations. The Chapman Report noted within its section on the subject that, during the initial DRB-Bristol development phase, 103 rockets were success-fully flown, most of which were *Black Brant IIA* models.[41] As well, there was no doubt that the *Black Brant* design would go on to serve both Canadian and commercial interests abroad in the future.[42] In 1966–1967 alone, the sounding rocket market generated roughly $25 million per annum in the United States, and the *Black Brant* family of rockets rapidly gained wide acceptance as a reliable launch platform for all manner of scientific research experiments. The *Black Brant* rocket had generated $500,000 in sales in 1966, and it was estimated that by 1969 direct sales would reach $1.3 million. When combined with sup-porting sales (additional user requirements, instrumentation, telemetry, etc.) that total amount would go much higher. It was already estimated by the Canadian government that the total income for *Black Brant* rockets and services from August 1966 to December 1967 was approximately $3 million. By 1969, that amount had nearly doubled.[43]

Financial success with the sounding rocket market, however, did not automatically translate into a similar financial success with larger launchers. Simply put, the larger the rocket the more infrastructures and support it required. While it was estimated that the minimum requirement for breaking even on the cost of establishing and maintaining a launch facility was four to five launches a year, it seemed unlikely at the time that Canada even had a requirement to deploy that many satellites in orbit ever, let alone in a single year. The Canadian policy, therefore, became launch on demand and by the late 1960s there simply was no longer any major demand from the Canadian military or defence sci-ence community for rockets. With the defence client out of the picture, that left only the Canadian universities and the NRC as probable regular customers, and the requirements of these clients could be successfully met with smaller *Black Brant* and *Arcas* rockets. For larger civilian satellite programs, the Canadian government decided instead to rely on its ability to purchase foreign launches as required.

A Shifting Space Culture

Canadian space activities in the 1950s and 1960s served as something of a crossroads where three influential Cold War communities converged. These were the government and the civil service, the defence community, and the scientific and engineering community. The first two had a long tradition of convergence and at times even conflicting with each other, but the scientific and engineering community, and in this case the space scientific community in particular, was a new party to the paradigm in that it only matured during the Second World War. The nature and status of the subsequent relationship between this triad influenced all space program development in Canada during the first three decades of the space age, and understanding the ideals and attitudes prevalent within each culture during the era of the *Alouette-ISIS* program is essential to further defining the relationships

---

[41] Ibid., p.62.

[42] The fact that Black Brant remains in use today is a clear testament to the success of the design.

[43] Chapman Report, pp.62–63, 102–105.

that influenced other developments in Canada's space program over the longer term. Finally, what one discovers is that, interestingly but perhaps not surprisingly, Canada during the early space age had experiences similar to those of its two main international partners, the United States and the United Kingdom.

Though the Second World War had brought soldier and scientist together, it appeared that after the war most scientists and engineers working within Canadian defence establishments preferred to return to their own research interests rather than directly support the goals of the organizations that employed them. Dr. Peter A. Forsyth was a talented defence scientist and one of Canada's space pioneers who later taught at the University of Saskatchewan and then served as one of four central figures in the 1966 National Space Study. In 2002, during a workshop on the history of early Canadian space science he described an environment within the DRB and DRTE during this period that was somewhat openly contemptuous of the defence establishment.[44] Defence scientists and engineers were increasingly concerned about becoming involved in classified projects that would hinder the sharing or publication of research, and they often held contempt for the higher politics that might impose specific tasks or objectives on the various staffs, especially those that had little relation to their own research interests.[45]

Fig. 3.9 Dr. Peter Forsyth served as a member of the 1966–1967 National Space Study committee to build Canada's space roadmap going into the 1970s

---

[44] *UWO Space Workshop.* Transcript – DRTE and UWO, November 25, 2002.

[45] Ibid. A lot of the discussion from this section of the workshop focuses on the period between 1955–1965, during the height of which Prime Minister Diefenbaker was influencing many aspects of Canadian space research and development.

The *Alouette-ISIS* satellite project therefore only served to entrench this attitude amongst the staff at the DRB. In the absence of a ratified national space policy or a clearly-defined space agenda during the 1950s, separate government defence and scientific establishments were largely given *carte blanche* to pursue their own research interests regardless of whether or not these projects coincided directly with any specific national defence or technology objective. Sor example, even the idea for the S-27 Topside Sounder that later became the *Alouette* satellite was born this way. When the DRTE was allowed to pursue the project directly with Canada's international partners and without much government oversight, its realization only served to validate the idea that the DRB's establishment's main objective was to not support defence science and technology specifically, but rather any project that might garner wider Canadian political and, hopefully, financial support. Though not necessarily a flawed or unrealistic business practice – after all, DND did perceive itself benefiting from the acquisition of new rocketry and space technology – it must be noted that the process created an undesirable precedent for the DND that it was later unable to influence politically or bring under strict departmental control.

Further exacerbating the issue was the fact there was also divergent opinion between government, defence, and the scientific community about the main purpose of the *Alouette* satellite project. To the government and to defence, the rationale for *Alouette* was based on the need to advance space technology in Canada for both security and industry. To the scientist and engineer, its purpose was pure research, design, and the desire to collect data for further studies. So much so, in fact, that this project took over the DRTE nearly completely, and effectively drew away most resources from other sections to make them available for upper atmospheric research. In a way, it ensured the untimely demise of the organization as the DRTE's near single focus on the development of satellite communications-related science between 1957 onwards ensured its subsequent transfer to the Communications Research Center under the newly-created Department of Communications in 1969.

To the engineer, *Alouette-ISIS* was about developing new technology, but the project ultimately exposed many of the weaknesses inherent in Canada's technological base. This is not to suggest that Canada's satellite-building abilities were weak; they were in fact very strong. That said, the base was also small compared to the juggernaut in the US and the USSR, and it often struggled to develop the advanced materials and electronics it needed for its projects. Often, the DRTE could not successfully draw much-needed resources from its civilian industry either, as the Americans could to sustain a long-term scientific program when expenses overtook all available government funding. Whereas in the United States private industry consumed much of the research and development cost associated with developing and testing new space technologies thanks to generous contracts from the US government and NASA, in Canada both the DND and the DRB were forced by circumstances to assume much of the cost internally.

Each of these groups felt very strongly about their respective roles in developing Canada's official rocketry and space program, and they regularly clashed throughout the *Alouette-ISIS* project, but ultimately it was the government in power that set the future course. As demonstrated in the following chapters of this book, both the DND and defence-oriented agendas were eventually dropped from most participation in national space projects heading into the 1970s; even the influential and successful coordinator of the *Alouette-ISIS* project, Dr. John Chapman, could not sustain further political support for a space science agenda in the wake of the government's desire to see commercial space application overtake pure research. Despite the challenges of political direction, however, the *Alouette-ISIS* project marked the convergence of these three cultures and remains to the present day one of Canada's most remarkable technological achievements at the dawn of the space age.

Unable, or perhaps unwilling, to generate new distinctive policy for science and technology during this period, both Conservative and Liberal governments defaulted to the traditional Canadian approach of specific objective-oriented science or the application of science to other political or social objectives.[46] This is certainly how most Canadian space activities were treated prior to 1967. One could argue, for example, that the *Alouette-ISIS* satellite program did not exist because Canada had a long-term goal to develop a domestic satellite communications capability, but instead because the Canadian government chose to support a single scientific application in order to gain other more significant political, technological, and defence benefits from its spacefaring American partner. True to form, rather than expend resources and effort on an elaborate and all-encompassing space program, the country sought instead to consolidate and concentrate where it could, trading scientific and technological innovation for more direct applications that provided other more immediate economically tangible benefits. Up until 1967 at least, a longer-term vision for outer space was abandoned in favour of short-term returns.

It is little wonder, then, that at the end of 1967 Canada still did not have an officially ratified space policy and the government did not seem prepared to embrace the creation of one. It seemed that Canada's government simply did not perceive any need for such policy as yet, even though its suborganizations desperately needed some more concrete direction in order to engage in longer-term planning. Even when strategic guidance for future space activities finally did arrive with the publication of Chapman's *A Space Program for Canada*, it was admitted at the same time by the Science Council that the space program's identification as a priority was, "partly determined by circumstances, and partly by present importance" rather than having a connection to the achievement of other national goals.[47]

Still, by 1968 there was finally a realization within Cabinet that from an economic perspective as well as a national welfare perspective, Canada could no longer afford to "expose herself to the degree of economic and technological dependence" that then existed within its national space program.[48] At some level or in some program, Canada needed to declare technological independence in support of its national interests. The first steps were therefore taken to realize some level of technological independence, as changing circumstances became important, if not critical, to new national science goals and a new agenda for Canada in outer space.

---

[46] G.B. Doern, *Science and Politics in Canada*, p.172.

[47] Science Council of Canada, *Towards a National Science Policy*, p.35.

[48] Science Council of Canada,. *A Space Program for Canada*, p.7.

Satellite Communications

With the results of the Chapman Report before them, in the summer of 1967 the government moved to formalize the country's space agenda for the next several years. Of all the recommendations found within *A Space Program for Canada*, the Liberal government was primarily interested in and supportive of the suggestion to create a domestic communications satellite capability for the country. Throughout its history, due to its massive territory and its small and widely dispersed population Canada's survival and prosperity often depended on the establishment and maintenance of strong lines of communications.[49] By the 1960s, both television and telephone were rapidly gaining on radio as a primary medium through which people communicated with one another both in Canada and internationally.[50] Possibly perceived as a modern-day version of Canada's transcontinental railways first built in the nineteenth century, the notion of linking Canadians from coast to coast was endorsed by Prime Minister Pearson and his successor, Pierre Elliot Trudeau, with little question.

On June 6, 1967, Cabinet met to establish government policy on satellites in the short term.[51] The government agreed to undertake negotiations at the earliest possible opportunity to register with the international community Canada's plans for the establishment of a communications satellite system for domestic use. As part of this process it also agreed to form a special task force under the auspices of the Science Secretariat to undertake a comprehensive study of all questions on communications satellite development that should be considered by the government of Canada in its pursuit of this goal.[52] The newly established Task Force on Satellites (TFS) was to draw its members from inside the government service, and a list of potential candidates was approved in principle. The Honourable Charles Mills 'Bud' Drury, a veteran politician in the Liberal government, was appointed chairman.[53] A graduate of the Royal Military College of Canada, Drury had served with distinction during the Second World War, reaching the rank of Brigadier. In 1949, he was appointed as deputy minister of national defence where he served until 1955. Elected to Cabinet 1962, he was first Minister of Defence Production and then later the

---

[49] Until the mid-nineteenth century, trade and communications in Canada largely depended on the establishment of means by which to navigate its rivers and inland seaways. Towards the end of the nineteenth century, a transcontinental railway was built to connect the country from the Pacific coast to the Atlantic. In the mid-twentieth century, Canadian communications took another step forward with the advent of airpower. The development of a space-based capability in the latter half of the century probably seemed the logical next step

[50] As previously discussed in this study, Canada was part of the electronics revolution in the West; however, it was limited in its capacity to exploit various technologies at more advanced levels. Still Canada's technological evolution in the post-war period remains to be explored in detail. The American experience is examined in D. Mowrey and N. Rosenberg, *Paths of Innovation: Technological Change in 20th-Century America*. Cambridge: Cambridge University Press, 1998.

[51] Record of Cabinet Decision – Meeting of June 6, 1967 – Government Policy on Satellites. Prepared for CDS by R.J. Sutherland. Vol.46. File 1150-110/M16 pt.1, Acc.83-84/232, RG 24, LAC.

[52] Ibid., p.1.

[53] Other members from the government included Mr. Martin, Mr. Winters, Mr. Pickersgill, Mr. Hellyer, Mr. Mellwraith, Miss LaMarah, Mr. Sauve, Mr. Benson, Mr. Trudeau, and Mr. Chretien.

country's first Minister of Industry.[54] He was more than suitable for the post, having both knowledge and experience of scientific and technological matters from both military and civilian perspectives.

Fig. 3.10 Charles Mills 'Bud' Drury, former army officer, veteran minister, and the driver behind Canada's strategy to focus on satellite communications going into the 1970s

Drury's Task Force on Spaceon Space was authorized to commission special studies as required and retain any subject matter experts it needed to complete its work, and was expected to report back to Cabinet within the year. Like Dr. Chapman's previous report, it was expected that Mr. Drury's White Paper would lay out the way ahead for putting

---

[54] Drury also later served as Treasury Board President.

Canadian communication satellites into space within the decade. But unlike Chapman's schedule, Drury was subject to a greater sense of urgency from Cabinet to accomplish his assignment. Perhaps concerned with establishing a leadership role in the field, the government indicated a desire to get into the business early before foreign competition seriously degraded the market and diminished Canadian opportunities to secure a good portion of the potential industry that would result from their efforts. It seemed that once the government had decided to pursue satellite communications, it wanted to proceed with that development as quickly as events would allow them to.

Research and consultation on satellite communications began immediately. A number of trips were undertaken by the TFS in the fall of 1967, including visits to Italy, Britain, and France to exchange views and knowledge on their respective domestic satellite communications research. Interviews were also arranged with key personnel at home both in Canadian government and industry. Similarly, the Air and Space Institute at McGill University's Faculty of Law was consulted, as were other academic and professional scientific organizations such as the Canadian Aeronautics and Space Institute. By early November, enough material on the subject was collected and analyzed to allow the TFS to submit an initial report on its findings and make recommendations back to Cabinet. Asked to deliver quickly, Drury obliged his employers in good order.

Ultimately, the efforts of the TFS proved successful. Cabinet was generally pleased with Drury's findings and proposal and accepted the task force's initial report in December 1967. Further consideration of the matter took place in early 1968, when Drury introduced a more finalized version of the White Paper to Cabinet. Titled *A Domestic Satellite Communications System for Canada*, the ninety-four-page bilingual document consisted of seven chapters and three appendices outlining current and future options for Canada in the field of satellite communications technology.[55]

The first half of the report summarized general techniques and achievements in satellite communications research and development to date, including Canada's own success with experimental satellite communications technology in its previous *Alouette-ISIS* project. The remaining four chapters dealt with the heart of the matter, most importantly providing a number of key decision recommendations to put Canada in a position to exploit domestic satellite communications as soon as the early 1970s.

The task force envisioned an initial space-based communications system of two synchronously positioned satellites in geosynchronous orbit over Canada providing country-wide coverage. Each satellite would have the capacity to transmit between four and twelve television channels. Alternatively, each channel could handle as many as six hundred two-way telephone circuits. One satellite would be designated as the primary while the second or 'back up' would act as a redundancy for continuity of service should the primary experience technical difficulties or fail. A third satellite would be built and held in reserve on the ground, ready for launch should a catastrophic event cause the premature failure of either of the two orbiting satellites. Finally, each satellite was expected to have an operational life

---

[55] Canada, *White Paper on a Domestic Satellite Communications System for Canada*. Ottawa: Queen's Printer, 28 March 1968; hereafter referred to as the Drury White Paper.

of between five and seven years.[56] In addition to the space segment, three major Earth stations were envisioned, one each for television, telephone, and data. As well, a tracking, telemetry, and command (TT&C) facility would be needed to operate and maintain the satellites in their positions.

The whole endeavor was expected to cost somewhere between $40 million and $75 million for the three satellites, including the research and development costs. The ground stations alone would cost in the range of $100,000 each, with an additional few million dollars for the TT&C station and staff. As with most space projects, however, without any precedent it was difficult to predict accurately in advance just what the final costs would be.[57] The White Paper recommended that the government was in a position to shoulder most of the cost, and that these expenses could be spread over a period of four to six years assuming that early development did not meet with any serious obstacles.

Perhaps the most interesting part of the proposal was the brief discussion on the importance of the satellite communications project to the future of Canadian television broadcasting. In the late 1960s, many parts of Canada still had no access to broadcast television due to the prohibitive costs associated with the emplacement of terrestrial microwave feeds in remote areas. As a result, only those people living in large urban centers or along the Canadian-American border where such systems were more common could easily and regularly receive television programs broadcast by satellite through local television stations. Canada's new domestic satellite communications plan proposed to eliminate this problem from space and with the installation of new locally placed receiving stations.[58] It was a small and yet remarkable demonstration of consideration for all Canadians, including its native populations, living in the remote north beyond urban and rural Canada, and showed that space technology could still support national welfare if not the national economy as well.

Finally, Drury recommended in the report that, given the size and scope of the entire effort, the new satellite communications system should be a true national undertaking under government jurisdiction and control. He suggested the formation of a new Crown Corporation, which would ensure that the system was Canadian made, that real competition would exist between suppliers and contractors, and that there would be some guarantee of the infusion of new technologies into the system as it matured. Also, a Crown Corporation could ensure an efficient sale of services while at the same time fulfilling what the report described as 'the minimum conditions for financial success.'[59] With a certain amount of government control at the very outset, Drury noted, he felt that Canada could guarantee some measure of regulation as well as profitability from the development of the new technology.

The Prime Minister and Cabinet agreed in general with this assessment. Shared public and private ownership of the corporation was considered a valid proposal given the magnitude of the cost, and Cabinet foresaw the government holding fifty-one percent of the new corporation's shares. Also, the government agreed that the competition brought by the new satellite system was good for Canada, but that it "would eventually be such as to

---

[56] Drury White Paper, p.42.

[57] Ibid., p.42.

[58] Ibid., p.32.

[59] Ibid., p.46.

induce the private sector not to use the new system unless it was given some participation in its ownership and operation".[60] Similarly, it was estimated that the private sector would demand that its own interests be protected, but the Cabinet cautioned that, "it should be made clear that there would be no such protection in the future as the monopoly on the diffusion of ideas was very illiberal".[61]

Thus the overall aim of the new domestic satellite communications effort was to eventually link all Canadians together from coast to coast, preferably in both of Canada's official languages, English and French. It was also seen, however, as an opportunity to stimulate Canadian industrial capability and participation in advanced technology concepts, as towards the end of the 1960s Canada's lag in adopting more modern space systems writ large was starting to reveal itself. While Canada had demonstrated success in space electronics and subsystem design over and over, its collective experience in directing large-scale projects such as launch and satellite systems was in fact waning. Both government and industry were regularly warning of the 'de-industrialization' of Canada, so this new project was conceived and designed to encourage new research and development, regulation, industrialization, and even international cooperation and investment. Still, much work lay ahead to put Canada's next generation of satellites into orbit and, with them, the Canadian Broadcasting Corporation (CBC) in everyone's living room.[62]

Towards a New Agenda for Space

The findings and conclusions of the Chapman Report produced in the mid-1960s served as a catalyst for the establishment of a new government agenda for outer space that would guide it through to the mid-1970s. After years of seeming political aloofness and indecision the country's space advocates finally received clear signals from the government that Canada's role in space was to be formalized to some degree. With plans for indigenous launch capability discarded for an emphasis on organization and communications, Cabinet directed the government to begin committing the resources necessary to build and deploy a domestic satellite communications capability for Canada by the early 1970s.

Such decisions were likely linked to the larger trends of the space race as realism replaced idealism. Not only had the United States won the race to place men in lunar orbit in December 1968, they soundly defeated the Soviet Union when *Apollo 11* landed safely on the lunar surface in July 1969. Soon after this climactic event, however, the romanticism of space travel appeared to recede from the public eye or interest, as did political support for large-scale expensive programs even as impressive as *Apollo*. The enthusiasm that had fuelled public support for the space program at the height of the Cold War had been dulled by the American experience in Vietnam, and some began to question the need of continuously beating the Russians in outer space when it appeared that the Soviet Union was taking the lead in military capability and influence on Earth.

---

[60] Proposed White Paper on a Domestic Commercial Satellite Communication System for Canada dated February 27, 1968. Privy Council Office, Series A-5-a, Vol.6338, RG 2, LAC.

[61] Ibid., p.1.

[62] An appropriate 'national' project for the centennial year.

Within a year of the first moon landing the United States government curbed funding and began eliminating subsequent lunar missions. With an initial plan to make ten lunar landings, the program subsequently ended in 1972 after only six missions to the surface.[63] Though the United States continued to pursue manned space flight in the 1970s, there was less apparent need for haste or unrestricted funding. Instead resources were turned towards other endeavors such Earth-orbiting satellites and less-expensive robotic interplanetary explorers.

Canada faced a similar evolution within the scope of its own rocket and space program. The initial growth that was spurred on by Cold War security concerns, defence, and a genuine attempt to innovate and create new technologies, had reached a plateau at the end of the 1960s. The civilian-oriented industrial policy designed to take over during that decade was ill defined and yet to mature, and ultimately failed to provide the impetus to sustain a high rate of technological development through into the 1970s.[64] Although there was little question that industrial competence in advanced technology (included space-related systems) was essential for the nation's future prosperity, the means simply did not exist to push Canada to sustain its own diversified and comprehensive rocket and space program.

The Canadian government chose to abandon those projects it felt were unbeneficial to the Canadian economy, already struggling to avoid recession, and focus instead on a select few areas where true political and economic returns could be realized in the short term. Of those space capabilities then easily within Canada's grasp, satellite communications became the priority. The country had achieved considerable success in the field thus far, had a small but highly trained and experience cadre of scientists and engineers dedicated to the field, and was capable of designing and manufacturing most of the technologies needed to put a communications capability into orbit. Those technologies and resources that Canada did not already have it would arrange to procure from other sources such as the United States.[65]

---

[63] *Apollo* 18, 19, and 20 were subsequently cancelled after cutbacks to NASA funding in the late 1960s/early 1970s. The lunar missions that landed men on the moon consisted of *Apollo* flights 11 through 17, with *Apollo* 13 not landing due to damage suffered from an explosion in its service module while en route to the moon.

[64] This fact is demonstrated, for example, by attempts in Canada to create an indigenous computer industry in the 1960s. Though both government and private industry were encouraged to buy Canadian, national economic policy at the time encouraged subcomponent manufacture and importation from the U.S. rather than developing complete systems at home. As well, the market for electronic goods and services simply did not exist to make large-scale technological development attractive or profitable. Attempts to use the government as a large 'customer' backfired when policy contradicted departmental directives. This was especially true with defence, designated as a prime advocate for advanced technological procurement while at the same time having its budget and purchasing power drastically reduced. For related notes see J. Vardalas, *The Computer Revolution in Canada*, Cambridge: MIT Press, 2001, 173–180.

[65] The American industrial base became a solid resource for Canadian technological procurement. Since Canada could not encourage or sustain its own indigenous space technological development, economic policy was designed around a type of moral persuasion that encouraged American technology firms to cross Canadian tariff barriers and invest in high-level research and development and manufacturing in Canada to a level commensurate with their sales. This gave Canada access to the space technologies needed to build its own satellite systems. See Vardalas, *The Computer Revolution in Canada*, p.173.

Reorganizing Canada's Space Departments

Seemingly endless government departmental reorganization remained the bane of Canada's space program throughout the 1960s. Without the oversight of a dedicated minister or space agency it was a constant struggle to set goals, secure resources, complete projects, and maintain a long-term vision. As with all other Canadian space projects to date, the satellite communications system project ultimately needed both government and private resources to be successful. The difference this time, however, was the fact that the Cabinet had endorsed the Drury White Paper directly.

In 1967–1968 Canada's Department of National Defence remained the largest owner of national space resources, material, and expertise. The Defence Research Board still oversaw the Churchill Research Range and its launch facilities and resources, while the Defence Research Telecommunications Establishment was actively engaged at the time in preparing for the launch of the first of the *ISIS* series satellites. Though the National Research Council and various civilian universities boasted a reasonable cadre of space scientists, the engineering component rested largely within the domain of Canadian defence research.

Rather than create new parallel civilian organizations to provide for the Drury White Paper plan, Cabinet proposed instead to simply transfer Canada's space expertise out of the DND and into the civilian sector. This would allow for civilian authority and control over the country's existing space resources, while not demanding any significant degree of new funds, staff, or effort. Of course, the Department of National Defence was not pleased with the decision. It was then struggling to consolidate and ratify its own military interests in space and the loss of the DRTE and its associated space resources came as a serious blow to defence space programming. The Minister of National Defence at the time, the Honourable Leo Cadieux, did not openly oppose the decision and as a result little could be done by the DND to stop the transfer of its resources. What was clear, however, was the political decision that satellite communications had priority over all other space projects or programs. Satellite communications was now *de facto* Canada's national interest in space.

In April 1968, as the Drury White Paper was making its way through various channels, the government created a new Department of Communications (DOC) and assigned it the responsibility of oversight over all Canadian domestic satellite communications development.[66] Soon afterwards, the DRTE initiated the transfer of all of its material resources and personnel from the Defence Research Board into the new department, and renamed itself the Communications Research Center (CRC).[67] As the best concentration of experienced satellite engineers within the federal government, the transfer of this group to the civilian sector reflected both the end of Canadian focus on defence and scientific-oriented satellite projects and its transition towards the new 'civilian' space agenda. It also, unfortunately, continued the trend of exclusivity within Canada's space program by isolating various

---

[66] Eric Kierans was appointed as Minister of Communications.

[67] If the scientists' general attitude towards defence was similar to that experienced during the *Alouette-ISIS* program, it is unlikely that many were disappointed by the move into a civilian department.

components of the country's space capabilities in various organizations rather than bringing all space endeavors under a single banner. The situation did not go unnoticed, and once again Canada's primary space advocates tried their hand at convincing Cabinet to bring all of the country's rocketry and space efforts under the aegis of a single civilian agency.

In July 1969, Drury, by that time Chairman of the Privy Council Committee on Scientific and Industrial Research, tabled a confidential memorandum before his committee and later to Cabinet that outlined the imperative for the creation of a central body to coordinate Canada's new space activities.[68] Drury noted how the rapidly expanding space projects within government, the scientific community, and industry all continued to grow in relative isolation from one another causing in some cases duplication of effort. Using the 1967 Chapman Report to further support his argument, Drury detailed how a single centralized agency could better coordinate a national space program such as the domestic satellite communications system, and how such a centralized body was the proper tool for oversight over the current transition from pure scientific research towards space applications.[69] More important, Drury argued, an amalgamation of all currently existing organizations into a single Canadian Space Agency would ultimately lead to savings in the space program budget overall.[70] Finally, the plan called for the immediate creation of an Interim Canadian Space Council (ICSC), composed of senior representatives of government, industry, and academia, to prepare the groundwork for the official Canadian Space Agency that would follow.

Despite a well-thought out and supported argument, Drury's efforts to create political traction for a new organization were to no avail. The newly installed Trudeau Cabinet was not convinced of the need for a full-up Canadian Space Agency and no government department would step forward to support the plan on its own.[71] Also, some ministers questioned if there was any need for yet another government organization. The recently established DOC was functioning well so far; it had successfully negotiated the purchase of a controlling stake in *Intelsat 1* from the United States the previous year, and appeared well prepared for the upcoming creation of a new Crown corporation – Telesat Canada – that autumn. Meanwhile, the National Research Council was already overseeing the majority of other Canadian space research now that the Department of National Defence had

---

[68] Cabinet Document No.719-69. (confidential) Memorandum to Cabinet – Canadian Space Program: The Need for a Central Body, dated July 9, 1969. DEA. Box 112, File 4145-09-1, RG25, LAC. This was likely the most opportune moment to make the proposal after past repeated failures, and demonstrated the aggressive and insightful nature of C.M. Drury on these affairs.

[69] Chapman Report, p.109–110. See also, *A Space Program for Canada*, main conclusions.

[70] Cabinet Document No.719-69. (confidential) Memorandum to Cabinet – Canadian Space Program: The Need for a Central Body, 3-6. DEA. Box 112, File 4145-09-1, RG 25, LAC.

[71] Part of the 'failure' may have been due to Drury's approach to selling technological policy. His memoranda to Cabinet often exposed glaring gaps in Canadian growth and development and warned of impending disaster if not acted upon immediately, and then in the same note advised caution in approaching the matter and recommend resolving each issue through a series of graduated steps. Examples of this are numerous. His 1969 memo on centralizing administration for Canada's space program is designed this way, as was a previous memo to government in 1968 outlining departmental policy towards computer technology. Though undoubtedly well thought out it did little to incite decisive action and advance technological agendas.

essentially been cut out of the space technology development business for the time being. Neither it nor the new DOC was overly supportive of any proposal to create yet another administrative space organization, especially if it meant taking budget and resources away from either of these new departments.

There was, however, Cabinet consensus towards the idea of greater coordination within existing federal agencies on space activities. Therefore, it was to this objective that political support was given. After much consideration, the Privy Council Committee on Scientific and Industrial Research met once again and agreed that:

> "An interdepartmental committee consisting of representatives of the Department of Communications, External Affairs, Industry, Trade and Commerce, Energy, Mines, and Resources, Transport, National Health and Welfare, Fisheries and Forestry and Agriculture, the National Research Council, and the Defence Research Board be [sic] formed to consider;
>
> a. The desirability of establishing an Interim Canadian Space Council and, if accepted,
> b. The terms of reference, organization, and reporting channels of the proposed council, and,
> c. The Chairman of the Interdepartmental Committee be designated by the Chairman of the Privy Council Committee on Scientific and Industrial Research."

This was not exactly what Drury and his colleagues had desired, but it reflected their staged approach to building policy and bureaucracy, and Drury himself appeared content to accept this decision for the present in lieu of fighting harder for a more formal agency. As well, the concept of an interim organization made logistical sense and it immediately suggested that the ultimate decision to create a Canadian Space Agency would logically follow eventually after the rest of the pieces were in place. It is also interesting to note that the Department of National Defence was not counted among those members assigned to the ICSC; instead, the DRB was chosen to represent defence interests. Given the schism between DRB and DND space agendas at the time, it is likely that defence received little real representation within Cabinet at all, and it is very possible that their lack of direct voice or presence at the table could only have further contributed to diminishing any concrete interest in defence space programs at the end of the decade.[72]

Finances also reflected the waning interest in initiating further defence space projects. In the fiscal year of 1969/1970 the Canadian government was spending roughly $17.3 million on space.[73] Of this amount, $7 million was budgeted to the NRC with another $6 million budgeted to the new Department of Communications. By contrast, only $4.2 million was assigned to the DND, of which $2.3 million went to the reorganized Defence Research Establishment (Ottawa) and the remaining $1.9 million to the Defence Research Board. Most of this funding, however, was devoted to the ongoing *ISIS* satellite project with very

---

[72] It is very likely, given the fact that DND was at the time being integrated and then politically unified, as well as massive budget cuts, that any defence space programs were considered by the study group as too expensive and therefore automatically unsupportable. However, there were also competing agendas at stake, as reflected in the final recommendations back to Cabinet.

[73] Confirmation of the Decisions of Cabinet Committees – An Interdepartmental Committee on Space, dated December 4, 1969. PCO, Series A-5-a, Vol.6340, RG 2, LAC.

little being attributed to other missile or defence space projects. Finally, there was no political indication that any new funding for defence space proposals was planned.[74] Clearly, Government priorities for outer space lay elsewhere.

The formalization of the ICSC began in earnest in late November 1969, when the Cabinet Committee on Science Policy and Technology met to coordinate its organization. Again time was a factor, as Canada had recently received official notification from the United States that NASA representatives wished to visit Ottawa during December to discuss possible Canadian participation in various American space projects proposed for the 1970s, and the country still did not have an official government body to act as overeer and to coordinate all Canadian space activities at the national or international level.[75] Quickly, the committee agreed that the ICSC, which by then had been renamed the Interdepartmental Committee on Space – ICS – be established immediately with the same membership and terms of reference as outlined previously in Drury's proposal to Cabinet for a centralized body for space activities.[76]

The ICS and America's Post-Program

The ICS met for the first time in January 1970, and consisted of representatives from eight different government departments. It was responsible for coordinating all government involvement in space programs both present and planned. Its mandate "was to advise on policy and planning for Canadian space activities…to ensure the coordinated development of government, university, and industrial activities, and international cooperation."[77] Many bureaucrats, however, felt the ICS to be a questionable mechanism for directing a national space effort, lacking both the executive authority and the dedicated financial authority needed to determine clear objectives and make concrete decisions with respect to national interests in outer space. Some even argued that it "could lead to tragic consequences for Canada in the loss of technological opportunity", and that "…a central coordinating and contracting agency for space research and development"[78] was what was really needed. Succinctly, even though no one department would take on the space portfolio, many officials felt the ICS was a poor substitute for a Canadian Space Agency and consequently gave it little serious respect or support.

---

[74] The absence of any DND space or missile reference in the 1971 White Paper on Defence may also be used to demonstrate the departure of these activities from Canada's defence agenda. This period marked the beginning of a long atrophy in Canada's military space program.

[75] Dr. Thomas Paine, NASA Administrator, visited Ottawa on December 15, 1969 where he invited Canada to participate in the next era of American space travel and exploration.

[76] LAC. RG25, DEA. Box 112, File 4145-09-1, Cabinet Document No.719-69. (confidential) Memorandum to Cabinet – Canadian Space Program: The Need for a Central Body; see also RG2, PCO, Series A-5-a, Vol.6340. Confirmation of the Decisions of Cabinet Committees – An Interdepartmental Committee on Space (Cabinet document 1150-69), dated December 4, 1969.

[77] Government of Canada, Interdepartmental Committee on Space, *Annual Report 1976*. Ottawa: Supply and Services, November 1977.

[78] Chapman Report, pp.109–110; see also conclusion of *A Space Program for Canada*.

Acting only as a consultative body with a policy-planning role and therefore no department or financial authority, it is not surprising then that the Interdepartmental Committee on Space ran into problems almost immediately.[79] Loosely organized, many of its members were inexperienced in space policy, science, and technology matters, indecisive about what Canada's future space policy should look like, and divisive over whether or not space activity was even an appropriate arena for the pursuit of Canadian national interests.[80] Meanwhile, other government departments quickly criticized the ICS for its general weakness as a mechanism for establishing national space policy or executing its decisions. The ICS had no funding authority of its own and members were unwilling to commit funds from their own respective departments; many had no power to do so anyway. Combined with a lack of political strength or consensus, the ICS was subsequently often ignored when it came to making serious decisions concerning Canada's major space projects, as it had no real power or authority to direct or oversee them.

The true organizational weakness of the ICS came to light when an official invitation from the United States to participate in its post-*Apollo* program was horribly mishandled.[81] Shortly after he came to power as President of the United States Richard Nixon established a special Space Task Group (STG) to advise the White House on options for U.S. space program objectives in the post-*Apollo* period. Completing its report in 1969, the STG recommended as a primary objective the development of "new systems and technology for space operations with emphasis upon the critical factors of (1) commonality, (2) reusability, and (3) economy".[82] At the time, the main manned programs in NASA included the Apollo Applications Program (known later as Skylab) and the Space Transportation System (or Space Shuttle). The STG also recommended that the United States promote international cooperation by inviting other nations to participate and cooperate in its future programs.[83] Canada was among those countries personally invited to participate, but was also asked for an official response as soon as possible.[84]

The ICS was tasked by Cabinet to coordinate the government's official response to NASA's invitation. It immediately established four additional subcommittees to address particular issues, whose reports would provide the basis for the ICS formal policy advice to Cabinet. The four subcommittees covered the topics of scientific research, space vehicles and propulsion, satellite applications, and international aspects. Each of these

---

[79] J. Ghent, *Canadian Government Participation in International Science and Technology.* Ottawa: Norman Patterson School of International Affairs, 1979, pp 44–45

[80] Ibid., pp.46–47.

[81] Cabinet Document No.924/72 (confidential), Memorandum to the Cabinet, *Development of Policy and Coordination of Activities in Space Technology.* Dated August 16, 1972. RG 25, LAC.

[82] USA. Space Task Group Report to the President. *The Post-Apollo Space Program: Directions for the Future.* Washington: Congressional Research Service, September 1969, p.iii.

[83] Some historians have suggested this was done also to reduce the possibility of Congress canceling funding or political support for future programs. In the early 1970s there was a large lobby in U.S. government which felt that landing on the Moon was a glorious end, but still the end, to U.S. manned space flight.

[84] Dr. Paine, NASA Administrator, made the invitation during his December 1969 visit to Ottawa.

committees spent nearly a year investigating the American proposal and collecting other data, after which all subcommittees returned a generally favorable recommendation of the United States' offer. There were, however, two particular issues of concern that subsequently derailed the entire deal.

The first issue concerned international cooperation. In late 1970, C.M. Drury, then serving as Treasury Board President, suggested in a public speech that linking international space activities with Canadian domestic priorities was desirable in order to promote Prime Minister Trudeau's vision for an independently-minded foreign policy that was "the extension abroad of domestic priorities".[85] While conceding that continued Canadian–American space cooperation "is undoubtedly desirable and probably inevitable", he also noted, "for this very reason there is a real political need to look beyond the continental relationships. Association with Europe offers such an opportunity and hopefully could be achieved at a tolerable cost."[86] Drury went on to suggest later that Canada might even consider seeking associate membership in the new European Space Agency (ESA) then being considered to replace its two existing institutions – the European Space Research Organization (ESRO), and the European Launcher Development Organization (ELDO).

Fig. 3.11 Established in May 1975, the European Space Agency granted Canada associate member status in 1979

[85] Government of Canada, *A Foreign Policy for Canadians*. Ottawa: DEA, 1970.

[86] C.M. Drury, 'International Aspects of Possible Future Canadian Participation in Space Programs', *Canadian Aeronautics and Space Journal*, February 1971, 33–35.

Drury's push for international space cooperation with the Europeans became strong. He noted that Canada, "would probably have more influence in the process of evolving an international institution" if they joined with Europe rather than maintaining the status quo and "try to influence the United States on the strength of what would necessarily be a relatively very modest contribution to the overall NASA program".[87] Further, Drury argued that entering into a relationship with the Europeans would give Canada access to new scientific and technological relationships with countries like Germany and France, as well as provide new commercial opportunities for the research and development Canada gained through the U.S. Post-Apollo Program. Finally, Drury stated that the government "intends to pursue a space policy consistent with Canadian resources and Canadian objectives".[88] Though all of this was to some degree a valid argument, the undertones of Prime Minister Trudeau's "third option" politics, i.e., alternatives to Canadian–American cooperation, could not be ignored.

Though the scientific lobby within the ICS tended to oppose defence-related or exclusive Canadian–American cooperative relationships, in the case of space policy and programs their attitude became almost the complete opposite. Aside from the fact that few felt space activities was the appropriate place for practicing foreign policy, the fact was that in 1970 the United States was the only country other than the Soviet Union with a mature launch capability. Neither Europe nor Japan, whom Drury had also suggested cooperating with, had their own assured access to orbit, and the European launch program was in fact notorious at the time for its technological failure and disjointed program management.[89] Others still warned against becoming involved with Europe's various space organizations which, at the time, lacked any cohesion or common sense of purpose and were instead constantly quarrelling with each other.[90] Finally, Canada had no history of cooperation with either Europe or Japan in space activities; such exchanges would only come later. In contrast, the United States had assured access to space, had some degree of space control, and had long been a cooperative partner to Canada in the domains of rocketry and outer space. Some members of the ICS felt that diluting Canada's already limited space resources by chasing unstable relationships with under-developed European space organizations just in order to satisfy a political agenda was irresponsible and wasteful.

Within a year of its standing up, the Interdepartmental Committee on Space began to implode. The pro-Europe lobby within the committee stated that the United States space program should not encompass the whole of Canada's future strategy, and that the current disarray in Europe's space planning was a temporary problem that would eventually be overcome. The anti-Europe lobby on the committee disagreed, and argued that there was little connecting Europe's space program to Canada's beyond a party politics. Unable to

---

[87] Ibid.,pp. 33–35.

[88] Ibid., pp.33–35.

[89] For a detailed analysis of European space planning difficulties see S. Johnson, *The Secret of Apollo: Systems Management in American and European Space Programs.* Baltimore: The Johns Hopkins University Press, 2002. Of particular note is Chapter 6, Organizing ELDO for Failure.

[90] D. Simonelli, 'Cooperation in Space', *European Community*, Jan–Feb 1978, 19; see also B. Valentine, 'Obstacles to Space Cooperation: Europe and the Post-Apollo Experience', *Research Policy* I, 1971–1972.

reach a consensus within the committee, the argument initially prevented the ICS from completing its task of providing Cabinet with an appropriate and timely response and agenda for participation in the U.S. Post-Apollo Program. By the time this issue was finally settled (but not resolved), Canada had missed the window of opportunity to participate directly in America's next major manned spaceflight program – the Skylab space station project.

Another serious problem confronting the ICS was that of financial authority. Any contribution to Skylab or the Space Shuttle had to be reviewed within the context of the entire Canadian aerospace program.[91] In 1970, almost the entire aerospace sector was committed to the development of Vertical Take-off and Landing (VTOL) and Short Take-Off and Landing (STOL) technology, and there was no new funding for Canadian space projects beyond that already committed. How then was Canada going to afford to participate in new international endeavors? Cost-sharing with the United States was out of the question; NASA had a firm policy that each contributing nation had to accept financial responsibility for their own involvement, and pay all of their own development and sustainment costs associated with each program or project.

Taken all together, the short timeline, the lack of an overall basic Canadian space policy, financial concerns, and the congestion caused by the debate over the merits of space cooperation with Europe all contributed to the ultimate demise of the ICS' ability to formulate an articulate and timely Canadian government response to the American Post-Apollo invitation. Though the inability of the ICS to react positively to the U.S. proposal was not entirely its own fault (the ICS was always perceived as an interim organization and it was attempting to find scientific consensus during the height of the national science policy debate) it never fully recovered from this initial folly. Though the ICS continued to exist for several years after, much of the government decision-making on science and technology was shifted to other parties and organizations.

The MOSST Takes Over

National reviews of science policy, including the critical review undertaken by the Lamontagne Committee – a Senate Special Committee on Science Policy – had nearly all advocated for the creation of a formal science portfolio within Cabinet. The existing policy structures such as the ICS were not truly effective in advising senior government decision-makers, and the new ministry of state concept employed in other portfolios, noted Canadian political analyst Jocelyn Ghent, "appeared to fit the need for centralization of science policy efforts."[92] Thus in October 1971, the Ministry of State for Science and Technology (MOSST) was created to oversee and coordinate, amongst other science and technological issues, Canada's official space program.

Amongst other things, the MOSST was responsible for working with other departments to ameliorate both the formulation and the execution of science and technology policy, especially when related to international cooperation. Having similar set of terms of

---

[91] J. Ghent, *Canadian Government Participation in International Science and Technology*, p.47.

[92] Ibid., p.18.

reference as the ICS, however, the MOSST was also only an advisory and coordinating body for government without any formal decision-making authority or mechanism and without dedicated funding. What it did have, though, was a clear mandate from Cabinet and a considerable degree of high-level political support. Still, the MOSST found itself challenged immediately by other venerable departments which already had considerable experience with international responsibility and were unwilling to yield it up to yet another new and unproven cog in the expanding Canadian government bureaucracy. In particular, the scientific and technology advisors were not keen to have their own subject matter expertise subjugated to more 'generalist' career public servants and bureaucrats who they felt could not fully appreciate the situation or the longer-term objectives of Canada's science and technology programs. As political scientist James Hyndman noted in an article in *International Journal* in early 1971, scientific-oriented departments "tend[ed] to question the wisdom of turning over science and technology matters from the hand of the expert to the hand of the generalist, and they [were] also apprehensive of the interference of foreign policy goals with the rationality of the missions."[93]

Acrimonious relationships between scientists and bureaucrats was nothing new, but it did underscore the difficult transition then taking place in the overall relationship between science and government in Canada at the beginning of the 1970s.[94] In the decades immediately following the Second World War, the need for technologically-enhanced security allowed science, engineering, and defence to play prominent roles in government decision-making. By the end of the 1960s those roles began to wane, and the standing of both scientist and soldier in policy-making decreased markedly. This transition took place in Canada for many reasons. Some argued that the warranted elitism of scientific advice in government which existed in the 1950s and early 1960s was very much out of place in the counter cultural attitudes of the 1970s.[95] Others still offered more philosophical explanations. As the scholar Yaron Ezrahi noted in his study between science and democracy, "the Icarian dream of flying on toward a 'knowledgeable society' in which ideology and politics are replaced by technically rational choices approved by an informed public, may have lost its earlier hold upon the political imagination".[96]

In the case of Canada's transition away from favoring scientific advisors, other factors were also likely important. Both Prime Minister Trudeau's style of modern government and his own political agenda were very different from that of his predecessor, and science and technology were simply not always considered priorities while he was trying to shape

---

[93] J. Hyndman, 'National Interest and the New Look', *International Journal,* Vol.26, Winter 1970–1971, 5–6.

[94] Nor was it a uniquely Canadian experience. The United States underwent a similar transition between 1968 and 1974. For an example see C. Twomey, 'The McNamara Line and the Turning Point for Civilian Scientist-Advisors in American Defence Policy, 1966–1968', *Minerva,* 37, 1999, 235–258; also C. Twomey, 'The Vietnam War and the End of Civilian Scientist Advisors in Defence Policy', *Breakthroughs: Security Studies Program MIT,* 9:1, Spring 2000, 12–20.

[95] Ibid., p.238.

[96] Y. Ezrahi, *The Descent of Icarus: Science and the Transformation of Contemporary Democracy.* Cambridge: Harvard University Press, 1990, p.15.

and retain his political party's voting base. Social and economic agendas were overtaking more traditional security and industrialization interests in Cabinet, and funding tradition-ally reserved for science, technology, and even national defence was being diverted to other initiatives such as international aid and development. As well, both military and scientific advisors were increasingly disillusioned with their roles in government. In 1970, neither the Defence Research Board nor the National Research Council had the same degree of access or influence that they had enjoyed only a decade before.[97]

The creation of the new Ministry of State for Science and Technology was also a signal that further change was coming in how Canada approached international scientific and tech-nology cooperation. Though perhaps unwelcome to some scientist advisors, government officials countered the scientist lobby's criticisms of who had authority and oversight and highlighted their apparent lack of appreciation for the larger political picture beyond scien-tific discovery for the sake of science. As one political scientist noted, "They [the scientist advisors] usually show little understanding for the argument that only ministries of foreign affairs have the necessary overview to evaluate the respective merits of competing priorities".[98] Despite the fact that some departments felt that MOSST's activities tended to transgress their own boundaries, the organization was perceived by Cabinet as an important component of improving the overall process of Canada's foreign policy decision-making.[99]

Interestingly, the creation of the MOSST was supposed to streamline government pol-icy and decision-making in Canada's international science and technological matters, but at first it produced very much the opposite effect. The ministry initially had considerable trouble developing productive working relationships with other government departments, as their role as "policy coordinators" seemed both threatening and unwarranted. At times, the MOSST even appeared as bullies, forcing their way into other departments' missions and agendas without invitation. Some argued that the high public profile of the MOSST tended to cause the ministry to try and take on everything at once and make hasty "deci-sions" without properly consulting the departments likely to be affected by such deci-sions.[100] This behavior "had the effect of diminishing the government's coordination of international science and technology cooperation instead of strengthening it as intended".[101]

It was in such a manner that the new ministry of state approached the issue of taking over the coordination of the country's future space program. Seizing on the failure of its predecessor, the ICS, to properly address the American Post-Apollo Program invitation, the MOSST submitted a memorandum to Cabinet in August 1972 offering alternatives to the current coordination of Canadian activities in space research and technology develop-ment.[102] Of course, these alternatives all consisted of the MOSST assuming control over

---

[97] There were likely other reasons as well; however, again this is an area of Canadian history of sci-ence and technology that has not been explored in any detail thus far.

[98] J. Hyndman, 'National Interest and the New Look', pp.5–6.

[99] J. Ghent, *Canadian Government Participation in International Science and Technology*, p.19.

[100] Ibid., p.19.

[101] Ibid., p.19.

[102] (Confidential) Memorandum to Cabinet – Development of policy and coordination of activities in space technology, dated August 16, 1972. Cab-Doc. 924/72, RG 2, LAC.

this portfolio. "Canada lacks an adequate mechanism for planning and coordinating the application of space technology in the national interest", the memo argued, and suggested three alternatives as follows:

a. The ICS could be reactivated in its present form but report to Cabinet through the Minister of State for Science and Technology;
b. The committee could become advisory to the MOSST and its responsibilities for formulating policy and coordinating activity could be assumed by the ministry; or,
c. The MOSST could take over the responsibilities of the ICS and the ICS itself could be dissolved.

The Minister initially advocated for option (b) as the MOSST lacked subject matter expertise in space technology within its own immediate staff and would need the ICS membership to advise on various courses of action as new space projects evolved.[103] This actually surprised some members of the ICS, who fully expected the demise of the committee once the MOSST had set up a space task force of its own.[104]

The reality was that, despite its own collective ego, the MOSST still needed the ICS to maintain oversight over current Canadian space activities and help integrate these projects into a more formalized national space strategy and policy. For example, the nature and complexity of Canada's domestic satellite communications project had increased tenfold when the U.S. announced plans to initiate its own domestic satellite communications program in mid-1970, and the Canadian government in turn opted for an American rather than Canadian prime contractor for building its own domestic communication satellites. This move was not very supportive of the 'made in Canada for Canada' pitch employed earlier to sell the government on the idea of pursuing domestic satellite communications, and would be difficult to place within the context of a space policy that was supposed to focus on putting Canadian national interests first.

The MOSST's own initial failure to connect with other government departments and agencies seriously hindered any and all space strategy and policy formulation throughout 1972 and 1973, while its over-commitment of its limited resources in trying to maintain a high public profile in all science and technology matters obstructed efforts to produce an effective working document before the summer of 1973. Still needing to consult heavily with the ICS members and their respective departments, the MOSST was only finally able to prepare the guidelines and recommendations for an official Canadian space policy and present them to Cabinet for final consideration in the spring of 1974.

Canada's First Civilian Space Policy

In April 1974, the Liberal government then in office published Canada's first official civilian space policy. Submitted by the Minister of State for Science and Technology, Mme Jeanne Sauve, and developed with the full participation of the ICS representatives, the

---

[103] Ibid., p.3.

[104] J. Ghent, *Canadian Government Participation in International Science and Technology*, p.48.

document finalized Canada's plan to focus on space applications over pure space science, with a special emphasis on developing Canada's satellite communications. Leaning heavily towards industry, the policy outlined four key points. First, the use of space applications had to contribute directly to the achievement of established national goals. Second, the policy had to ensure the growth of Canadian space industry by moving government space research into industry in accordance with a 'Make or Buy' policy.[105] Third, the policy directed that Canada's satellite systems must be designed, developed, and constructed in Canadian industry,[106] and finally, the policy called for the improvement of domestic space industry to meet domestic needs while continuing to rely on foreign launch services.

The policy made explicit what had been implicit within Canada's space program since the mid-1960s – a transition from purely scientific research towards space applications that contributed to the increase of national welfare. Satellite communications, then Canada's main space effort, clearly fit the bill to achieve this goal. Yet the policy recognized that diversification was also required, and it reflected the Cabinet decision to expand Canada's space technology capabilities beyond just ionosphere research and telecommunications.

More important, perhaps, the decision to pursue a 'Make or Buy policy', a strategy that encouraged joint government–industry capability to develop an indigenous productive capability, was aimed at satisfying domestic space system requirements, providing high-technology employment opportunities and enhancing Canada's aerospace industry's ability to penetrate additional export markets, particularly those in the United States. It was also estimated that such an approach would provide the opportunity to enhance Canadian knowledge of space systems and create opportunities to acquire new space systems hardware.

The caveat regarding foreign launches marked the continuation of Canada's dependency on international cooperation to achieve its space goals. Until the mid-1960s, albeit on a smaller scale, Canada's space program was largely self sufficient with the exception of heavier launch capability. Though recommendations had been put forth to government to build and operate a launch vehicle pad at Churchill for the American-designed *Scout* rocket, this plan never went beyond the concept stage and Canada continued its reliance on the United States for access to space. This issue became a real concern as the number of other nations also seeking access to launch services increased during this period, and additional customers meant increased competition for priority and preference. In order to ensure that access continued throughout the 1970s, Canada's space policy specifically directed that it should consider participating in the supplying nation's (in this case the United States) space program in return for guaranteed access to space. Fortunately, all the mechanisms were already in place to allow exactly that to happen.

---

[105] 'Make or Buy policy' was a term employed throughout Cabinet level memoranda defining the choice involved with technology acquisition.

[106] This line was obviously included as a result of the very bad publicity received from the government's previous decision to select an American company, Hughes, over RCA Victor Ltd. of Montreal, to provide the satellites for the country's communications project. For more details on this satellite provider dispute see D. Dewitt and J. Kirton, *Canada as a Principal Power* Toronto: John Wiley and Sons, 1983, pp.330–345.

Looking to the Future

The domestic satellite communications project was essentially initiated to finally organize and focus Canada's space efforts. It was largely successful in some regards, producing the world's first domestic communications satellite system within five years, and providing enough attention and impetus to Canada's space program to have some of its needs recognized within the highest levels of government. Still, in other ways even this project failed. Rather than providing the catalyst for the creation of an official civilian space agency with a broad mandate for science and exploration, it instead secured a specialized and niche-oriented future for Canada's space program which put commercial application permanently ahead of science and technology research and development. The transformation was so complete, in fact, that it resulted in Canada not launching another scientific satellite of its own for almost thirty years.[107]

Change, however, was both necessary and inevitable. Canada's space program had reached an impasse in 1972. Despite the early success of the country's launch and satellite projects, the Cabinet decision to concentrate most efforts solely within semi-privatized satellite communications during the early 1970s had greatly reduced Canada's overall space potential. Without a national policy and an effective centralized authority, neither defence nor the civilian sector writ large was able to advance their own interests on a broad scale let alone take part in further international efforts. Realizing its own paralysis, the MOSST sought a plan that harnessed the assets of the country's existing aerospace industry and combined it with a political and economic agenda for expansion. Expansion, in turn, suggested a plan for sustained international cooperation. The space 1974 space policy therefore focused on this theme.

Changes in the focus of the Canadian space program at the end of the 1960s also influenced the dynamics of the relationship between science, engineering, defence, and government. Though scientist-advisors played a central role in the initiation of the Canadian satellite communications program, they did not retain their influence long enough in decision-making through to the point where Canada's official space policy began to take shape. The machinery of government solidified in the latter days of the national science policy debate of the 1960s, so that in the end it was the bureaucracy, not the science community itself, that controlled the final stages of national space policy development. As well, social and economic goals rather than purely scientific objectives remained at the heart of the plan. Lastly, the 1974 space policy was designed to foster much-needed advanced industrial development. It was clearly designed to serve specific national interests.[108]

---

[107] ISIS-2 was finally launched on March 31, 1971. Canada's next domestic scientific satellites, MOST and Scisat-1, were launched in 2003.

[108] Namely, those interests such as economic and labor force development, but not necessarily the scientific or military potential of space.

4

Ad Astra: Establishing a Permanent Presence in Space, 1974–1984

Once the dust of Canada's massive government reorganization had settled, the country's space advocates turned their attention towards maturing those capabilities that the government had chosen to focus its efforts on. As highlighted in the recently-released 1974 national space policy, central to these priorities was the further development of the country's remote sensing and telecommunications industries, while at the same time initiating the relationships and processes that would eventually lead to the launch of Canada's first astronaut into space sometime during the mid 1980s.

The new civilian space policy made explicit what had been implicit within Canada's space program since the mid-1960s – a transition from purely scientific and experimental research in support of defence space programs, towards more space applications that contributed directly to the socioeconomic growth of the nation. Domestic satellite communications, then Canada's main space technology development effort, clearly satisfied the government's requirement. Yet the new strategy and policy recognized that diversification was also required in its program, and it reflected the Cabinet's subsequent decision to expand Canada's space capabilities beyond just ionospheric research and telecommunications into other domains.

More important, perhaps, the decision to pursue a 'Make or Buy' policy, a policy that encouraged joint government–industry capability to develop an indigenous productive capability, was aimed at satisfying certain domestic space system requirements, providing high-technology employment opportunities, and enhancing Canada's aerospace industry's ability writ large to penetrate additional export markets in an increasingly globalized world. It was also estimated that such an approach would provide the opportunity to enhance Canadian knowledge of space systems, and create new opportunities to acquire new space systems capabilities, hardware, and expertise.

The caveat regarding foreign launches marked the beginning of Canada's dependency on international cooperation to achieve its own space goals. Until the late 1960s, albeit on a smaller scale, Canada's space program was largely self sufficient, with the exception of heavier launch capability for its experimental satellites. But with a desire to launch larger satellites of its own with increased capabilities, heavier launch capability was a mandatory requirement. In order to enhance Canadian access to such services, the Sauvé space policy

© Springer International Publishing AG 2017
A.B. Godefroy, *The Canadian Space Program*, Springer Praxis Books,
DOI 10.1007/978-3-319-40105-8_4

directed that Canada should consider participating in the launch-supplying nation's own space program in return for a degree of guaranteed access to space for Canadian payloads. For Canada, the foreign launcher for the time being was most likely to remain the United States, a country with whom Canada had been cooperating in space projects since the very beginning of its space program.

Yet despite the hopes of some Canadian space advocates, the ratification of the 1974 space policy still did not provide a sufficient catalyst to entice the government to create a single national space agency to act as oversight and execute that policy. It remained a frustrating state of affairs for all of those involved in the country's space-related policy and management, especially when Canada's two main space partners had recognized the obvious need for permanent national space agencies of their own. The United States had created NASA in 1958, and the European Space Agency was created in 1975, but Canada's own space program remained organizationally and financially fractured, with several separate and sometime competing departments, agencies, and institutions continuing to play their own roles. The result of this forced decentralization, not surprisingly, was a degree of government incoherence and, worse, weakened links between government and the industry that would be needed to drive the main effort in Canadian space program development.[1]

At the highest levels of Canadian government the Ministry of State for Science and Technology (MOSST) and its suborganization, the Interdepartmental Committee on Space (ICS), remained the two official oversight bodies for Canadian space-related activity, especially that associated with external affairs and bilateral cooperation. Theoretically at least, the ICS reported to the MOSST, which in turn reported directly to the Minister and Cabinet. But as political scientist Jocelyn Ghent noted in her 1979 study of the MOSST-ICS relationship, "Since 1975, ICS has been reporting to the Department of Communications rather than MOSST", as most of Canada's program at the time dealt with domestic satellite communications, while space policy planning and other organization continued to suffer from bureaucratic fragmentation.[2]

Within this framework, unfortunately there was no establishment of a single vision for Canada's space program or maturation through a series of mutually inclusive goals. Instead, individual exclusive efforts continued to direct the nature of the national program, though fortunately by the mid-1970s the number and nature of these efforts had expanded considerably. By 1975, Canada was spending approximately $50 million annually on space projects and had negotiated a dozen international space agreements in support of these ventures. The country had also developed a mature satellite electronics and subsystem design and construction industry, and was aggressively contributing to a number of other high-profile international projects such as the planned American space transportation system program. It was disconnected perhaps, yet still a substantial effort as the country continued to evolve as a serious space-faring nation.

Similarly, Canada's space program had fostered the development of a healthy domestic aerospace industry despite its fragmented and inconsistent nature. Bristol Aerospace Limited, responsible for the development of the Black Brant rocket, was one of the country's first space industries but it was soon joined by others as satellite design and construction transitioned out of Canada's defence research laboratories and into the private sector. Both RCA Victor Limited of Montreal and the Special Products and Applied Research

---

[1] J. Ghent, *Canadian Government Participation in International Science and Technology*, p.53.

[2] Ibid., 53.

(SPAR) Division[3] of de Havilland Aircraft were contracted to support the Alouette-ISIS satellite project. With the creation of the Crown corporation Telesat Canada in 1969, Spar and another new company, Northern Electric[4] of Lucerne, Quebec, served as subcontractors for Hughes Aircraft of California, the company that was contracted to build the Anik series of domestic communications satellites for Telesat Canada. Meanwhile, Spar Aerospace Ltd. continued to play a leading role in many other Canadian space projects, notably the next-generation Communications Technology Satellite (CTS) project as well as the space shuttle Remote Manipulator System.[5]

Satellite Communications

With the cancellation of the *ISIS-C* satellite project in 1969, the Canadian government began its shift in effort away from experimental space science projects and towards commercial space application. Acting upon recommendations from both the Chapman report and the Drury White Paper, Cabinet directed its various space organizations to focus on the development of new advanced technologies for domestic satellite telecommunications. The newly-created Telesat Canada was already heavily engaged in preparing for the upcoming launch sometime in mid- to late-1972 of a new venture, *Anik-A1*, the world's first domestic communications satellite to be placed in geostationary orbit. This left the Department of Communications (DOC) to initiate a series of new conceptual studies for other satellite projects to contribute to the overall Canadian telecommunications strategy and policy. Dr. John H. Chapman, now one of Canada's most experienced satellite engineers, was selected to lead the DOC study team.[6]

Not wanting to completely abandon the effort put into the conceptual design of the ISIS-C satellite, Dr. Chapman's group examined options for continuing experimental communications research that would build on the success of the previous projects. Telesat's *Anik* series satellites at the time were based on well-tested existing technologies and delivered low-powered radio frequency capabilities in the 6/4 GHz band.[7] The group therefore advocated the design and construction of a new satellite that could test much higher power, high-frequency signals ranging somewhere in the 14/12 GHz band (KuBand). This experiment would also allow for the testing and evaluation of direct-to-home broadcasting, another area of great interest to Canada's communications strategy. The final concept for this new project, completed by the DOC in mid-1970, was for the time being given the simple functional title of 'Communications Technology Satellite' or CTS. It would be rededicated with a new name, *Hermes*, at a later date.

---

[3] SPAR Division became Spar Aerospace Products Limited in Toronto on January 1, 1968 and later spun off from de Havilland as a separate company.

[4] This company was later renamed Northern Telecom Limited and eventually became Nortel Networks Corporation.

[5] For an overview of this company, see Lydia Dotto. *A Heritage of Excellence: 25 Years at Spar Aerospace Limited*. Missasauga: Spar Aerospace Ltd. 1992.

[6] Hartz and Paghis, *Spacebound*, p.110.

[7] 6/4GHz band meant that the satellite communications system uplinked at approximately 6GHz and downlinked at approximately 4GHz.

Fig. 4.1 A 1975 CTS information brochure included an illustration demonstrating its intended mission of helping people separated by great distances to stay in touch with one another

The CTS project was a particularly ambitious and important endeavor for Canada. In the early 1970s, a significant minority of Canadians, perhaps more than five million, continued to live outside major urban centers and in remote regions across the country with little or no access to rapidly-evolving communications, television, and information-processing technologies enjoyed by those living in the country's major towns and cities. From the Canadian perspective, the CTS project, if successful, would lead to the future development and deployment of satellites that would forever change this equation. Some day all Canadians, regardless of where they lived in the country, would enjoy equal access to the ongoing communications and information data processing revolution.[8]

The major challenge then facing the DOC was how to turn the CTS concept into a reality. During its earliest stages, many criticized the project as too far-reaching and unrealistic. The size of the proposed satellite, its design and manufacture, and its launch and operations requirements all demanded resources beyond the scope of Canadian indigenous capabilities. No satellite had ever operated before in its proposed orbital radius or at that frequency band. New space systems and components would have to be developed by industries that were non-existent in Canada at the time. The new technologies needed for the satellite would very likely add to its overall weight, and therefore increasing considerably its overall cost. Canada also had no means of its own to launch medium or large payloads into space and again would have to rely on foreign assistance. Despite the hope that Canadian aerospace industries would lead its development, the private sector was extremely reluctant to assume the technical and financial risks associated with the development of the CTS project. This forced the government, through the DOC, to once again assume the lead in Canadian satellite research and development, thus reversing a trend it had started with the *ISIS* satellite series where it acted only in a supervisory capacity while industry led.[9]

---

[8] Hartz and Paghis, *Spacebound*, p.105. See also CRC, *Communications Technology Satellite.* CRC Serial Document 06. Ottawa: Research Publications and Documentation Services, May 1973.

[9] Arthur Cordell and James Gilmour, *The Role and Function of Government Laboratories and the Transfer of Technology to the Manufacturing Sector.* Science Council of Canada, Background Study No.35. Ottawa: Information Canada, April 1976, p.249.

Given the requirements to make the CTS a reality, the government quickly assessed that cooperation with the United States was necessary for success. On April 20, 1971, the Department of Communications signed an official agreement with NASA to undertake the Communications Technology Satellite program.[10] Canada would of course conceive, design, and build the satellite as well as operate it once in orbit. NASA agreed to provide specialized facilities and satellite components, as well as arrange for the provision of a launch vehicle to carry CTS into orbit. Once in space, both countries would share equal access to the satellite for their own experimental research purposes.

In Canada, the Communications Research Centre (CRC), a suborganizational element of the DOC, acted as the government lead and was made responsible for the management and systems engineering of the CTS project. It in turn assigned the veteran *ISIS* series satellite developer and director of the Canadian National Space Telecommunications Laboratory, Dr. C. David Florida, as the first CTS program manager.[11] Colin A. Franklin, previously head of the space electronics laboratory at the Defence Research Telecommunications Establishment, was tasked as the project manager.[12] Together, Florida and Franklin assembled an experienced team to support them including Dr. Leroy Nelms serving as Franklin's deputy manager, Dr. Mac Evans overseeing satellite systems, Harold Raine overseeing satellite integration and testing as well as the development of a new satellite assembly and test facility (SATF), and Bob Gruno overseeing satellite subsystem development.[13] The Canadian program of experiments that would travel aboard CTS was developed by Mr. Bert Blevis and implemented by a team led by Mr. George Davies.

In addition to this core leadership, several other government and private industry offices supported the CTS project. From industry, Spar Aerospace was assigned as the prime contractor and supplied the satellite structure and mechanical subsystems. RCA Limited built the electrical and electronic systems for the satellite as well as the satellite antennae and 18 small Earth-receiving stations, while SED Systems Limited provided three larger Earth-receiving stations as well as computer software support. In addition, a number of other smaller Canadian companies participated in the project, including Bristol Aerospace Limited, Canadian Astronautics Limited, COM DEV Limited, and Digital Devices Limited.[14]

---

[10] From an historical perspective, the CTS project remains the most well treated Canadian topic other than human spaceflight. For a good overview of the program's history and evolution see Hartz and Paghis, *Spacebound*, pp.104–170.

[11] Sadly, David Florida passed away in 1971 and was temporarily replaced by John N. Barry from the CRC. He was subsequently succeeded by Irvine Paghis. The Satellite Assembly and Test Facility was later renamed the David Florida Laboratory in his honor.

[12] Colin Franklin was one of Canada's space pioneers with extensive experience in space sciences and programs. For a detailed biography of his career see Andrew Godefroy, *A Collective Biography of the Canadian Space Program.* St. Hubert: Canadian Space Agency, 2007.

[13] Harold Raine, "Satellite Assembly and Test Facility", accessed on the World Wide Web at http://friendsofcrc.ca/Projects/DFL/satf.html on March 30, 2007. Hereafter referred to as CRC [HR] memoir.

[14] A complete list of participating companies is noted in Hartz and Paghis, *Spacebound*, p.111.

Domestic Communications: The *Anik A* Series Satellites

As work on the CTS project progressed, Canada prepared for the deployment of its first domestic communications satellites. The recently-established Telesat Canada Corporation, specifically created by an Act of Parliament to act as the main engine for the development of Canadian telecommunications, was initially tasked with the oversight, design and manufacture of these new satellites. The company soon suffered some controversy, however, when the government subsequently decided to award the satellite contract to a foreign bidder, the Hughes Aircraft Company. The political decision to award the contract for such a symbolic satellite program to an American firm seemed to run counter to the very reason that Telesat Canada was created in the first place – to support Canadian industry. When it was revealed that perhaps less than twenty percent of the contract's total monies would be allocated to subcontracting Canadian firms for various subcomponents of the satellite program, Canadian political and public opposition to the whole program quickly increased. The government defended its decision as Hughes had met all the requirements to be prime contractor and had submitted the lowest bid. Opposition parties decried the decision nonetheless, as it meant less jobs and less money going into the economy at home. For a short period of time, there appeared to be the possibility that the new domestic communications satellites might never get off the launch pad.[15]

As the politicians in Ottawa battled back and forth over the aim and scope of the project, the design engineers at Hughes Aircraft Company got down to work. Whatever hurt feelings existed over the government's choice of an American firm to build Canada's first domestic satellite communications system, Hughes ultimately proved to be a wise and practical decision. Founded in 1932 by the legendary business tycoon and aviator Howard Hughes, by the early 1960s the company had entered the space business and had already achieved considerable success building geosynchronous orbit satellites as well as the famed lunar lander *Surveyor 1*.

The new Canadian project consisted of a three-satellite constellation providing maximum coverage over Canadian territory, especially to those sparsely populated areas in Canada's northern regions. To meet the requirement, the Hughes Aircraft Company Space and Communications Group opted for its recently-designed HS 333 satellite, a second-generation standardized satellite design based on the *Intelsat IV* satellite frame that it planned to market worldwide. The HS 333 was a spin-stabilized, 300-watt, 12-channel, single antenna satellite that took advantage of its cylindrical shape to maximize its payload within the confines of the rocket's payload bus. Still, even contained within a tube, the HS 333 satellites were 1.8m in diameter and nominally 3.3m tall. Deploying a new 'see-through' parabolic antenna reflector designed to minimize the tipping effect of solar pressure, in orbit the HS 333 was capable of providing 6/4 GHz (C band) frequency to all of Canada. As mentioned above, the first HS 333 satellites included twelve transponders each

---

[15] For a detailed analysis of this controversy see Howard Fremeth, 'The Creation of Telesat: Canadian Communication Policy, Bell Canada, and the Role of Myth (1960–1974)'. Masters Thesis, Simon Fraser University, 2005.

with five watts output power – essentially enabling them to transmit twelve television programs or the equivalent of 11,520 one-way telephone channels. Anticipated to have an operational life of five to seven years, the new satellites would depend on an extensive network of over one hundred ground stations across Canada to transmit new TV and phone service to populations all across the country.[16]

Fig. 4.2 Canada's *Anik-A* satellite undergoing inspection by Hughes technicians c.1971. Its solar panel exterior and parabolic antenna design were the result of significant advances in the development of satellite communications technology

The first of the three HS 333 satellites dedicated to the Canadian domestic satellite communications program was ready for mating with its launcher in the autumn of 1972. The 577kg satellite was transferred from the Hughes Aircraft Company to Cape Canaveral,

---

[16] The *Anik A* satellite master telemetry, tracking, and command station was located at Allen Park in northwest Toronto, Ontario.

where it was stacked onto an American Delta 1914 rocket[17] and then moved out to launch complex 17B. As the first satellite in the series prepared for launch, back in Canada a public contest was held to name the new series, and thousands of entries were received. The three-member judging panel eventually selected 24-year-old Julie-Frances Czapla's entry. A supermarket bookkeeper working in Montreal, she had proposed the name 'Anik' – an Inuktitut word meaning 'brother'. According to one of the panel judges, songwriter Leonard Cohen, "the choice of Anik reflected a desire felt by many Canadians to pay homage to one of Canada's native peoples". Ultimately, the name fit both the government's own political agenda as well as being uniquely Canadian.

After a relatively short design, testing, and deployment phase, the first of the *Anik* satellites was ready to fly. Lifting off from the Cape at 01:14GMT on November 10, 1972, *Anik A1* enjoyed a flawless entry into orbit and began its operations soon after. A second *Anik* satellite, *A2*, was launched from the same location a few months later at 23:47GMT on April 20, 1973, followed by *Anik A3* at 23:35GMT on May 7, 1975. This trio of spacecraft marked another first for Canada, as the *Anik* constellation made it the first country in the world to own a domestic communications satellite system in the geostationary orbit.

The CTS Assembly and Launch

With the deployment of the *Anik* series satellites well under way, Canada's space sector now turned its attention to the Communications Technology Satellite program. The first order of business after ratifying the joint development agreement and assigning the team leadership was the development of new facilities for the engineering design, construction, and testing work that followed. At the time, the CRC lacked any specialized buildings for this type of work, and therefore Harold Raine was tasked with rapidly completing a needs assessment for the whole CTS project. During early 1971, Raine and his staff toured a number of satellite test facilities in the United States, including the General Electric test facility at Valley Forge as well as the Jet Propulsion Laboratory in Pasadena, California. That same year, Raine later noted in an online memoir, "NASA designated the Lewis Research Center in Cleveland, Ohio, to collaborate with DOC on the CTS project. This center was equipped with thermal-vacuum and vibration test facilities that could be used in the environmental testing of the CTS."[18]

As a result of this agreement, the Canadian team focused on the construction of a facility that could adequately test various subsystems rather than a completely finished satellite. Still, there was a requirement for an integration area sufficiently large enough to accommodate the assembly and operation of the full satellite including the deployment of its folded solar arrays, and this integration room became the center of the new Satellite Assembly and Test Facility at CRC. at CRC. In addition to this area, the SATF contained

---

[17] *Anik A1* was launched by a launch vehicle configuration Delta 1914 580/D92.
[18] CRC [HR] memoir.

a radio frequency anechoic chamber and antenna range, an environmental test area including three vacuum chambers and a vibration cell, and an office area for the project staff. A small meeting room and observation area comprised the second floor. Again, in a later memoir Harold Raine described the evolution of the SATF is some detail:

> "To cope with the work load expected in the establishment of the SATF, two supervisors were hired. Nick Steinmetz (formerly from RCA in Montreal) undertook the development of the specifications for the vacuum chambers. Bernie Kinney with a great amount of test experience from the United States took on the specifications for the vibration facility. Others from the Communications Research Centre programme office, from Site Services, and the support groups helped in defining the building and its services. Visits were arranged to the sites of potential vendors including HiVac and Unde-Holtz Dicke who eventually supplied the original vacuum chambers and vibrator, respectively. While equipment for performing solar simulation was not incorporated for a number of reasons, but especially because of the anticipated cost, the 8′ × 8′ chamber was designed so a limited capability could be provided by modifying the end bell and adding a xenon lamp. As operations began in SATF, it was realized that a data recording system would soon become a necessity. Since a Digital Equipment Corporation minicomputer that had been "retired" from CRC was available, it was put into use in SATF, with some misgivings. It proved to be a mixed blessing × good because the price was right, but cumbersome to use with out-of-date user interfacing. Ultimately, in later years in the upgrades that eventually occurred, it was replaced with a more modern easy-to-use, more powerful machine."[19]

The new building took about a year to construct and was officially opened for business on September 29, 1972, the tenth anniversary of the launch of Canada's first satellite *Alouette*. Known affectionately as the SATF to those who were part of its construction, the building was later renamed the David Florida Laboratory (DFL) in honor of the CTS program manager C. David Florida, who passed away suddenly in 1971 shortly after the new satellite program began.

In fact, work on the CTS project had already begun by the time the SATF was officially opened. Being an experimental satellite testing new concepts, there was a requirement for several new technologies and components to be developed, none of which yet existed and all of which had to be invented, built, tested, and integrated together whilst respecting the rigid limitations on the spacecraft's total weight. The Canadian CTS team originally wanted to employ a larger rocket to allow them some freedom to design a larger satellite, but NASA had already assigned a Delta 2914 rocket to the project and was not prepared to spend an additional $12 million to replace it with a larger launcher. Weight therefore became the central focus of all concepts and design work on the CTS project, though this restriction only served as a catalyst for considerable innovation amongst the Canadian engineers and technicians tasked with solving the problem.[20]

---

[19] Ibid.

[20] Hartz and Paghis, *Spacebound*, pp.111–120.

The end result of this innovation was a remarkable machine. The CTS had a launch weight of 676kg that would be more than halved to 346kg once it reached orbit. It was only 1.8m high and 1.7m wide, though once the solar arrays were extended it would reach out to a total length of 16.1m.[21] These arrays generated the satellite's power, along with body-mounted solar cells to give the platform a total of 1465 watts of energy. The communications subsystem contained a receiver in the 14–14.3 GHz range and a transmitter in the 11.8–12.1 GHz range, each divided into two 85 MHz channels.[22] The planned operational life of the CTS was just two years, but as with all Canadian satellites launched thus far they often exceeded their designed life expectancies and it was anticipated that, once in orbit, the CTS would be no different.

Fig. 4.3 The CTS completes vibration testing on its x-y axis at the David Florida Laboratory, May 1975

---

[21] For details on this technology see S. Ahmed et al., 'Canadian Solar Array Developments for Space Applications', *Canadian Aeronautics and Space Journal*, 30:1, March 1984, 3–14.

[22] CRC, *Communications Technology Satellite*. Serial Document 06. Ottawa: CRC Research Publications and Documentation Services, May 1973, pp.2–10.

Fig. 4.4 The CTS/[Hermes] spacecraft undergoing solar array deployment testing at the David Florida Laboratory, July 1975

The CTS payload was launched from Cape Canaveral's Launch Complex 17B at 23:27GMT on January 17, 1976 aboard a four-stage Delta 2914 rocket. All stages of the launch went smoothly and the CTS payload was spun up for stability at the conclusion of its initial staging. As the payload made its way into its transfer orbit, NASA handed control of the CTS over to the Department of Communications, which slowed the satellite spin and guided its $60 million investment through its nine-day drift phase and two-day attitude acquisition phase until the spacecraft reached its final destination in geosynchronous orbit over the equator. Then it went through a series of start-up tests to check out the spacecraft systems and certify the payload as ready for operations. In the end everything worked as it should have, much to the relief of both Canadian and American teams who had worked so hard on the project over the past five years. The satellite was ready to begin its work.

Once the operational status of the CTS was confirmed, it was then felt appropriate to give the satellite a proper name. On May 21, 1976 the Honourable Jean Sauve, then federal Minister of Communications, renamed the CTS as the *Hermes* satellite during an in-orbit inauguration ceremony held jointly between the CRC in Ottawa and the NASA Lewis Research Center in Cleveland. It was a fitting name, as Hermes was the son of Zeus and messenger of the Olympian Gods. He was also the God of Science and Invention, very appropriate considering that the satellite proved to be a new milestone in the evolution of satellite communications and direct broadcast.

Fig. 4.5 Another view of the CTS solar array deployment. Note the support mechanism to keep it in place in Earth's gravity

With the launch and orbiting tasks completed, a new operational team was brought in to succeed the initial *Hermes* organization and staff. Mr. John N. Barry, the acting *Hermes* program manager, handed over his responsibilities for the satellite to Mr. N.G. Davies, the director of the newly-created Space Communications Program Office (SCOPO). Over the next two years this office executed the Canadian experimental program, which included dozens of investigations including two-way radio and video for teleconferencing, telehealth, tele-education, and telemedicine. Communications experiments were conducted to connect ground stations in one hemisphere to another, as well as test components associated with specific space technologies and ground technologies. When the *Hermes* satellite exceeded its operational life expectation in early 1978, a decision was made to extend the project for another year so that a focused experiment evaluating direct-to-home television broadcasting could be completed before the cessation of the satellite's operations sometime in late 1979.

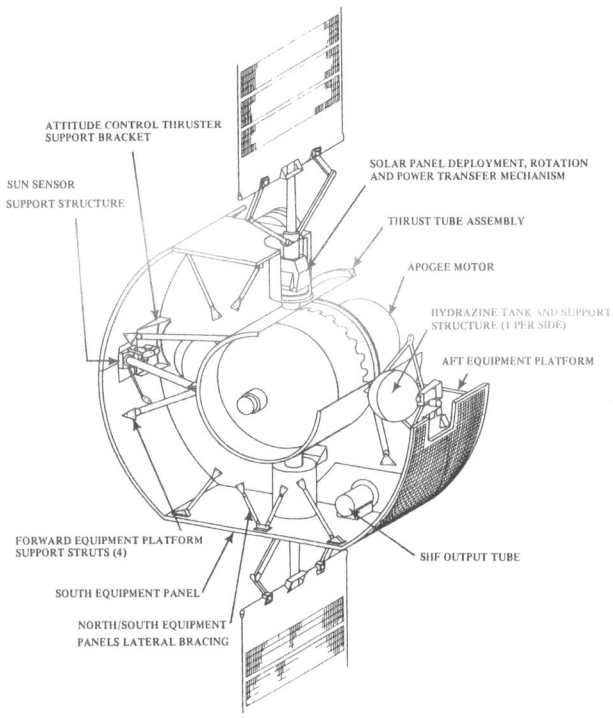

Fig. 4.6 Technical diagram of the CTS and its central components 1

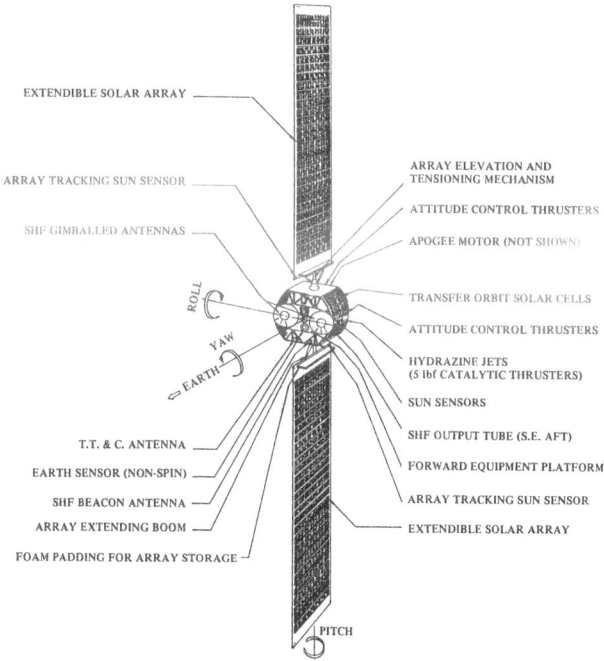

Fig. 4.7 Technical diagram of the CTS and its central components 2

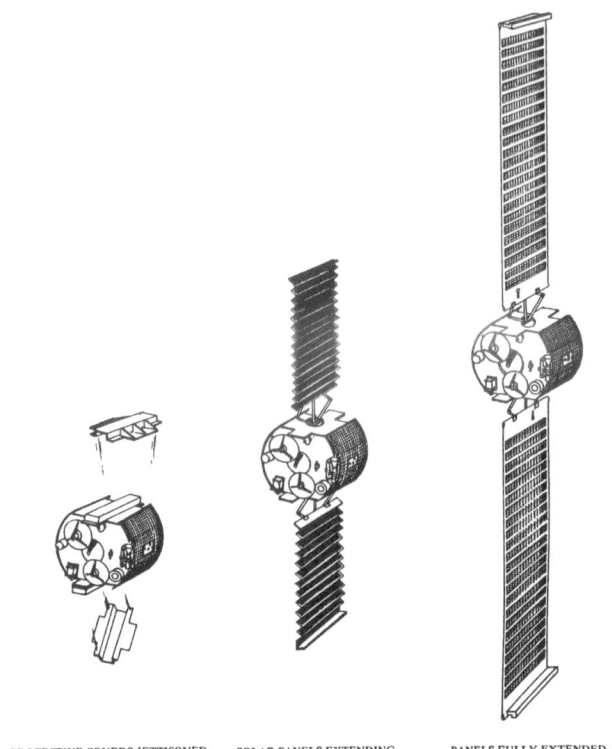

PROTECTIVE COVERS JETTISONED   SOLAR PANELS EXTENDING   PANELS FULLY EXTENDED

Fig. 4.8 CTS solar array deployment sequence

When the *Hermes* satellite project finally concluded, Canadians could be satisfied with themselves for having achieved yet another series of global firsts in satellite communications. More importantly, however, the project served as the catalyst for the development of a second generation of telecommunication satellites that would eventually have the potential to link all organizations and individuals throughout Canada through two-way voice and video communications. For a country whose citizens were used to being dispersed and only intermittently in contact with one another, the arrival of such communications was considered a great achievement. The *Hermes* satellite showed tremendous Canadian innovation, whilst at the same time again exposing the various weaknesses of its fragmented space program. Coordination between the many organizations, departments, contractors, and clients associated with the *Hermes* satellite remained a constant challenge throughout the project, and the development of a coherent departmental policy for the eventual diffusion of the *Hermes* satellite technology out to private industry was lacking. Once again,

Fig. 4.9 A U.S. Delta 2914 rocket launching the CTS from Cape Canaveral's Launch Complex 17B on January 17, 1976

the project inspired its leaders to advocate for the creation of a centralized government space agency with real budgetary authority and central management capacity, but it would still take many more achievements in space before such a dream would finally take hold for good on the ground.[23]

---

[23] For recommendations for the creation of a space agency during this period see Jocelyn Ghent, *Canadian Government Participation in International Science and Technology.* Ottawa: Science Council of Canada, 1979; and Hartz and Paghis, *Spacebound.* Ottawa: Department of Communications, 1982.

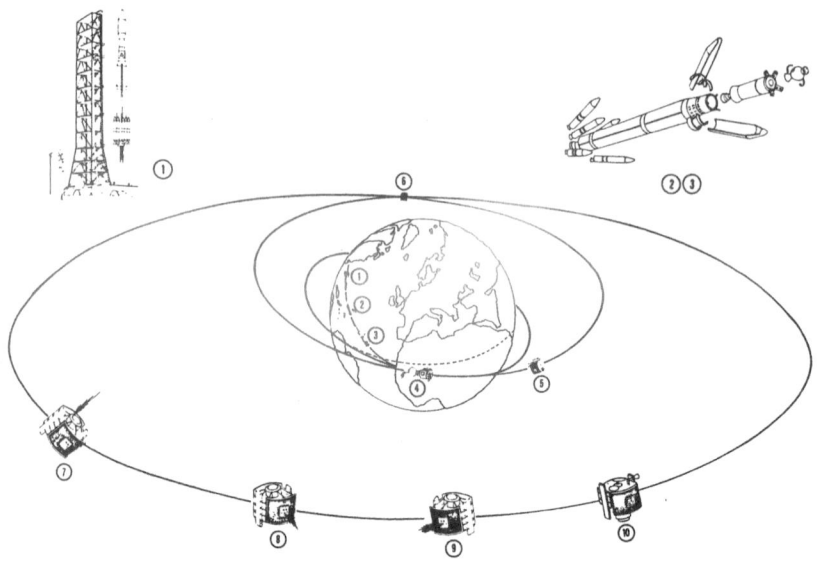

1 – LIFT OFF FROM EASTERN TEST RANGE (CAPE KENNEDY)
2 – SECOND STAGE IGNITION
3 – SPIN-UP TO 60 RPM
4 – THIRD STAGE IGNITION
5 – ALIGNMENT OF SPACECRAFT FOR APOGEE BURN

6 – APOGEE MOTOR FIRING
7 – ALIGNMENT OF SPACECRAFT FOR POSITIONING MANOEUVRES
8 –
9 – POSITIONING OF SPACECRAFT ON STATION
10 – SPACECRAFT CORRECTLY ORIENTED

Fig. 4.10 The CTS launch, orbit insertion, and deployment sequence 1

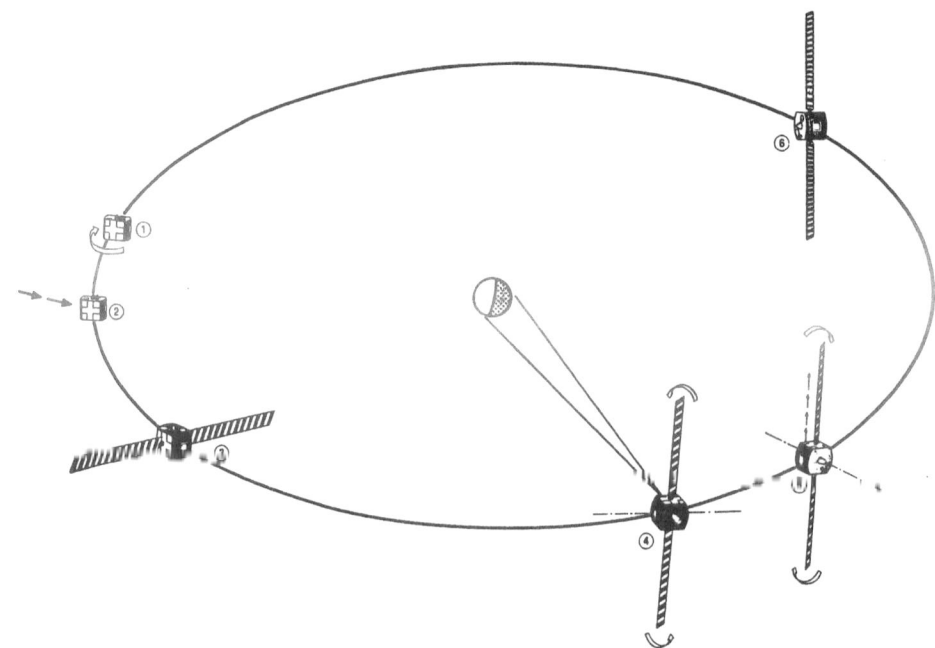

1 – DESPIN FROM 60 RPM
2 – STOP ROTATING WITH SUN SENSOR POINTING TO SUN
3 – DEPLOY SOLAR ARRAYS TO SUPPLY POWER
4 – ROLL ABOUT SPACECRAFT SUN-LINE TO LOCK ON EARTH WITH NON-SPINNING EARTH SENSOR
5 – YAW MANOEUVRE ABOUT SPACECRAFT EARTH-LINE USING SUN SENSOR DATA TO ALIGN ROLL
    AXIS WITH ORBITED VELOCITY VECTOR
6 – SPACECRAFT OPERATIONAL; ATTITUDE DETERMINED BY EARTH SENSOR AND SUN SENSOR

Fig. 4.11 The CTS launch, orbit insertion, and deployment sequence 2

Search and Rescue Satellite (SARSAT) Project

Canada's first civil–military space project since the days of the *ISIS* satellite series came when the government entered into an agreement with the United States and France to design and launch a new satellite system known as the Search and Rescue Satellite Aided Tracking, or SARSAT. Initiated in 1979, the role of the new SARSAT program was to simplify the means by which planes, ships, vehicles, and people activating emergency locator transmitters (ELT) were pinpointed through the use of satellite technology. The SARSAT concept involved placing a satellite in a polar orbit at 850–1000km altitude with a receiver tuned to the international distress frequency to intercept ELT signals and relay them back to Earth, thus pinpointing the location of the signal quickly and passing this information on to search-and-rescue teams.[24] Interestingly, thanks to some thawing of the Cold War in 1980, the Soviet Union formally joined the SARSAT program – after which it then became known internationally as COSPAS-SARSAT – and even took the lead briefly with its launch and deployment of the program's first satellite in 1982. Designated *Cosmos 1383*, the initial COSPAS-SARSAT spacecraft consisted of a Soviet navigation satellite fitted with an ELT signal repeater device that was capable of covering the Earth's entire surface twice a day. Canada followed soon after with the design and manufacture of the local user terminals that received and retransmitted SARSAT signals, and later headed the SARSAT mission control center located at Canadian Forces Base Trenton, Ontario.

In 1984, a second SARSAT was launched from Vandenberg Air Force Base, California, attached to a United States National Oceanic and Atmospheric Administration (NOAA) meteorological satellite using a retired Atlas ICBM as the launch vehicle.[25] Although piggybacking the SARSAT to other satellites helped advance the aims of the project, the Canadian government soon felt that SARSAT would only be truly successful if it was not tied so rigidly to American satellites and their launching schedules. Mr. Rod Hafer, who was the SARSAT project manager at Defence Research Establishment Ottawa (DREO) in 1984, was noted to say, "We have to launch our SARSATs on the NOAA schedule and this doesn't always conform to our schedules. It can mean that we can't always get our satellites up when we want to. This means we wouldn't get maximum value out of the SARSAT system."[26] Though SARSAT was a true joint venture, it was clear that some partners wanted more independence from the United States in the conduct of the SARSAT mission.

---

[24] The international distress frequency is 121.5 MHz. See Lt. A.B. Macpherson, 'The COSPAS/ SARSAT Program: An Example of Successful International Cooperation', *Canadian Defence Quarterly,* Autumn, 1984, 40.

[25] Macpherson, *The COSPAS/SARSAT Program,* p.41.

[26] Ibid., p.41.

Fig. 4.12 An early period COSPAS-SARSAT emblem

Fig. 4.13 Artist's illustration of a SARSAT on station

Yet despite some differences in program ideas and schedules, all four countries in the SARSAT partnership reached agreement on the development of the program and its continued evolution. In the end, the COSPAS-SARSAT program was a tremendous success between these four cooperating nations, a happy achievement in the midst of Cold War-era animosity. Later joined by another pair of satellites, the four-satellite COSPAS-SARSAT constellation continued to operate through to the 21st century and as was employed in 10,385 search-and-rescue events assisting in the recovery of 37,211 people as of 2013.[27]

Expanding Horizons

In addition to its focus on satellite telecommunications, the Canadian government was engaged in a number of other space science endeavors throughout the 1970s. One area of particular interest to Canada was the emerging field of space-based remote sensing. As the second largest country on Earth, after Russia, in terms of physical geography, Canada had a great need for timely and accurate data concerning its own territory, its environmental needs, and its future potential. The Canadian Centre for Remote Sensing (CCRS) was therefore created in 1971 within the Department of Energy, Mines, and Resources (EMR) with the objective of acting as the government's center of excellence for space-based remote sensing. It began constructing the facilities needed and developing the skills required for processing data acquired from American ERTS/LANDSAT satellites with the aim of eventually introducing the technology to Canadian users.[28] Specifically, LANDSAT data were extremely valuable for earth resources management, such as forest, water, and wildlife management, land use mapping, ice reconnaissance, and mineral and petroleum exploration, and this was of particular interest to the Canadian government.[29] Its close space partner, NASA, had launched the first LANDSAT in July 1972, with two additional LANDSATs following it into orbit in January 1975 and March 1978 respectively. A fourth satellite, LANDSAT-D, was subsequently launched in March 1981, this latest version giving Canadian researchers their best color and spatial resolution of the Earth to date.

Beyond earth remote sensing, the Canadian government also explored options to pursue similar data from American ocean monitoring satellites. In December 1982, for example, Canada joined 118 other countries in ratifying a new United Nations Convention on the Law of the Sea, and the surveillance and management of Canadian coastlines took on increased importance. The following year, maritime policy analyst Frank E. Bunn and others completed a major report for the Canadian Institute for Research on Public Policy concerning space-based approaches to comprehensively manage Canadian coastal zones.[30] The Interdepartmental Committee on Space had already negotiated Canadian

---

[27] Figures based on those reported at http://en.wikipedia.org/wiki/Cospas-Sarsat.

[28] Ghent, *Canadian Government Participation in International Science and Technology*, p.41.

[29] DOC, The Canadian Space Program: Five-Year Plan (80/81-84/85). Serial no. DOC-6-79DP Discussion Paper, dated January 1980, 10–11.

[30] Frank E. Bunn et al., *Oceans From Space: Towards the Management of Our Coastal Zone.* Montreal: Institute for Research on Public Policy, 1983.

access to American SEASAT-A data after its launch in June 1978, but the satellite suffered a catastrophic failure after only four months in orbit.[31] Still, before its loss SEASAT-A had provided very useful data and was also noted for being the first satellite to employ synthetic aperture radar (SAR) technology to monitor the world's oceans. The McDonald Dettwiler Company, an emerging aerospace industry in Vancouver at that time, saw opportunity and developed unique capabilities in the processing of SAR data from SEASAT, and this technique subsequently proved so useful to researchers that it was later evolved much further, in fact to the point where Canada would eventually pursue its own space-based SAR satellite systems.[32]

Canada also continued to expand its international partnerships through space development cooperation. The STEM technology developed by Canada for spacecraft communications in the 1960s was later exported to France, West Germany, the United Kingdom, and Japan during the 1970s. In 1975, Canada began exploring its opportunities for cooperating directly with Japan's national space agency. That same year it was also granted observer status in the newly-created European Space Agency (ESA). In 1976, the ICS and the ESA cooperated on both the *Hermes* and *Aerosat* projects, the success of which by January 1979 led to the signing of the first of several five-year cooperation agreements between the two organizations.[33]

Securing a cooperation agreement with the ESA was a considerable achievement for Ottawa and marked another space milestone for Canada, further reflecting the growing maturity of the country's official space program. By 1979, Canada was spending – not including dedicated expenditures to Telesat – just under $96 million on the country's space program, a very modest sum compared to other spacefaring nations considering the number of activities that its various organizations were then engaged in. The lion's share of the annual space budget, approximately 43 percent, went to support major projects such as Hermes, the space shuttle remote manipulation system (SSRMS), the SARSAT program, the follow-on *Anik-C* and *Anik-D* telecommunication satellite programs, as well as the expansion of the David Florida Laboratories (DFL) at Shirley's Bay in Ottawa. A quarter of the remaining federal budget dedicated to its space program went to government operations and support activities, and another quarter to developing new communications services to take advantage of the *Hermes* satellite and *Anik-B* telecommunications satellite experiments. The remaining nine percent of the budget served Canada's bilateral and multilateral cooperation initiatives, such as LANDSAT and SEASAT, as well as the country's growing relationship with its newest international partner, the European Space Agency.[34]

---

[31] Launched June 28, 1978 Seasat operated until October 10, 1978 when a massive short circuit critically damaged the satellite's electrical system and effectively ended the mission.

[32] DOC. The Canadian Space Program: Five-Year Plan (80/81-84/85). Serial no. DOC-6-79DP Discussion Paper, dated January 1980, p.11.

[33] For a preliminary history of Canada–ESA relations see Lydia Dotto, Canada and the European Space Agency: Three Decades of Cooperation. ESA History Project Report HSR-25, dated May 2002 accessed on the World Wide Web at URL http://www.esa.int/SPECIALS/ESA_Publications/SEMQPTZ990E_0.html on March 30, 2007.

[34] DOC. The Canadian Space Program: Five-Year Plan (80/81-84/85). Serial no. DOC-6-79DP Discussion Paper, dated January 1980, p.8.

Canada and Manned Spaceflight

Beyond the development of new facilities and new and more ambitious satellite projects, the Canadian government also sought to renew its involvement in the direct human exploration of space. Having missed the recent opportunity with NASA's Apollo Applications Program (Skylab) in 1970, the ICS endeavoured to be prepared to take advantage of whatever opportunity may be presented next. At the time, only the two superpowers had the capability to launch human beings into space. As the Soviet Union remained a political and strategic adversary, any effort put towards sending Canadians into orbit would be directed towards the United States. Fortunately for the committee it did not have to wait long, for work was already under way in the United States on the next space program that would eventually replace Skylab and see the end of the employment of the venerable Saturn series rocket as America's primary means of putting human beings into orbit.

Though the concept of a reusable space vehicle could be traced through science-fiction literature back prior to the Second World War, the American space transportation system (STS), or space shuttle as it became commonly known, evolved quickly during the latter half of the 1960s. A major proponent of the concept was George E. Mueller, then head of the Office of Manned Space Flight at NASA headquarters. Though at the time he oversaw Apollo operations and was very focused on the ongoing exploration of the Moon, Mueller was also keenly interested in NASA's next project, a space station, and had the foresight to appreciate that the development of reusable low-cost launch vehicles would be critical to making it a reality. In January 1968, he invited industry leaders to meet and discuss their current thinking on reusable launch vehicles with both United States Air Force (USAF) senior officers and NASA officials, with the hope of igniting more serious attention to what would be the next great American adventure after the Moon.[35]

As talks progressed it soon became clear that, while the Americans could potentially go it alone in the development of a space station, both the technological and financial burden might be eased somewhat if they were to opt for the pursuit of international partnerships to support the concept. Canada was at the top of the list. A legacy of Canadian-American human spaceflight cooperation had already put Ottawa in a favorable light as a potential partner for future American-led manned spaceflight systems. NASA had already once before been the beneficiary of Canadian assistance when the Royal Canadian Air Force (RCAF) had sent a number of its best officers with engineering backgrounds on exchange to various American spaceflight establishments across the United States as part of a project then known as the Space Indoctrination Program (SIP).[36]

In early 1961, the RCAF selected twelve officers with advanced technical backgrounds and experience, and assigned them for a period of three to four years to various USAF space programs and projects, though on the odd occasion the tour of duty may have been

---

[35] T. A. Heppenheimer, *The Space Shuttle Decision: NASA's Search for a Reusable Space Vehicle.* NASA SP-4221. Washington DC: NASA, 1999, p. 86.

[36] Correspondence from Air Marshal Hugh Campbell, Chief of Air Staff to Air Marshal C.R. Slemon, Deputy Commander in Chief of NORAD, dated March 28, 1961, 3. Vol.17829, File 840-105-001.8, RG 24, LAC.

shorter.[37] Most, if not all, of the original contingent of RCAF officers fell under the command of the USAF Headquarters Space Systems Division (SSD) in Los Angeles, California, though some members worked at other facilities throughout the United States.[38] The SSD was tasked with the planning, programming, procurement, development, and management of dozens of new space projects and systems, and it also acted as a primary center for capabilities and future systems research, including weapons systems concepts and development.

From the first group of RCAF officers, Squadron Leader (Sqn Ldr.) John Webster and Sqn Ldr. Allan Pickering were assigned first to the Mariner and then later to the Atlas-Agena Program Office, while Flight Lieutenant (Flt. Lt.) Andy Thoma was assigned as a project officer in the Engineering Division of the Standard Launch Vehicle II (SLV II) Directorate.[39] Another officer, Sqn Ldr. Jack Henry was assigned to work in the Aeronautical Systems Division at Wright Paterson Air Force Base and did extensive work alongside Flt. Lt. J.H. Lathey, who was engaged on the high-technology X-20 Dyna-Soar Program. Although the X-20 program was cancelled in December 1963, it was at the time one of the most advanced spacecraft programs in the United States and a precursor to the space shuttle program itself.[40]

Sqn Ldr. Robert White had perhaps the most interesting, if not peculiar, assignment to the Mercury and Gemini Launch Vehicle Directorate. This office was at the heart of the space race, and was directly responsible for getting American astronauts into orbit and eventually to the moon, preferably ahead of the Russians. "I spent most of my first year seconded to the Mercury Program Office", White later recalled, "traveling with the program director, accepting launch vehicles at General Dynamics Astronautics, San Diego, and participating in the final two launches of the Mercury Program at Cape Canaveral. My basic task had been to bring the lessons learned on Mercury back to Gemini."[41] Interestingly, White's duties did not end there. With the ramp up of the Gemini Program in 1966, White noted, "As the GLV [Gemini Launch Vehicle] Pilot Safety Operations Officer, I became responsible for all acceptance and flight safety and pre-launch review board operations, and also for propellants, loading, and engine operations at Cape Canaveral."[42] It was a considerable degree of responsibility and trust given to a young foreign air force officer who shocked his NASA colleagues the first day he arrived at work wearing a Canadian uniform.

---

[37] For the only known published Canadian source on the SIP see Anon. "Canadian Missile Men", *Sentinel*, (September 1969), 32-33. Interviews with RCAF officers in the SIP have revealed that they were never informed that their 'loan' to the United States fell under any organized plan, though official documentation and the above-mentioned article make reference to SIP.

[38] Reports and Returns – RCAF Personnel on Exchange Duties – USAF – HQS Air Force Space Systems Division – Los Angeles – Calif. 1963-1965, File 813-89/3-42, RG 24, LAC.

[39] Exchange Officer Report Flight Lieutenant A. Thoma, Headquarters Air Force Space Systems Division, Los Angeles, California, dated 15 March 1963. File 813-89/3-42, RG 24, LAC.

[40] D.R. Jenkins. *Space Shuttle: The History of the National Space Transportation System*, 30–31.

[41] Email interview with S/L (ret'd) Robert 'Bud' White, March 20, 2003.

[42] Ibid.

Fig. 4.14 RCAF exchange officers routinely worked at various USAF and NASA organizations during the 1960s. Here Captain C.S. Lines (left) and Major B.C. Dimock smile for the camera at the USAF Space and Missile Systems Organization Headquarters

The level of responsibility afforded the Canadian exchange officers throughout their tours was considerable, and their success in their respective assignments was further demonstrative of the quality of RCAF officers in service at that time. Flt. Lt. Lathey made a considerable name for himself as a result of his contributions at the Air Force Flight Dynamics Laboratory (FDL) to the guidance and control and 'fly-by-wire' systems employed in the X-20 Dyna-Soar spacecraft. Likewise, Sqn Ldr. MacFarlane did considerable work on hypersonic ramjet research and development at Arnold Engineering Development Center, Tennessee, and Flt. Lt. T.M. Harris worked on spacecraft re-entry at the FDL.[43] By the end of the 1960s, Canadians had distinguished themselves in many U.S. manned spaceflight projects, helping set the stage for later consultations that would eventually lead to Canada's direct involvement in space shuttle operations.

The question therefore became, what opportunity and potential role was there for Canada in the American space shuttle program? The issue and desire of a manned spaceflight program was discussed, but NASA was not seeking this level of involvement from other countries at the early stages of its own program. Canadian industrial representatives traveled to Washington D.C. in October 1969 to attend a technical conference for the

---

[43] RCAF Exchange Officer Report – F/L T.M. Harris and F/L J.H. Lathey dated February 28, 1965. File 813-89/3-42, RG 24, LAC.

purpose of assessing shuttle concepts and laying out design considerations for the next steps of space shuttle development. Canadian industry was certainly interested, of course, but also somewhat concerned that unless Canada participated on a national basis there would be little hope or opportunity for individual companies to compete successfully against American organizations for contracts. It was this advice that industry leaders gave back to the ICS after the conference for further consideration.

As the ICS deliberated the potential role for Canada through 1970 and early 1971, Spar Aerospace Limited began developing its own package that envisioned a major contribution to shuttle development. It was noted during the 1969 Washington conference that early space shuttle concepts showed a spacecraft armed with a crane of sorts, but that the initial conceptual drawings forwarded by industry did not include anything resembling a remote manipulator system. It seemed instead that more effort was placed on determining the design of the airframe itself and the main engines of the shuttle than the payload bay or how exactly it was to be utilized. A subsequent proposal forwarded by North American Rockwell in 1969, for example, showed a satellite being released into orbit from a set of docking clamps in the orbiter's payload bay. Another design advanced in 1970 by another American space company, Grumman, also showed a satellite deploying unassisted. Even more interesting is a 1971 concept artist illustration showing a NASA straight wing orbiter servicing a space station via astronaut extra-vehicular activity (EVA) and a tethering system. While later shuttle operations did in fact employ both astronaut EVAs and a self-ejected satellite deployment mechanism, one can easily surmise how much the value of a shuttle remote manipulator system (SRMS) to assist orbital payload operations was underestimated at the time.[44]

It was from this initial situation that the Canadian Spar Aerospace Ltd. shuttle arm proposal grew. Mr. Lloyd Secord, an engineer working for the DMSA Acton Limited company had suggested after the 1969 Washington conference that the future space shuttle might employ a remotely controlled robotic arm similar to one then being designed for remotely controlled refueling of CANDU nuclear reactors. Frank Thurston, a National Research Council of Canada engineer later tasked to coordinate Canadian industrial options for the space shuttle, and John McNaughton, an engineer with Spar, supported Secord's idea and carried the proposal even further. A number of feasibility studies were conducted in collaboration with North American Rockwell, the space shuttle prime contractor, and NASA's Goddard Space Flight Centre, throughout 1971.[45] The results of these efforts were promising, and a remote manipulator system was added to space shuttle conceptual designs from 1971 onwards after a number of mission requirements highlighted both the necessity and value of equipping the orbiter with such a system. As director of Spar's Market Development, Mr. Terrence Ussher worked on the development of the initial proposal. By early 1972 the Spar Aerospace arm proposal was complete, and this submission subsequently became the basis for possible Canadian participation the space shuttle program.

---

[44] For a comprehensive analysis of early space shuttle concept design and definition see Dennis R. Jenkins, *Space Shuttle: The History of the National Space Transportation System – The First 100 Missions*. Hong Kong: World Print Ltd., 2001.

[45] Ghent, *Canadian Government Participation in International Science and Technology*, p.48.

The SRMS 'Canadarm'

A remarkable piece of technology, the SRMS design itself consisted of a remotely controlled six degree-of-freedom payload handling robotic arm with a shoulder, elbow and wrist joint separated by an upper and lower arm boom giving it shoulder pitch and yaw, elbow pitch, and wrist pitch yaw and roll. At a length of 15.2m and a total weight of approximately 431kg, it was designed to maneuver payloads of up to 14,515kg at a rate of .06m/sec with a maximum contingency operation payload weight of 265,810kg. Under unloaded conditions the SRMS could achieve a maximum translational rate of 0.6m/sec. Ironically, the final SRMS design was incapable of supporting its own weight on Earth and had to be supported by specialized ground handling equipment during its acceptance testing and later shipment to NASA. Designed to handle very heavy payloads in space, however, the SRMS also demonstrated an impressive degree of finesse that allowed very precise handling of delicate cargos, including the astronauts themselves.[46] A computerized control system commanded the arm and it could deploy payloads to a positional accuracy of +/- 2.0-in and +/- 1.0-degree of a preprogrammed target zone at the rates and load conditions detailed above. Interestingly, if required the SRMS could also be operated manually by an astronaut onboard the shuttle to the same accuracy with the use of hand controllers and closed circuit televisions (CCTV) mounted on the manipulator arm itself.[47]

Since the SRMS would spend most of its existence exposed to the vacuum of outer space, it was designed to have an operational life of ten years or one hundred space shuttle missions. The entire arm itself was covered over with a multlayer insulation thermal blanket system, which provided passive thermal control to the whole arm. This material consisted of alternate layers of godized Kapton, Dacron scrim cloth, and a Beta cloth outer covering. In extreme cold conditions, thermostatically controlled electric heaters (resistance elements) attached to critical mechanical and electronic hardware could be powered on to maintain a stable operating temperature throughout a longer task.[48]

The complexity of the SRMS often reflected how impressive a technological innovation it really was. Each subassembly component of the SRMS, for example the shoulder, elbow, or wrist, was made up of a basic element called a joint one-degree-of-freedom or JOD. The JODs were simply motor-driven gearboxes that allowed the basic structure of the arm to articulate much like the human arm. There were two JODs in the shoulder joint that allowed the whole arm to pitch and yaw. One was located in the elbow joint to allow the lower arm to pitch, and three were placed in the wrist joint to allow the tip of the arm to pitch, yaw and roll. The SRMS was much more articulating than even the human arm and could therefore accomplish very complex maneuvers. The JOD motors were equipped with their own brakes and joint motor speed control. Each JOD also incorporated a device called an encoder, which accurately measured the joint angles. Thus, each joint was

---

[46] Anon. The Shuttle Remote Manipulator System – The Canadarm. McDonald Dettwiler Space and Advanced Robotics Ltd. Accessed on March 30, 2007 on at http://ieee.ca/millennium/canadarm/canadarm_technical.html

[47] Ibid.

[48] Ibid.

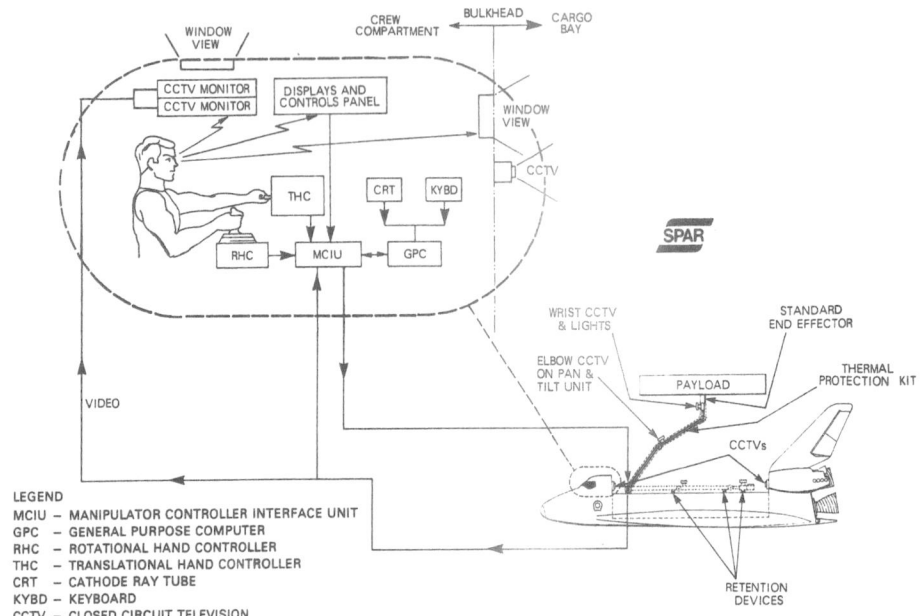

LEGEND
MCIU – MANIPULATOR CONTROLLER INTERFACE UNIT
GPC – GENERAL PURPOSE COMPUTER
RHC – ROTATIONAL HAND CONTROLLER
THC – TRANSLATIONAL HAND CONTROLLER
CRT – CATHODE RAY TUBE
KYBD – KEYBOARD
CCTV – CLOSED CIRCUIT TELEVISION

FIGURE 1   REMOTE MANIPULATOR SYSTEM

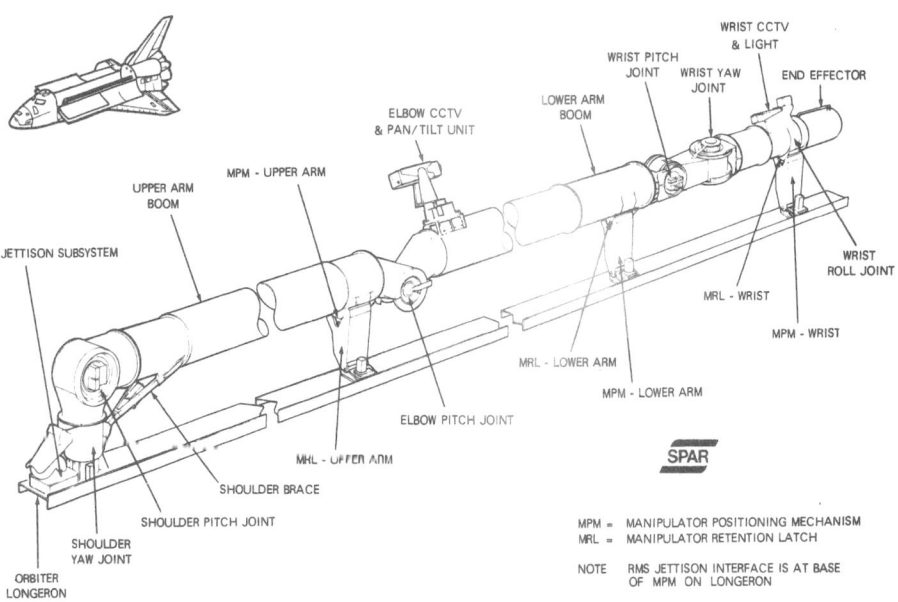

MPM = MANIPULATOR POSITIONING MECHANISM
MRL = MANIPULATOR RETENTION LATCH

NOTE   RMS JETTISON INTERFACE IS AT BASE
OF MPM ON LONGERON

FIGURE 2   MECHANICAL ARM ASSEMBLY

–5–

Fig. 4.15 An SRMS assembly and operation schematic from a Spar Aerospace Ltd. manual c.1981

capable of moving independently at different speeds and in different directions with respect to any or all the other JODs. Every possible range of motion was conceived for the SRMS, giving it a degree of reach never before witnessed in robotic technologies.[49]

Two booms constructed of graphite-epoxy linked the shoulder, elbow, and wrist joints. The upper arm boom was approximately 5m long by 33cm in diameter comprised of 16 plies of graphite-epoxy (each ply was .013cm thick) for a total weight of just under 23kg. The lower arm boom was approximately 5.8m long by 33cm in diameter comprising of 11 plies of graphite-epoxy for a total weight of just over 22.7kg. Each boom was further protected with a Kevlar bumper to preclude the possibility of dents or scratches on the carbon composite. Just as the arm booms linked the shoulder, elbow, and wrist joints mechanically, the wiring harness (electrical cabling) accomplished the same thing only electrically. The wiring harness provided electrical power to all the joints and the End Effector (mechanical hand) as well as data and feed back information from each of the joints. This link continued from the SRMS in the payload bay and carried on into the cabin of the space shuttle where astronauts controlled the actions of the arm remotely. [50]

Perhaps the most impressive part of the SRMS was the End Effector or mechanical hand at the end that allowed the arm to capture stationary or free-flying payloads by providing a large capture envelope (a cylinder 20.3cm in diameter by 10cm deep) and a mechanism/structure capable of soft docking and rigidizing. This action was accomplished by a two-stage mechanism in the End Effector which closed three cables like a snare around a grapple probe (a knobbed pin) bolted onto the payload and then drew it into the device until close contact was established and a load of approximately 499kg k was imparted to the grapple probe. All of this occurred under the observation of two closed circuit televisions (CCTV), one at the elbow joint and one at the wrist joint. The CCTV units were by the astronauts to help the positioning of the arm for payload capture, retrieval, and deployment.[51]

From inside the space shuttle, astronauts originally used a general-purpose computer and two handgrips to give the SRMS instructions. Built-in software examined the inputs commanded by the astronauts and calculated which joints to move, what direction to move them in, how fast to move them, and what angle to move to. As the computer issued the commands to each of the joints it also looked at what was happening to each joint every 80 milliseconds. Any changes inputted by the astronauts to the initial trajectory commanded were immediately re-examined and recalculated by the computer and updated commands were then sent out to each of the joints. The SRMS control system was continuously monitoring its "health" every 80 milliseconds and, had a failure occurred, the computer would have automatically stopped all joints and noted a failure. The control system provided a continuous display of joint rates and speeds, which were displayed on monitors located on the flight deck in the orbiter, allowing the astronauts to closely monitor every aspect of the SRMS task. As with any control system, the computer could be over-ridden and the joints operated individually from the flight deck by the astronaut if needed.[52]

---

[49] Ibid.

[50] Ibid.

[51] Ibid.

[52] Ibid.

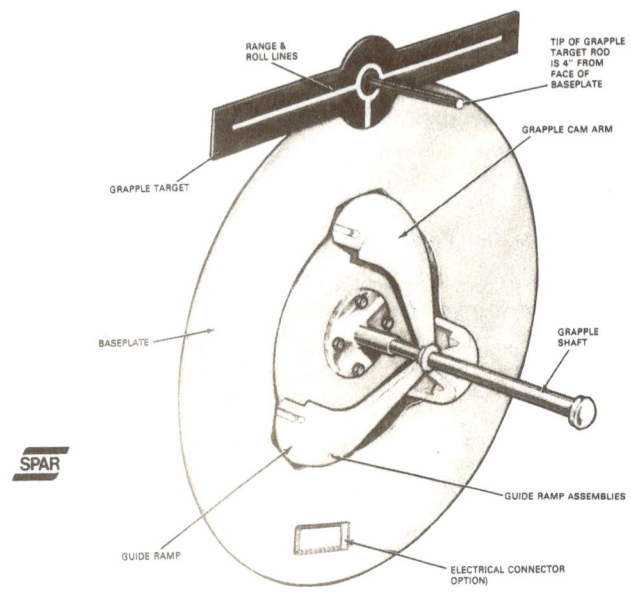

FIGURE 3   GRAPPLE FIXTURE AND TARGET ASSEMBLY

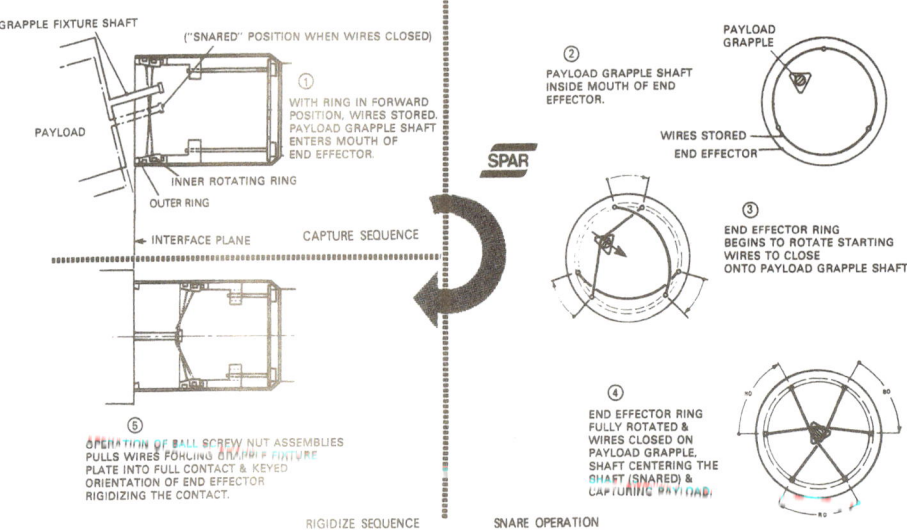

FIGURE 4   END EFFECTOR OPERATION

Fig. 4.16 Diagram of the SRMS End Effector operation

The Canadian SRMS project was later identified by the President of the National Research Council's Annual Report as "one of the most important developments for the future of space research in Canada", but in 1973 acceptance of the the technology company's proposal still remained in some doubt.[53] The new national space policy had not yet been revealed, and in the absence of a centralized government space organization there was no political champion to sponsor its development. The ICS, being only an advisory committee, had neither the budget nor the authority to supervise the SRMS, and neither the Department of Communications nor Telesat Canada could justify the SRMS within their own mandates despite both being obvious Canadian space advocates. Needless to say, SPAR and other Canadian aerospace industries were frustrated. It appeared for a time that Canada would once again miss an opportunity to become engaged in human spaceflight endevors.

Fortunately, space advocates prevailed and a directed government policy appeared the following year from Cabinet. There was still some concern over whether or not the high investment made into developing the SRMS technology would in the end pay dividends in the Canadian aerospace and robotics technology sector, but with a policy in place it seemed likely at least that the company's proposal would receive the support it needed. An official invitation to participate in the American space shuttle program was received from NASA Administrator Dr. Thomas Paine in July 1974, and Canada's Ministry of State for Science and Technology announced that the National Research Council had opened discussions with NASA for the proposal of official Canadian participation in the American space shuttle program soon after. This was also followed by an approval from the Treasury Board for $1 million to fund further studies of the SRMS technology project.[54] Under the 1974 space policy's "make-or-buy" conditions, the Treasury Board also later committed $75,787,000 to the project with an additional $12 million contingency funding in case of any development uncertainties. This may seem like a huge figure, but it is necessary to consider the difficulty budget planners experienced at the time trying to estimate the cost of something that had never been done before.

The potential policy implications of the whole project,, plus its total cost of nearly $90 million required the government to seek a formal treaty with the United States. On July 9, 1975, the National Research Council concluded a Memorandum of Understanding (MOU) with NASA for a cooperative program concerning the development and procurement of a space shuttle attached remote manipulator system, followed by another official Exchange of Notes on June 23, 1976.[55] This MOU also laid out the initial details of "the provisions for access by Canada to the use of the space shuttle and for the procurement by NASA of additional RMS units".[56] Last but not least, the MOU also effectively further shortened the project's acronym of SRMS to just RMS. NASA was planning for a total fleet of five space shuttles, meaning a requirement for at least five RMSs, and in return Canada sought some of the payment in the form of direct access to spaceflight.

With the agreements in place, more than 25 Canadian companies went to work on the SRMS project. Spar Aerospace Limited served as the prime contractor, with CAE Electronics Limited as well as several other companies providing various parts of the project.

---

[53] NRC. Report of the President, 1975–1976. Ottawa: Information Canada, 1976, p.102.

[54] MOSST. News release dated July 16, 1974. The definition phase of the SRMS project eventually cost $2.5 million.

[55] Canada Treaty Series MOU 9 and July 18, 1975, Exchange of Notes June 23, 1976. No reference number.

[56] Ibid.

Dr. Garry Martin Lindberg, a Cambridge-educated mechanical engineer who joined the National Aeronautical Establishment (NAE) of the NRC in 1964, was brought in as the project manager for the RMS and guided it from conception in 1974 to completion in 1979. Dr. Karl-Heinrich Doetch, previously the assistant director of the NAE, joined him in 1976 as the Simulation Systems Manager and closely assisted Dr. Lindberg throughout the many phases of the project. Dr. Doetch was also responsible for the analyses and simulations necessary in the design and verification of the RMS and its potential developments, a critical step in ensuring that the final arms would be ready for space operations.

Fig. 4.17 The *Canadarm* takes shape. Engineers conduct vibration testing on the RMS joint at the David Florida Laboratory

Soon after, Spar Aerospace Limited appointed John D. MacNaughton as vice president and general manager of the company's RMS Division. The new division was responsible for the overall design, development, and manufacture of the space shuttle's remote manipulator systems. An obvious choice for the post, Mr. MacNaughton had graduated in Aeronautical Engineering in 1954 from the de Havilland Aeronautical Technical School and Hatfield College in the United Kingdom, and had previously worked on a number of Canadian space-related projects, including the famous STEM antenna technology. As chairman of the AIAC Space Committee in the late 1960s and early 1970s, he played a central role in informing Canadian space policy and seeking opportunities for Canadian industry. Terrance Ussher, the gentleman who produced the initial RMS proposals in 1972,

Fig. 4.18 A technician makes an adjustment on the RMS End Effector. Only one small piece of the whole device, this photo provides a good sense of the overall size of the *Canadarm*

assisted MacNaughton in a number of management roles. He was responsible for the testing of the individual RMS subassemblies, the design and procurement of all ground support equipment for the project, and the delivery of the first flight system to NASA.

As the RMS was too heavy to test completely on Earth, a specially-designed general-purpose manipulator system simulation facility was created at the Spar plant in Weston, Ontario, to conduct the majority of the testing under simulated zero gravity conditions. The facility was completed in October 1976 and was employed throughout the critical design phase of the project. In April 1978, NASA conducted a review of the Canadian progress and was suitably impressed, satisfied that the first RMS would be ready and delivered on schedule. The Americans were not to be disappointed, with qualification and flight test units ready in July 1979.[57] The first space-ready RMS was officially released to NASA on February 11, 1981, having been driven from the Spar Aerospace plant in Toronto down to Kennedy Space Centre on specially-designed trucking.[58] During the official handover

---

[57] A letter contract for a follow-on production (FOP) program was signed between NASA/JSC and Spar on May 1, 1979, with the definite contract signed in April 1980.

[58] One account indicates that the driver for this task was the same man who drove the famous King Tutankhamon exhibit across North America. His name remains unknown. For reference see NRC. "Canadarm: Background", *Science Dimension*, 14:2, 1982, 2.

Fig. 4.19 SPAR technicians prepare a section of the RMS for vacuum chamber testing

ceremony in Toronto, Dr. Larkin Kerwin, then president of the National Research Council, officially christened the remote manipulator system as the *Canadarm*, the name by which the system has been commonly known since.[59]

---

[59] National Research Council of Canada. Canadarm STS-3 Press Kit. Undated, p.11.

Fig. 4.20 A full-length view of the RMS undergoing wiring testing at the Spar facilities

The first *Canadarm* was integrated into space shuttle Columbia spacecraft in June 1981, and flew into space for the first time that November aboard the second space shuttle mission (STS-2).[60] NASA astronauts Joe Engle and Richard Truly flew a near flawless mission, and the newly-christened *Canadarm* performed admirably to their expectations. Perhaps the most memorable part of this mission in fact was the *Canadarm's* first unveiling to the world. Prior to the mission, as the RMS was being prepared for spaceflight, NRC program manager Art Hunter and his assistant manager, Bruce Aikenhead,[61] officially arranged that a beta cloth Canada wordmark and Canadian flag be attached near the base of the arm. When astronaut Richard Truly unberthed the arm during the second day of the spaceflight, television cameras caught a majestic sight of the RMS with its prominent Canada

---

[60] For an overview of this mission see NASA History Division. History of the Space Shuttle. Accessed on the World Wide Web on March 30, 2007 at URL http://history.nasa.gov/shuttlehistory.html

[61] Bruce Aikenhead was one of the many skilled engineers who, having started at A.V. Roe Canada on the Avro Arrow, left to work for NASA after the program's cancellation, where he subsequently trained American astronauts for three years. He then returned to Canada and became one of the pioneers of the Canadian space program. He worked on Dr. Gerald Bull's controversial High Altitude Research Project, on Canadian satellites, and on the Remote Manipulation System (RMS or Canadarm) as the Deputy Manager of the National Research Council's Remote Manipulator System program.

Fig. 4.21 Testing the RMS End Effector against a mounted grapple fixture and target assembly at the Spar plant

wordmark clearly visible with the Earth rolling by in the background. The scene was truly breathtaking for Canadians and instilled tremendous pride all across the country, and has since become one of the most iconic images of the space shuttle era as well as the Canadian space program writ large.

The Space Shuttle Columbia returned to Earth after the completion of its second mission on November 14, 1981, landing on runway 23 at Edwards Air Force Base, California. In post mission reports and statements, the appreciation of the performance of the Canadarm was universal. Prime Minister Pierre Trudeau congratulated the NRC and Spar, "whose talents were directly responsible for the design and construction of the Canadarm, which lived up to all expectations during its first test in space".[62] John Roberts, Minister of State for Science and Technology and also Minister responsible for space policy, echoed Trudeau's sentiments. He noted, "the effectiveness of the arm not only demonstrates the world-class technology of which Canadian industry is capable, but also is testament to the benefits of an international collaboration, which advances the interests of all mankind."[63]

---

[62] National Research Council of Canada. Canadarm STS-3 Press Kit. Undated, p.12.

[63] Ibid., p.13.

Very satisfied with the Canadian technology, NASA quickly put in place an agreement for the delivery of a further three RMSs to arm additional space shuttles, one each scheduled for delivery in May 1982, November 1983, and November 1984. In addition, it asked the NRC to pursue the design and development of a space vision system (SVS) that could help shuttle astronauts use the Canadarm to grapple moving objects on future missions. The long-term plan was to employ the space shuttle for the deployment, retrieval, and redeployment of satellites; thus it would need a sophisticated means of both capturing orbiting objects as well as returning them safely to their stations. It was also expected by some senior Canadian bureaucrats that, if successfully developed, eventually perhaps a Canadian astronaut might test the final product aboard a future space shuttle flight.

Fig. 4.22 This National Research Council Presskit for the space shuttle's third mission proudly displays the new logo for the Canadarm

Fig. 4.23 A Canadian launch pass pin for the STS-41C mission highlights the use of the *Canadarm* to deploy the LDEF (Long Duration Exposure Facility)

The Canadian Astronaut Program

Such a major technological contribution to the American space shuttle program by Canada also gave the country an opportunity to join the shuttle crews that would fly aboard the orbiter. In 1979, NASA sent a letter to John Chapman, former chairman of the ICS, suggesting it was open to allowing Canada to fly its own astronauts aboard the space shuttle at some future date as two European astronaut candidates had already entered the process and were expected to fly. Unfortunately, Chapman died later that year and so the invitation was not initially acted upon. Three years later, in September 1982, the issue resurfaced. Lieutenant General James Abrahamson, then serving as NASA associate administrator, publicly issued a formal invitation to Canada to join the shuttle astronaut corps during a celebration in Ottawa to mark the twentieth anniversary of the launching of *Alouette 1*. Needless to say, those politicians and bureaucrats in attendance to receive the news were delighted, and in turn delegated the NRC to seriously study the offer with a view to returning a recommendation to the government as soon as possible. The Canadian space research community was widely canvassed by the NRC about potential experiments that could be conducted aboard the shuttle, and it received numerous submissions from which it selected two – one dealing with space physiology and the other being proposal for the development of a space vision system.

At the beginning of the space shuttle program, NASA designated two categories of crewmembers for the new orbiters: pilot-astronauts and mission specialists. To account for other potential crewmembers joining missions from outside the traditional astronaut corps, however, NASA created a third designation known as payload specialists. Originally, these add-ons to the shuttle crews were not even referred to as astronauts, as they were not employed by NASA and were not expected to fly into space more than once. In fact, the only reason a payload specialist might join a crew was because that person had specialist knowledge or skills to carry out a specific task related to the client's payload. Otherwise, the professional astronauts then in NASA saw them as something of a burden, or worse, an obstruction to those anxiously waiting for their own assignments to a shuttle crew and mission.

Regardless of the designation, the Canadian government was pleased to have the opportunity to fly its own citizens aboard the shuttle. Having officially endorsed the American offer by the end of the year, the NRC was directed by the government to officially stand up its own Canadian Astronaut Program and begin soliciting the Canadian population for potential candidates as soon as possible thereafter.[64] Karl Doetsch from the NRC's Space Division was selected to direct the initial program, and he decided to initially recruit as many as six people to become payload specialists, not just to ensure that there would be two people ready to carry out the initial experiments, but also to leave the door open for other opportunities that may arise. As Clive Willis, director of the NRC public relations and information services office, commented, "we are not picking six people to go out and do experiments alone; we are picking an astronaut corps."[65]

On July 14, 1983, a unique "help wanted" advertisement appeared in Canadian newspapers all across the country. Its message, however surprising it must have been to many who read it, was straightforward. The National Research Council was looking for Canadian men and women to join them for up to three years, develop two sets of scientific experiments to be performed aboard the U.S. space shuttle, help inform the Canadian public about human spaceflight and Canadian space activities, and lastly, be prepared to carry out and assess one of the experiments. In essence, the NRC was looking for someone to become the first Canadian to fly in space.[66]

---

[64] The history of Canada's astronaut program is perhaps the best-developed aspect of the country's space history to date. Two popular books on the subject exist by the same author. For a detailed popular account of Canadian astronauts see Lydia Dotto, *Canada in Space*. Toronto: Macmillan, 1987; and Lydia Dotto, *The Astronauts: Canada's Voyageurs in Space*. Toronto: Stoddard Limited, 1993.

[65] Dotto, *Canada in Space*, p.56.

[66] According to the ad, the starting salary for the astronaut candidates was to be between approximately CDN$40,000 and $55,000 depending on previous qualifications and experience.

Fig. 4.24 The Canadian Astronaut Program was launched in 1983

The NRC left the deadline for application open until August 8, 1983 with the expectation of receiving maybe as many as a thousand applications. In the end they received well
over four thousand responses, far beyond what the application office had originally prepared itself to handle. Despite the stipulations outlined in the call, applicants ranged in age
from 6 to 73 and came from all walks of life. Those who had applied with less serious
intentions were immediately weeded out of the pile, with the remainder being thinned
down by two separate committees and a selection panel over a six-month period. The
screening committee was chaired by Madeleine Hinchey, then serving as Secretary-
General of the NRC, and included Lorne Kuehn, director of the science and technology
(human performance) division at the Department of National Defence, Ray Marchand of
the ICS, and Karl Doetsch and Clive Willis.[67]

---

[67] Op. Cit. Dotto. *Canada in Space*, pp.56–57.

Four rounds of elimination took place between the initial call for candidates in July and the final selection in December. As with any astronaut candidate program the selection process was both meticulous and merciless, with many seemingly suitable applicants being cut for one reason or another. By November 1983, the NRC produced a shortlist of 19 potential persons, from which the final six were chosen on December 3. Out of more than 4400 applicants, Ken Money, Steve MacLean, Bjarni Tryggvason, Robert Thirsk, Roberta Bondar, and Marc Garneau were ultimately selected to be Canada's first astronaut group. They each received the news by telephone, with orders to be prepared to report to their new offices in Ottawa shortly after.

Space Team Canada   L'Équipe spatiale canadienne

Canada

National Research   Conseil national
Council Canada   de recherches Canada

Fig. 4.25 The first group of Canadian astronauts, 1983. Standing left to right – Ken Money, Roberta Bondar, and Bjarni Tryggvason. Seated left to right, Robert Thirsk, Steve MacLean, and Marc Garneau

Roberta Lynn Bondar was born in Sault Ste. Marie on December 4, 1945. She received her early education at the University of Guelph, earning a Bachelor of Science degree in zoology and agriculture in 1968. She went on to earn postgraduate degrees in pathology, neurobiology, as well as a doctor of medicine from McMaster University in 1977. She became a fellow of the Royal College of Physicians and Surgeons of Canada in neurology in 1981. A qualified pilot since age 23, her knowledge and expertise in medicine, and in particular neurology, made her an obvious choice for Canada's new space program.

Born in Quebec City on February 23, 1949, Marc Garneau received his early education at the Royal Military College of Canada and went on to serve as a naval officer with the Canadian Forces. An expert in engineering and naval weapons systems, he spent his early career troubleshooting new systems being fitted on Canadian warships. Promoted to the rank of Commander while attending Canadian Forces College in 1982, he was assigned to Ottawa the following year where he worked on naval communications and electronic warfare systems. Garneau was selected as a Canadian astronaut candidate in December 1983.

Steve MacLean was born in Ottawa, Ontario, on December 14, 1954. He received his early education at York University, earning degrees in physics and astrophysics. An expert in lasers and electro-optics, the future president of the Canadian Space Agency would first lead critical work on the Advanced Space Vision System (ASVS), (ASVS), a computer-based camera system designed to provide enhanced guidance to the control of the Canadarm in space. When not engaged in world-class physics research, MacLean pursued his passion for gymnastics, competing at the Olympic level with Canadian national teams. This combination of fitness and scientific mind made him a solid choice for selection as an astronaut in December 1983. MacLean later admitted in an interview with journalist and author Lydia Dotto that when he received the phone call at home with the good news he did a backflip right then and there.[68]

Kenneth (Ken) Money was born in Toronto, Ontario on January 4, 1935, making him the oldest member of the first group of Canadian astronaut candidates. That said, he was perhaps the most likely candidate to be chosen from all applicants, having already had extensive experience working with NASA scientists since 1962 in the field of space motion sickness and orientation. Money had earned degrees in physiology and biochemistry while training in the Canadian military to become a pilot. Later graduating from the National Defence College at Kingston, Ontario, he went on to become the senior scientist at the Defence and Civil Institute of Environmental Medicine in Toronto, having logged over 4000 hours flying time on a number of jets and fixed-winged aircraft. Perhaps surprising no one, Ken Money was selected in the first group of Canadian astronauts in December 1983.

Bob Thirsk was born in New Westminster, British Columbia, on August 17, 1953 and later earned degrees in both mechanical engineering and medicine. An expert in biomedical engineering, he was serving as Chief Resident in Family Medicine at the Queen Elizabeth Hospital in Montreal in 1983 when he was selected as one of the first astronaut candidates. Having received the good news of his selection, he took his then girlfriend to one of Montreal's most exclusive restaurants and later proposed to her. They were married in January 1984, but the honeymoon would have to wait after Thirsk was selected as Garneau's backup for the first Canadian spaceflight.

Bjarni Tryggvason was born in Reykjavik, Iceland on September 21, 1945 and later earned degrees in engineering physics, applied mathematics, and aerodynamics. An expert in meteorology and the effects of weather and climate on structures, in 1982 he joined the Low Speed Aerodynamics Laboratory at the National Research Council of Canada as an associate research officer. At the same time, he served as a lecturer in building aerodynamics at the University of Ottawa. This combined experience and expertise led to his selection as an astronaut candidate in 1983.[69]

---

[68] Lydia Dotto, *Canada in Space*, p.67.

[69] Contemporary biographical information for the original six astronauts derived from National Research Council of Canada. *Canadian Astronaut Program – Space Team Canada, January 1986.*

Fig. 4.26 Ken Money was the most obvious choice for selection as an astronaut candidate, but the Challenger accident in 1986 derailed his chances of getting on a mission, and he subsequently resigned in 1992 without ever having flown in space

In early 1984, Canada's astronaut candidates pulled up stakes from their former lives and reported to the National Research Council east of Ottawa, where they were installed in rather nondescript offices at the back of building M-60. Their workplace was defined solely by a Canadian Astronaut Program decal on the office door, and the six candidates shared three tiny offices with face-to-face desks. Yet despite this rather unglamorous introduction to the astronaut business they did not have to wait long for things to get busy. At a meeting with Larkin Kerwin in January 1984, the candidates were informed that NASA had offered a spot on a space shuttle mission scheduled for that October – a mere ten months away. It was a full year earlier than expected or planned, but the government had officially accepted the offer so it was time to get on with things. Still, the astronaut candidates were excited. Yes, it would mean even tighter training deadlines, but the addition of a flight sooner than anticipated increased the chances that each of them would ultimately have their own chance to fly. As MacLean later noted, "the program has barely begun and already it's expanded".[70]

---

[70] Dotto, *Canada in Space*, pp.80–81.

The original plan for Canada's astronaut corps had been to fly two missions in 1985 and perhaps one more in 1986. The first mission, dedicated to physiology experiments, would be flown by one of the "life sciences" candidates, and the second, dealing with the development of the ASVS, would be flown by one of the "engineering" candidates. A third mission offered the opportunity to fly a follow-on experiment, but this would be decided later. For now, the six candidates needed to do everything they could to get ready. Among the tasks on their plate – pilot's licenses. The three who didn't have one – Garneau, Thirsk, and MacLean – decided that they would try and check this box off in early 1984. Another of the candidates, Tryggvason, agreed to serve as their instructor. Interestingly, Garneau did not actually get his first solo flight in before his first spaceflight, possibly making him the only astronaut up to that time to record shuttle flight time in his log book before Cessna-172 flight time.

At the weekly staff meeting held on March 13, 1984, Doetsch announced with little preamble that Garneau had been selected to fly first with Thirsk chosen to serve as his backup. A round of handshakes followed by a bottle of champagne was quickly followed in turn by briefings from the NRC public affairs officer, Wally Cherwinski, on the questions they could expect to receive the next day at the scheduled press conference. Once this was done, the candidates went about their business. Garneau decided to break the good news to his family in person rather than phone. His wife, it was later reported, "went into shock".[71] Garneau himself admitted to the press on March 14 that his new situation "hasn't quite sunk in yet" but there was not much time to reflect on such things. Both he and Thirsk had just six months to get ready for Canada's first spaceflight and there was still so much work ahead of them.

The schedule was demanding. In addition to the training and study, both Garneau and Thirsk were expected to maintain an active public appearance across Canada. Eventually, however, their training took them more and more often to the Johnson Space Center (JSC) in Houston, Texas. In early August, the pair moved to Houston to train with their U.S. crew in preparation for the upcoming flight. Garneau and Thirsk were given offices in a wing of Building 32, located on the outer fringes of the JSC, which was the newly-designated area for all payload specialists. Needless to say, it was a less than warm welcome for them at JSC, as most of the "career" astronauts were located over in Building 4, which was much closer to the heart of space operations at the Center. Both, however, understood to some extent the general ambivalence that was shown towards them. In 1984, open crew slots were still rare and many NASA astronauts quietly resented the fact that these temporary newcomers were being added to crews when they themselves had yet to fly.

Over the following weeks, the pair slowly integrated themselves into the NASA crew that Garneau would eventually fly with. Thirsk was especially pleased to learn that Robert L. 'Bob' Crippen, a veteran astronaut who flew on the very first space shuttle mission, would command their flight. He made them both feel welcome, and according to Doetsch, "was a big factor in helping Marc be ready".[72] After months of training, simulations, and other preparations, everyone was about to find out.

---

[71] Ibid., p.93.

[72] Ibid., p.116.

Canada's First Astronaut: Flight STS-41G

Spaceflight STS-41G was the thirteenth American space shuttle mission and the sixth launch of the space shuttle Challenger. Like its predecessors it too was a mission of many firsts. It was the first mission to consist of a crew of seven astronauts.[73] It was the first flight to include two female crewmembers – Mission Specialists Kathryn D. Sullivan and Sally K. Ride. It would witness the first spacewalk by an American woman – Kathryn Sullivan. The crew also included the first NASA astronaut to fly a fourth space shuttle mission – Commander Robert L. Crippen. The mission would undertake the first demonstration of a satellite refueling technique in space, and would also be the first shuttle flight to employ a reentry profile crossing the Eastern United States. Finally, STS-41G would include the first flight of a Canadian payload specialist, naval officer and astronaut Marc Garneau.

Fig. 4.27 The STS-41G crew. Canadian Forces Naval Officer and Canadian Astronaut Program Payload Specialist 2 Marc Garneau stands to the right of the mission Commander Robert L. 'Bob' Crippen at center

---

[73]The crew of STS-41G comprised Robert L. Crippen (commander), Jon A. McBride (pilot), Kathryn D. Sullivan (Mission Specialist 1), Sally K. Ride (Mission Specialist 2), David C. Leestma (Mission Specialist 3), Paul D. Scully-Power (Payload Specialist), and Marc Garneau (Canadian Payload Specialist).

Fig. 4.28 The Earth Radiation Budget Satellite – ERBS – being deployed by the Challenger crew, October 1984

Space Shuttle *Challenger* lifted off the pad at Kennedy Space Center (KSC) in the early morning hours of October 5, 1984, carrying its crew to orbit in what appeared to be another flawless launch from Florida. Garneau steadied himself for the short 8.5-minute ride to achieve orbit, but any anxieties he may have had were soon washed away once he had the opportunity to first look out at the Earth. "That first view is spectacular," Garneau remembered in a later interview. "That's something that you've been dreaming about seeing for the longest time. The thing that struck me was how crystal clear the colors were. I was expected there would be a sort of haze due to pollution or moisture in the air." Garneau, like many of his fellow astronauts before him, was deeply moved by the experience of seeing for the first time the Earth as a whole planet, the home to all human beings.[74]

As much as Garneau could have spent the entire mission looking out the window, he had many tasks before him that had to be completed in a few short days. The STS-41G mission carried a number of major experiments aboard. NASA flew its Earth Radiation Budget Satellite (ERBS) as well as its third Office of Space and Terrestrial Applications (OSTA-3) payload, along with the Orbital Refueling System (ORS) Experiment. As a direct participant in this mission, the National Research Council of Canada flew its own package of ten investigations known as the Canadian Experiments (CANEX) payload.[75]

---

[74] Op. cit. Dotto, *Canada in Space,* p.17.

[75] NASA, Space Shuttle Mission 41-G Press Kit, dated October 1984, p.22.

Fig. 4.29 Payload Specialists Marc Garneau [left] and Paul D. Scully-Power, a civilian oceanographer with the U.S. Navy, carry out one of the Canadian experiments on Challenger's middeck during STS-41G

The CANEX package covered various fields, including space technology, space sciences, and life sciences. Key amongst these experiments was further testing of the NRCC's Space Vision System (SVS). During the mission Garneau conducted a number of recordings of stationary and moving test targets with the RMS closed-circuit television cameras, collecting data that would assist in the development of a future system to be used in space for rendezvous, inspection, and assembly tasks beginning in 1986.[76] In addition, CANEX included an advanced composite materials exposure experiment. This test consisted of different material samples attached to the RMS and exposed to the direction of the shuttle spaceflight for thirty-six hours. Upon return to Earth, the materials were evaluated for any deterioration that may have occurred during the short duration flight. This information assisted with the ongoing development of new materials needed to build future spacecraft, satellites, and space stations.[77]

---

[76] Ibid., p.22.
[77] Ibid., pp.22–23.

The CANEX package also contained two important space science experiments. The first was a hand-held Canadian-made sun photometer instrument which when pointed at the sun obtained readings of solar radiation at several wavelengths in the visible and near-infrared region of the spectrum. This data assisted the Canadian Atmospheric Environment Service in its measurements of local atmospheric constituents and the monitoring of acidic haze. The second experiment, Atmospheric Emission and Shuttle Glow Measurements, was a response to the discovery of a phenomenon whereby during spaceflight some surfaces of the space shuttle orbiter developed a faintly visible reddish glow. The causes and characteristics of the glow were not fully understood at the time and there was some concern that it might affect the optical instruments of payloads. The experiment employed a very high-resolution optical filter and image intensifier to obtain photographs that were later analyzed to identify the reactions which produced these emissions.

Finally, Garneau conducted a series of studies known as the Space Adaptation Syndrome (SAS) Experiment, in which he isolated and measured several of the key adaptation processes which occurred in people during the first days in space. Being a first-time space traveler himself, Garneau was an ideal test subject. Over a period of several days, he examined a number of human conditions including vestibulo-ocular reflex, sensory function in limbs, proprioceptive illusions, awareness of external objects, space motion sickness, and taste in space. Each investigation yielded invaluable data about how the human body coped with the conditions of space travel, data that would be necessary for the preparation of astronauts who would eventually stay in space for much longer periods of time.

The eight-day mission successfully concluded on October 13 after a rather routine re-entry and landing at the Kennedy Space Center in Florida. Garneau exited the space shuttle when his turn arrived having become part of history, the first Canadian to travel in space, and for the time being holding a national flight record for traveling just over 5.3 million kilometers in only eight days.

The STS-41G mission was also a notable in that it provided the first popular view of NASA's shuttle space program to the public. Interestingly, much of Garneau's mission was captured in spectacular visual record, as by 1984 space shuttle equipment routinely included a new device known as an IMAX motion picture camera. This super-high-definition film camera allowed for the massive projection of movies on multiple stories-tall screens on Earth, and it proved especially worthy to capture the splendor and majesty of spaceflight. The first IMAX film about the flights of the space shuttles *Columbia*, *Challenger*, and *Discovery* played around the United States and Canada in 1985 to fantastic review and praise. In addition to already being so many other things in Canadian popular culture, Marc Garneau had also become an accidental movie star.

More important, perhaps, Marc Garneau's successful spaceflight secured an open door for more Canadian participation in future NASA space shuttle missions. As the Canadian backup astronaut for STS-41G, many assumed that Bob Thirsk was slated as the next likely to go into space. The program, however, was not structured that way. The Canadian astronauts were selected based on their viability to effectively carry out the two main experiments that Canada sought to undertake in orbit, the SAS and SVS experiments. As a result, the NRC chose three life science candidates (Bondar, Money, and Thirsk) and three engineering candidates (Garneau, MacLean, and Tryggvason). Thirsk was paired up with Garneau partially to ensure that both specialist groups gained experience, as it remained uncertain which of the two main experiments would receive priority for inclusion on future American space shuttle flights.

The Five-Year Plan

By the early 1980s, the growth of Canada's space program from its modest origins after the Second World War had been spectacular. The government's space budget had tripled from just over $30 million in 1972–1973 to $95.7 million in 1978–1979. In the fifteen years prior to 1974, the Canadian space industry had grossed roughly $150 million in sales. By 1980, however, the industry had reached an annual sales volume of approximately $140 million. Canada's space program had effectively evolved from a single office at the Defence Research Board to an operation including five major facilities and laboratories across the country developing seven major space projects. More than fifteen major Canadian companies were also directly involved in the space industry sector in some way. Few other Canadian high-technology sectors had fared as well as the space sector during this period.

Yet despite this tremendous growth there remained a number of challenges. A space agency had yet to be created, and even with a strengthened interdepartmental committee network, Canada's space projects remained spread across nine different departments of government. There was still no move towards a concentration of space effort even though success had revived a government interest in a single space budget and agency. As well, while international space cooperation was encouraged there was a limit to the government's interest in expensive space programs, not to mention a great deal of debate over which direction Canada should ultimately follow. The older operations and research-oriented bodies – the National Research Council [NRC], Energy Mines and Resources [EMR], Department of National Defence [DND], etc. – favored a strictly bilateral relationship with the United States, while the newer policy and coordination-oriented organizations – the Department of Communications [DOC], Ministry of State for Science and Technology [MOSST], the Science Council, etc. – favored links as well with European and other international partners. To address this and many other issues, the membership of the ICS convened a major meeting in 1979 to prepare a strategic framework for the next decade of Canadian space exploration. Their aim was to create the conditions for the development of Canada's space program that would eventually see it evolve into a sustained national program under the direction of a centralized civilian agency with real budgetary authority.

In January 1980, the Minister of Communications, as the Minister responsible for space, tabled a discussion paper and financial envelope to Cabinet detailing a plan for the research and development of Canada's space program over the next five years. The proposal, based on extensive analysis by the Interdepartmental Committee on Space members and endorsed by the four major project departments, was designed to further enhance both public and government awareness of space as well as assist in filling the gap in medium-term planning in Canada's space program.[78] The five-year plan put before Cabinet reinforced the pragmatism of Canada's space policy and formally recognized international space cooperation as part of its central agenda. Well-thought-out and crafted, the government's acceptance of the plan led not only to the swift approval of several of its proposals, but also to more formalized government space budgets and program direction. The plan was designed to ensure that Canada continued to have access to key aerospace technologies and also contribute enough to international space cooperation to have a say in the

---

[78] Canada, Department of Communications, *The Canadian Space Program: Five-Year Plan (80/81– 84/85)*. Serial no. DOC-6-79DP Discussion Paper, dated January 1980.

setting of international standards for space, particularly in bodies such as the International Telecommunications Organization [ITU] and the Commonwealth Telecommunications Organization [CTO]. Like the previously-released 1974 space policy, the focus of the 1980/81–1984/85 Five-Year Plan remained the use of space to enhance national objectives, only now mainly through international forums.

Overall, the new agenda was generally a success. It gave the ICS both the longer-term vision as well as the procedural power to develop Canada's space program through the early 1980s, providing industrial and economic details of the country's seven major space projects then under way as well as details for fifteen future proposals. The current projects included: a space science program, the shuttle Remote Manipulator System [RMS or Canadarm] program, the CTS [Hermes] and *Anik-B* experimental satellite program, the *Anik-C* and *Anik-D* domestic communications satellite program, the Search and Rescue Satellite [SARSAT] experimental project, the remote sensing satellite projects, and environmental satellite development. The new agenda proposals included the expansion of Canada's remote sensing and satellite communications programs, as well as an increase in government aerospace industry support. All of the projects were research and development intensive, yet now well within the ability of the ICS membership to coordinate and fund.

Fig. 4.30 Teleset *Anik C* Press Kit cover

The ICS wasted no time in exploiting its new window of opportunity. With the satisfactory delivery of the first RMS to NASA in 1980, it secured an order for another three RMS for the expanding fleet of American space shuttles. As well, during the same timeframe Canada successfully deployed five new communications satellites between 1982 and 1985, one of which – *Anik C3* – was deployed during the first commercial flight of the space shuttle *Columbia* in 1982. Canada's astronaut corps was created in 1983, and quite literally got off the ground within a year when Marc Garneau made history as the first Canadian into space on shuttle mission 41-G on October 5, 1984. Canada signed on as a full member of the SARSAT program in 1979, and by the early 1980s was made responsible for the design and construction of the satellite system's repeaters and Local User Terminals [LUT]. A number of other cooperative ventures were also launched during this period, including a Canadian aurora experiment aboard the Swedish experimental satellite *Viking* in February 1986. Whenever possible, the ICS exploited any opportunity to expand Canada's national space program through international endeavors.

The Catalyst

A consistent string of Canadian successes and achievements in the space sector since the announcement of the 1974 space policy had finally made the political impact the country's space advocates had desired. After many attempts to get Cabinet's attention, the government was finally taking serious notice of the advantages of direct investment in space science and technology, and expressed publicly its desire for an organized approach to further Canadian research and development. With so many diverse space programs to manage, the choice of creating a centralized body to coordinate these efforts became increasingly logical if not essential. Combined with the advent of Canadian human spaceflight, and the proposed way ahead for the American shuttle program that included a proposal for an international space station to challenge the Soviet *Mir* space station, it appeared to many involved that Canadian government-sponsored space activity would only increase both in scope and complexity over the coming years. Happily, space advocates in Canada would not have to wait long now for a centralized agency to lead these efforts.

As the space shuttle program progressed with launches occurring on a regular basis, it seemed to many that human spaceflight would soon become a routine affair. Access to space through the American space shuttle, as well as the construction of a new space station, would open up a number of new opportunities for countries ready to take advantage of what outer space had to offer. Seeing such opportunities to further its own national interests, Canada decided during the mid-1980s to institutionalize its own government space program so that it too could take full advantage of what appeared to be the next chapter in the ongoing exploration of outer space. First, however, it would need to learn how to deal with sudden change and then adapt to rapidly changing circumstances surrounding international access to space over the next few years and beyond.

5

Maple Leaf in Orbit: Institutionalizing the Canadian Space Program, 1984–1995

The success of Canada's space program in the first half of the 1980s was largely the result of the country's increased bilateral space cooperation with the United States. High-profile projects such as the shuttle's Canadarm and the flight of Marc Garneau brought considerable public attention and praise to the program, in turn boosting government support for Canadian space activities at a time when the criticism the government faced over other American bilateral space cooperation may have seriously curbed such endeavors. Public opposition to Canadian participation in American military space efforts such as the Strategic Defense Initiative (SDI), for example, had little impact on other non-military cooperative ventures. While the government seriously debated Canada's position and policy towards the American SDI in 1985–1986, the ICS was at that time busy finalizing plans for Canada's future participation in the American Space Station Freedom project now scheduled to begin in the early 1990s. Such initiatives were a demonstration of how both Canadian nationalism and internationalism could work in outer space, despite the presence of some political friction between the two partners on Earth.

Though the grand vision for Canada's space program remained more or less the same throughout this period, the mechanism through which it was focused had greatly changed. The vision itself could still be defined by three key factors. First, projects undertaken had to have practical objectives. Canada simply could not afford space stunts or activities that offered little beyond a news headline. Second, space activities had to be directed at developing a competitive space industry in Canada. The Canadarm was an impressive start, but future efforts had to follow a similar pattern of industry development and offer the same level of returns for the investment made. Third, Canada attempted in its space science to continue its contribution to global knowledge through a reasonable level of applied research activity. The mechanism to strive towards this vision, however, had evolved. Following the success of the initial space shuttle flights, the United States made a strategic decision to begin moving away from the employment of expendable launch vehicles (ELV) in favor of using the space shuttle as its main vehicle to deploy and maintain satellites. As a country that relied almost exclusively on America for space access, this meant that future Canadian satellite launches would also migrate from ELVs over to deployment by space shuttle.

© Springer International Publishing AG 2017
A.B. Godefroy, *The Canadian Space Program*, Springer Praxis Books,
DOI 10.1007/978-3-319-40105-8_5

Further to this, Canadian participation in both future satellite operations as well as human spaceflight meant that its own vision for outer space would very likely now be fulfilled through its partnership role in such endeavors as the American space station project. This shift created tremendous challenges for the current status quo programming, but if engaged, offered tremendous new opportunities for future Canadian space activities.

With new requirements and demands being created in Canada's space activity as a result of this evolution in the U.S. space program, in March 1985 the MOSST published a very brief document simply titled, *Interim Space Plan 1985–1986*, to focus short-term research and development, as well as set the stage for the first Long Term Strategic Plan (LTSP) due to be published the following year.[1] As with previous government statements, the new interim plan issued by the Honourable Tom Siddon, Minister of State for Science and Technology, sought to reaffirm Canada's future commitment to space exploration and international cooperation as well as identify a roadmap for short- and medium-term development. Specifically, the new plan identified four main lines of effort – technology development, remote sensing, communications, and space science – committing to a budget of just over CDN$194 million.[2] Interestingly, especially for Canada's space advocates, was the plan's formal acknowledgment that the government would commit to three new areas of focus. The first new domain was the formal acceptance of the American invitation to join its space station project and the firm commitment of an initial $8.8 million in funding to conduct project design and preliminary definition studies for Canadian participation. Second, the plan confirmed government intentions to continue its development of a new advanced earth remote sensing satellite, to be known as RADARSAT. Third, continuing on from its earlier successes in satellite communications projects, the plan also called for the government to support the implementation of a new commercial mobile communications satellite system that would be known simply as MSAT.[3]

Though barely a dozen pages, *Interim Space Plan 1985–1986* carried considerable punch. Including new projects, the new space budget of CDN$194.1 million constituted an increase of nearly thirty percent over that of the previous year. A large amount was being spent on the new space station project, of course, and was expected to consume CDN$60.3 million of the budget during the next year alone, and potentially as much as CDN$219 million by the years 1990–1991. Only Canada's new remote sensing projects were expected to cost more in the short term, spending roughly CDN$73.5 million during the 1985–1986 fiscal year and as much as CDN$240 million by 1990–1991.

The rapid expansion of Canada's space budget, from roughly CDN$96.7 million during 1981–1982 to the estimated CDN$194.1 million for the 1985–1986 fiscal years again raised questions concerning the future coordinated management of the space program writ large, and the advantages that creating a centralized agency could bring to such management. David Low, who had at one time served as Chairman of the government's Interdepartmental Committee on Space, noted in a presentation at a conference on Canada and the American space program at the University of Toronto in 1985 that if Canada

---

[1] Canada, MOSST, *Interim space Plan: 1985–1986.* Ottawa: Canadian Space Program, March 20, 1985.

[2] Ibid., 2.

[3] Ibid., pp.3–5.

continued to expand its space program at such a rate then a centralized managed approach would only seem logical if not absolutely necessary. He did not go so far as to suggest, however, that an official civilian space agency was the best option, but rather that the long-term future of Canada's role in space exploration would depend greatly not on what kind of program it currently was, but on the kind of program it sought to become. As the title of the conference suggested, Canada's program had become increasingly dependent on close cooperation with the United States.[4]

Canada and Space Station Freedom

Canada's involvement in low Earth orbit space station development officially began in March 1985 when the Canadian government accepted a formal invitation from the United States government to participate in its new Space Station Freedom program.[5] As mentioned above, Ottawa initially allocated CDN$8.8 million towards project design and preliminary definition studies during the following fiscal year to develop specific proposals for Canadian participation. Of this funding amount, approximately CDN$4.7 million went to Spar Aerospace Ltd. and its subcontractors to continue with ongoing studies for a proposed servicing facility aboard the station.[6] The success experienced with the design and operation of the Space Shuttle's Remote Manipulator System influenced Spar to continue lending its robotics expertise towards potential space station development, and it was felt that a mobile servicing facility would meet several Canadian space objectives in a single effort.

Still, further studies were conducted by both the Canadian government and industry, which subsequently produced three potential options for its involvement in the U.S./International space station project. The first option was to provide one of the two free-floating remote sensing platforms planned for placement in polar orbit to complement space station operations. Canada already had a strong interest in Earth remote sensing and was making investments in this domain. A second option explored the construction and provision of solar power arrays for the unmanned polar-orbiting platforms associated with Space Station *Freedom*. This was an area in which Canadian researchers had developed considerable experience with previous satellite programs, and it offered some opportunity for Canadian companies as well. The third option was to provide the servicing facility platform that Spar was already examining options for. Now known as the Integrated Service and Test Facility (ISTF) concept, this mission-critical technology was conceived and designed to provide a centralized core capability for the space station to service and repair satellites, scientific instruments mounted on the station, space vehicles visiting the station, and even the station itself as needed.[7]

---

[4] CIIA, *Canada and the American Space Program.*

[5] Canada, MOSST, "Interim Space Plan for Canada Announced", *News Release.* Ottawa: MSST, March 20, 1985.

[6] Subcontractors included CAE Electronics, Canadian Astronautics, DMSA Atcon, SED Systems Inc., and Dynacon.

[7] L. Dotto, *Canada in Space.* Toronto: Irwin Publishing, 1987, pp.259–262.

Fig. 5.1 Insignia of the Canadian Space Station Freedom Program

As a result of its investigations, the Canadian Ministry of State for Science and Technology opted to stick to its proven technology in the short term, announcing in mid-1985 that Canada would offer to take a lead in building the future space station's mobile servicing capabilities while it continued to examine options for other contributions to the station, notably its life sciences and medical science modules.[8] As Canada had not yet formed an official space agency to oversee such efforts, Canadian space projects at the time remained an interdepartmental affair assisted by the private sector.[9] At the highest levels, Mac Evans represented Canada in policy discussions with NASA. The MOSST and the NRC, meanwhile, working in close cooperation with Canadian industry, were tasked to manage the initial Canadian space station program effort on behalf of the government. Dr. Karl Doetsch directed the space station program at the NRC, while Mr. Jim A. Middleton of Spar Aerospace Ltd. led the half dozen companies on the industrial team.

---

[8] Canada, MOSST, *Interim Space Plan, 1985–1986*. Ottawa: MSST, 1985; and Canada, MSST, *The Canadian Space Program: New Initiatives*. Ottawa: MSST, 1986.

[9] Ibid., pp.2–4.

Fig. 5.2 Space Station concepts – Artist Paul Fjeld's rendering of the ISTF on a future Space Station *Freedom*

The government's announcement to pursue the design of the ISTF as well as provide increased funding for its development was welcome news to the country's space advocates, who previously feared that the newly-elected government's commitment to

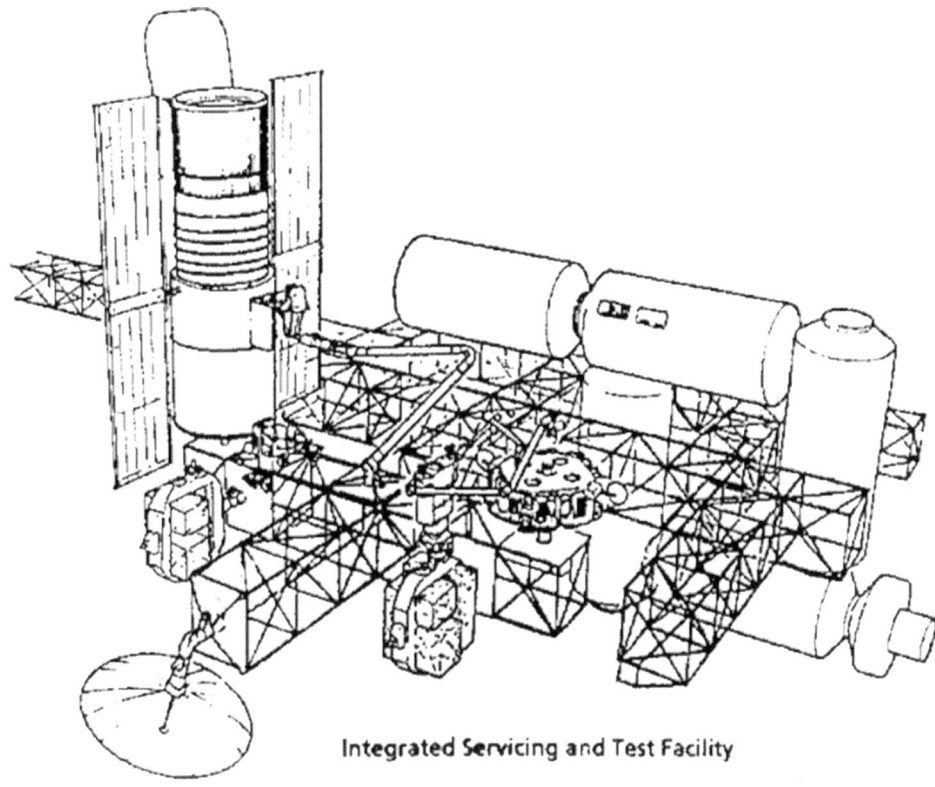

Integrated Servicing and Test Facility

Fig. 5.3 Technical drawing of the ISTF on Space Station *Freedom* repairing a Hubble Space Telescope-like satellite

reducing the deficit might cause it to step back from the space program. Instead, the government appeared to double down on its investment, giving the MOSST, NRC, and others, the financial support necessary to see Canada's space station program get off the ground.

At first inspection, the ISTF was a challenging proposal that consisted of a centralized, fixed service bay augmented by an advanced mobile inch-worming Canadarm. The new Canadarm was meant to move about the exterior of the station bringing payloads into the bay for servicing as well as service any part of the station itself if needed. Canadian designers presented their initial concept to NASA during the summer of 1985, delicately emphasizing that the design was to be an integrated facility. "It wasn't a bunch of subsystems like an arm here and a screw there", Jim Middleton noted in a later interview, "It all had to be in a neat package that could have a Canadian flag on it and had a unique task."[10] High visibility was critical from Canada's point of view, as it would help drive technological development as well as future business. As a result, much work was done to offer as complete a

---

[10] Op cit., L. Dotto, *Canada in Space*, p.262.

concept as possible to NASA, and the hard work subsequently paid off. Mac Evans commented, "At that point, I think, NASA came to life on this servicing issue. They realized that they had to come to grips with some of the questions we were asking and they were going to go into high gear and try to catch up."[11]

At first, NASA was not wholly comfortable with the idea that a foreign country might have such a central role in the development of their space station. The ISTF, what some Canadian designers were then simply calling "the space garage", was a piece of mission-critical technology. Without it, the space station would not be able to sustain itself let alone carry out one of its proposed core functions. As such, NASA had its own ideas concerning design. While the agency approved in principle the basic Canadian design, it wanted to alter it to include a thermal shelter around the bay and a flatbed carrier to improve the mobility of the arm along the central truss of the station. Though Canada was interested in building the advanced arm, it had no desire to build the thermal shelter. Dr. Doetsch noted that, "we felt that the number of satellites that would really require shelter of that nature was too small to justify the fairly large expenditure on our program."[12] The suggested cost, possibly as much as US$120 million, would present a serious pressure on Canada's proposed space station budget. At the same time, other projects, such as the RADARSAT program office, saw the cost increase as an opportunity to cut back on or abandon altogether the ISTF project in favor of its own proposal. Then in October 1985 there was another major development. Just a few short months before NASA intended to freeze the design of the space station, it introduced some radical changes to the basic configuration of the structure as a result of the outcome of various studies on how to best utilize the structure indefinitely.

The Canadian space station team responded to each of these challenges in turn, reaching an agreement in principle with NASA on December 19, 1985 regarding the basic configuration of the ISTF. In the agreement both parties spelled out in general terms their expectations of the partnership, with a view to resolving further more technical details regarding the Canadian contribution by the end of January 1986. All appeared to be on track until the arrival of the infamous "Boland letter" at NASA headquarters in mid-January, a strongly worded statement of concern from Congressman Edward Boland, chairman of a powerful subcommittee that oversaw part of NASA's budget, regarding the space agency's temptation "to make compromises with potential foreign partners in order to gain a short-range financial benefit that may have a long-range negative impact on U.S. high-technology development."[13] At the working level, Doetsch later acknowledged that U.S. resistance to Canada's vying for such a central role in space station technology development was to be expected. Certainly, various US interests in robotic development would challenge NASA's attempt to farm out this domain to a foreign interest. That said, Doestch was also passionate about putting Canada front and center in the space station program, and thus he made every effort to find accommodation amongst both American government and private interests regarding the servicing facility role. At a meeting in Washington D.C. on February 4, 1986, just a week following the *Challenger* disaster, disaster, NASA

---

[11] Ibid., p.262.

[12] Ibid., p.264.

[13] Ibid., pp.268–269.

officials put forth a counterproposal to him and his team outlining how the two countries would now share the space station's servicing function. While the counterproposal introduced certain changes to the original Canadian design that could not be avoided, Canadian officials left the meeting overall satisfied with the new arrangements. NASA's counterproposal kept with its core philosophy of functional allocation, and Evans, Doetsch, Middleton, and others felt that the agency had given the Canadians enough room to maneuvre while meeting its own objectives. Doetsch later commented that, "if we had not found this path that allowed both parties to sort of move in the directions they wanted to, I think the agreement would have been in jeopardy."[14] Fortunately, NASA was now offering a new way ahead, although the time available to respond to their new offer was incredibly short.

Amazingly, American officials asked their Canadian counterparts to provide a formal response to the new NASA proposal within a week. Jim Middleton later wrote, "I sat in my office and thought about the problem and an old idea I had came to mind. Instead of a garage to do servicing and repair, maybe a 'tow truck' could drive along a railway going from spot to spot on the station. It would have the big arm and the little robot mechanic. I decided to call it the 'Mobile Servicing Centre' (MSC). I got the group of engineers working on the program, told them about the idea and got them putting meat on the bones".[15] Thus the revamped Canadian Mobile Servicing Center concept was born.

After a hectic week of revising their plan Evans, Middleton, Doestch, and others briefed the revised MSC concept first up to Ottawa, then to NASA, the following week. The new servicing system was designed to perform a wide range of functions from maintenance tasks to the movement of materials around the station. Ultimately, the repackaging of Canada's proposed contribution to the space station proved a success. After further meetings and briefings, on February 12, 1986 Canada secured an agreement in principle that it would design and construct the new space station's mobile service center, equipped with among other tools a special purpose dexterous manipulator (SPDM). Still, further challenges from other contenders for Canada's space budget had to be overcome before the government formally announced on March 18, 1986 that Canada would commit to the space station program through its MSC project. Surprising even the most enthusiastic Canadian space advocates, Frank Oberle, then serving as Canada's science minister, announced that the government planned to commit CDN$800 million over fifteen years to develop and deploy the MSC. Of this budget, CDN$220 million would be spent in the first five years to ensure that Canada had "a highly visible role…in the most important international civilian space program of the century".[16] At the time, it was estimated that the commitment would create nearly 80,000 person years of work and ultimately generate CDN$5 billion in revenues.[17]

The Canadian government's commitment to the space station would both garner support and endure further criticism, but Cabinet held firm on its decision even during a time of national fiscal constraint. There was always a persistent concern that the country's

---

[14] Ibid., p.270.

[15] Jim Middleton, "Space Stations – Part 2: The Canadian Contribution", *King Weekly Sentinel*, April 6, 2016. Accessed online on March 31, 2016 at http://kingsentinel.com/?p=7655

[16] Op. cit., L. Dotto, *Canada in Space*, p.276.

[17] Ibid.

fledging space program might land on the budgetary chopping block before it had the opportunity to mature, so it offered tremendous reassurance to those in the space sector, especially at a time when even America's own future in space appeared uncertain.

The Accident and Aftermath

The tragic destruction of the American space shuttle, *Challenger*, on January 28, 1986, exposed once again the risks involved in both pushing the boundaries of human exploration of outer space and the precarious nature of Canada's increasing dependence on the U.S. space program for its own access to space. With the sudden indefinite halt to all space shuttle missions until the accident could be fully investigated and the problems solved, Canada's astronauts found themselves immediately grounded. MacLean, who had been selected to fly just two months before, had to accept that his own flight would not only be delayed for perhaps a year or more until shuttle operations resumed, but even when they did, it was likely that he would be bumped off his scheduled flight until NASA was satisfied it could return to safely hosting international partners aboard missions. Even then, he faced the additional possibility that the sequencing of the missions could change. If this occurred, Canada's life sciences experiments might fly first, meaning that the astronaut office would have to put him on hold while one of his team mates flew first. Roy VanKoughnett, then serving as head of the Canadian astronaut program, noted in an interview in mid-1986 that he still hoped Steve MacLean would get onto a mission within the first year after shuttle operations resumed. Regardless, he was forced to admit that "until we see something in the form of a manifest from them, some kind of schedule", it was uncertain when NASA might schedule the next Canadian to fly.[18] Even if NASA resumed shuttle missions by 1988, it was unlikely that MacLean would get into space before sometime in 1989.

Similarly, upcoming Canadian satellites, originally scheduled for future deployment by the shuttle, also now had to be reconfigured for launch using expendable rockets as it was unknown when the shuttle might resume offering commercial flights for international partners. The U.S. government had been phasing out ELV production as part of its transition to making the STS its sole launch provider, however, and by the mid-1980s there were simply not enough ELVs remaining in stock to satisfy all immediate demands. In protecting its own access to space, Canada and other nations were bumped from their scheduled flights.

Political problems, budget overruns, and delays that were further compounded by the *Challenger* disaster disaster as well as technical problems with America's ELV fleet in fact delayed all space endeavors, including the initiation of the U.S./International space station project, now projected for a start sometime in the mid-1990s. Canada's ISTF design underwent considerable alteration throughout the whole period, faced further challenges from aggressive American industry lobbying wanting to secure the core station technology development for themselves, and eventually emerged into a new mobile servicing system (MSS) that would allow the future station's astronauts greater flexibility to perform activities on location as opposed to being fixed at a certain location.

---

[18] Ibid., p.147.

The Canadian Space Agency

Throughout the early evolution of Canada's space program, the creation of a central civilian-led space agency remained a contentious issue for the government. Repeated attempts by various space advocates within government to convince Cabinet to create an agency in the 1960s and 1970s had all resulted in failure, but by the 1980s other actors were beginning to assert their own influence on the process, and this would ultimately force the government to commit to definitive action. The existing advisory body, the Interdepartmental Committee on Space, was once dominated by the Department of Communications, which also chaired the committee. By the mid-1980s, however, chairmanship had shifted to the Ministry of State for Science and Technology, while a third major player, the Department of Energy, Mines, and Resources (EMR), also became increasingly important to the space program. The result of this shift in power further exacerbated the problems of what many saw as a completely dysfunctional committee. As one official noted, "In my mind, it has always been a funny committee. It is advisory in a sense and a coordinating body in a sense, and it can't really take decisions … it wasn't a forum that could rank or rate projects. It was too large and cumbersome for that. It really couldn't focus and get agreement."[19] As a result, departments often worked around the ICS to advance their own space agendas.

Similarly, the Canadian aerospace industries that produced the majority of Canada's space technology were likewise beginning to assert their own influence. Having achieved considerable success both at home and abroad, the consortium of aerospace industries in Canada wanted a greater role in shaping the policy under which they all operated. Among their many concerns was the issue of coordination. Future growth in Canada's space industry depended greatly on good alliances and international partnerships. Yet because there was little coordination between departments when concluding international agreements, it often left the impression with foreign customers that the Canadian government did not have its own house in order. This, many companies felt, was adversely affecting business and the space sector wanted to see an improvement to the situation. In January 1985, the Aerospace Industries Association of Canada (AIAC) drew up an extensive proposal for the creation of a national space organization and submitted it to the government for consideration. The plan was very well received by the recently-elected government, and they announced during a throne speech later in the year the government's intention to create a new space agency.[20] The reaction to the announcement was one of promising interest but also caution. As one study of the agency's establishment noted, "…the two actors with the least knowledge of government operations in the space sector, the Prime Minister's Office [PMO] and the [Department of] Industry, were the most enthusiastic proponents of the agency concept".[21] Any new agency would have to gather under its own aegis both the political power as well as the corporate knowledge that previous space bodies had, otherwise it would be ineffectual in convincing government to commit funding and support to the space sector. Doing so would be no easy task, as the existing departments that currently

---

[19] Michael M. Atkinson and William D. Coleman, 'Obstacles to Organizational Change: The Creation of the Canadian Space Agency', *Canadian Public Administration*. 36:2, Summer, 1993, 137.

[20] Ibid., pp.137–138.

[21] Ibid.

oversaw Canada's various space programs were reluctant to give up those projects and activities they already owned or controlled. As the weeks and months passed following the announcement, various departments involved in the ICS reoriented themselves to downplay the centralizing function of space in their activities in order to safeguard their own resources against being sequestered for the new agency. For example, at the time of the government's announcement about the new space agency the DOC had four directors overseeing space technology divisions. By 1986, these had been reduced to two, and two years later the word 'space' was removed altogether from the job titles of the DOC directors involved in satellite communications development.[22] As agency planning continued, such resistance to the concept appeared to grow. Aside from protecting its own assets, departments involved in space activities seemed willing to transfer to the new space agency only those divisions it could do without.

In early 1987, the Privy Council Office began its search for the person who would lead the agency. In the meantime, the government appointed Art Collin, a veteran public servant with extensive experience in Canada's space sector, to lead the transition team and refine the agency concept. He envisioned an agency which would be organized in a fashion similar to that of a crown corporation, with the ability to report to parliament through a minister but not necessarily under the control of the minister. This would prove to be a point of contention from the start. The Privy Council Office did not relish the idea of a highly autonomous agency it could not politically control, fearing that in the future such an agency could potentially embarrass the government on issues such as science and technology policy. Disagreement over how the agency would be led and controlled soon led to Art Collin resigning from the transition team in February 1988. Even after his departure, it took another two years to finally resolve matters of reporting and financial authority, more than five years from when the government first announced the agency would be created.

While the debate over organization and authority raged on, the physical location of the new agency also became highly politicized. While many simply assumed that the country's future space agency headquarters would be located in the nation's capital, the Conservative government's Quebec interests mounted an aggressive campaign to have the new agency located in the Montreal area instead. Such a move made little practical sense, as the majority of the country's space expertise and laboratories were already situated in and around Ottawa. Only a few aerospace industries were in Montreal, with the majority elsewhere in the country. Yet despite this fact, the Quebec lobby persisted in its efforts and by the time opponents to such a move in Ottawa got organized to mount their own campaign, it was too late. The Prime Minister and his Cabinet agreed that the space agency would be located somewhere near Montreal. Both the AIAC and many other space research organizations were simply flabbergasted by the decision. As one industry senior executive later commented:

> 'We thought in our naiveté that by creating a space agency and showing that we were creating valued jobs and all these sorts of things that that would matter to politicians … When I say naiveté, one of the problems with the space industry is that it is run primarily by persons whose training has been in engineering. Engineers are notoriously naive. They believe that if this is logical, then you, as a reasonable human

---

[22] Ibid., p.140.

being, are going to see this is logical. We thought our vision of the universe was reasonable and that the politicians would have to see that too. But they didn't. We opposed in every way we could moving the agency to Montreal. You won't find any of the space companies that ever spoke in favour of it.'[23]

Still, despite the challenges before it, the government remained committed to eventually bringing all of its space activities under the political and financial oversight of a single organization. Yet even the decision to relocate to Montreal could not effectively accomplish this. For a number of reasons the federal government was unable to build the agency in Montreal itself, with the headquarters finally being constructed in St. Hubert on the far side of the city. After further delays the Canadian Space Agency headquarters were officially opened for business on March 1, 1989.[24] The choice of such an isolated location continued to frustrate many in the space community, and several of the first-generation Canadian space scientists and engineers opted for transfer to another organization or seek retirement over relocation.

Fig. 5.4 The John H. Chapman Space Centre, St. Hubert, Quebec

[23] Ibid., p.149. This said, some Canadian aerospace companies were more adaptable to 'illogical' government decision-making than others, and those that did tended to fare better in the long run.

[24] Though the agency began operations in 1989, it took a while longer to get the charter through all of the bureaucratic gates. The Canadian Space Agency Act received royal assent on May 10, 1990, and came into force on December 14, 1990.

Fig. 5.5 Details of the John H. Chapman Space Centre, St. Hubert, Quebec

Fig. 5.6 Dr. Larkin Kerwin served as first President of the Canadian Space Agency from 1989–1992

Perhaps the only aspect of establishing the new agency that did not draw significant criticism was the search for a suitable candidate to lead it. After much consideration of many possible candidates, the government offered the new post to Dr. John Larkin Kerwin, an accomplished physicist who had previously served as President of the National Research Council of Canada from 1980 to 1985. Kerwin was a well-known and respected figure and had been closely involved with Canada's space policy and projects for some time. At the time of his selection, he was serving as President of the recently-created Canadian Academy of Engineers, a non-profit organization of top talent dedicated to providing leadership and consultation on engineering matters of national importance. Kerwin left this appointment in 1989 to take up his new assignment in St. Hubert. Dr. Kerwin directed the new organization through its crucial first years, establishing with difficulty (due to the loss of some of the corporate knowledge and talent) a core capability for the country's space program by the time of his retirement from the appointment in 1992. Still, it would take about another decade before the agency finally achieved a degree of political and bureaucratic stability.

As the agency slowly organized, it continued a program of space science, including the launch of an imagery experiment aboard NASA's Upper Atmospheric Research Satellite (UARS).[25] During this time the agency also assumed responsibility for the training of Canadian astronauts, moving their offices from the NRC in Ottawa to the new CSA headquarters before preparing two members of the existing corps for the next Canadian shuttle flight opportunity expected sometime in 1991 or 1992. The long wait was understandably hard on the Canadian astronaut corps, all of them eager to fly but also wondering if they ever would. Regardless, the hiatus in spaceflight gave each of them the opportunity to pursue other career goals while they waited, in addition to continuing their work on space-related projects. MacLean, Tryggvason, and Garneau continued preparing for MacLean's anticipated flight, the primary goal of which was to test the prototype space vision system (SVS) that would improve the speed, accuracy, and safety with which astronauts would be able to employ the space shuttle's Canadarm. MacLean himself was also already at work on a second-generation SVS and, as a result, he was appointed manager of a new program to develop it for use on future missions. As the assigned backup to MacLean, Tryggvason closely mirrored him in all his training; he was also made responsible for the design of the Canadian Target Assembly (CTA), a specialized metal plate that could be captured by the Canadarm during the SVS tests in space. Garneau, meanwhile, was responsible for shepherding the Canadian SVS equipment through the bureaucracy of NASA's flight safety review in order to get it approved for use in the space shuttle crew compartment. He noted in a later interview, "you could write a book just to clear a pencil for flight. The system thrives on paper".[26]

The three life sciences members of the astronaut corps – Thirsk, Bondar, and Money – each returned to their own fields of work during the hiatus in NASA spaceflight. Thirsk took up clinical medicine again while continuing his research and development of a new antigravity suit for astronaut use during shuttle re-entry. Essentially designed to protect astronauts against faintness or blacking out on returning to Earth by stopping a sudden

---

[25] The UARS was deployed by space shuttle *Discovery* on September 15, 1991 during mission STS-48.

[26] Lydia Dotto, *The Astronauts: Canada's Voyageurs in Space*. Toronto: Stoddard Publishing Company, 1993, p.54.

flow of blood to the lower half of the body, he was hoping to test a prototype on a future mission. In addition, Thirsk was also made a member of two international working groups that examined various aspects of space station crew selection, training, and health. Bondar had similarly returned to her medical roots, beginning a research program with colleagues at three separate institutions examining the effects of spaceflight and microgravity on the human body. Ken Money, meanwhile, simply returned to his ongoing space-related work at the Defence and Civil Institute of Environmental Medicine (DCIEM) in Toronto, continuing his research into motion sickness. Yet he also picked up extra duties, as all astronauts did, being assigned to a NASA-led international study examining future manned missions to Mars.[27] At age 51, however, Money grew increasingly concerned that his chances of a mission were once again passing him by, as no one knew for sure when Canadians would finally return to spaceflight despite being promised at least two more opportunities by NASA.

The space shuttle program officially resumed operations on September 29, 1988, with the launch of the *Discovery* on her seventh flight. STS-26 was formally known as the 'return to flight' mission, and *Discovery* carried an all-veteran crew into orbit. After a successful mission and a safe return to Earth, NASA shuttle flights continued although it would be some time before Canadian astronauts would also return to the mission assignment roster. After further years of waiting and postponement, Canadian astronauts were finally reassigned to crews in early to mid-1991, with two projected flight opportunities scheduled for the following year. Due to a rescheduling of priorities, the first new payload specialist assignment slated for STS-42 went to Roberta Bondar with Ken Money acting as her backup. MacLean, originally selected to undertake Canada's second spaceflight, was then reassigned to fly Canada's third mission later the same year, serving as a payload specialist on STS-52.

Canada's Second Astronaut: Mission STS-42

Space Shuttle mission STS-42, NASA's 45th space shuttle orbital mission and the fifteenth flight of *Discovery*, marked the continuation of worldwide research efforts in the behavior of materials and life in weightlessness. Scientists from NASA, the European Space Agency, the Canadian Space Agency, the French National Center for Space Studies, the German Space Agency and the National Space Development Agency of Japan cooperated in planning a broad range of experiments that were conducted aboard a spacelab module stationed in the cargo bay of *Discovery* and designated the International Microgravity Laboratory (IML-1). As with previous shuttle spacelabs, the IML-1 was connected to *Discovery's* middeck via a tunnel and provided a crew with a spacious (by shuttle standards) fully-pressurized shirt-sleeve environment in which to conduct round-the-clock monitoring of experiments. More than 200 scientists from sixteen different countries participated in developing and analyzing the investigations, which made it one of the largest international space research endeavors held to date.

---

[27] Ibid., p.52.

Fig. 5.7 Dr. Roberta Bondar flew as payload specialist on mission STS-42. She was Canada's first female astronaut and the second Canadian to fly in space

Fig. 5.8 Roberta Bondar's personal spaceflight mission badge

Fig. 5.9 The Canadian logo for the IML-1

In addition to the IML-1, mounted in its cargo bag, *Discovery* carried twelve Getaway Special containers, each containing experiments examining materials processing, the development of animal life in weightlessness, and many other similar investigations. Another familiar sight onboard was the IMAX camera, now seemingly standard equipment to capture a visual record of the flight. As well, on *Discovery's* lower deck further experiments examining Polymer Membrane Processing investigated possible advances in filtering technologies in microgravity, while the Radiation Monitoring Equipment-III recorded radiation levels in the crew cabin. A number of other smaller experiments were also included in the flight, requiring eight days of intensive work on what some described as the most demanded scientific research mission conducted on the shuttle to date.[28]

NASA astronaut Colonel Ron Grabe commanded the shuttle on this mission with Steve Oswald serving as his pilot. Dr. Norm Thagard, MD, Lieutenant Colonel Dave Hilmers, and Bill Readdy served as mission specialists, while CSA astronaut Dr. Roberta Bondar, MD, Ph.D., and Ulf Merbold of the European Space Agency, served as the mission's payload specialists.[29] After more than seven years of waiting, Roberta Bondar could barely

---

[28] Ibid., p.73.

[29] NASA. News release 92-211 dated December 27, 1991. Space Shuttle Mission STS-42 Press Kit (January 1992), 5.

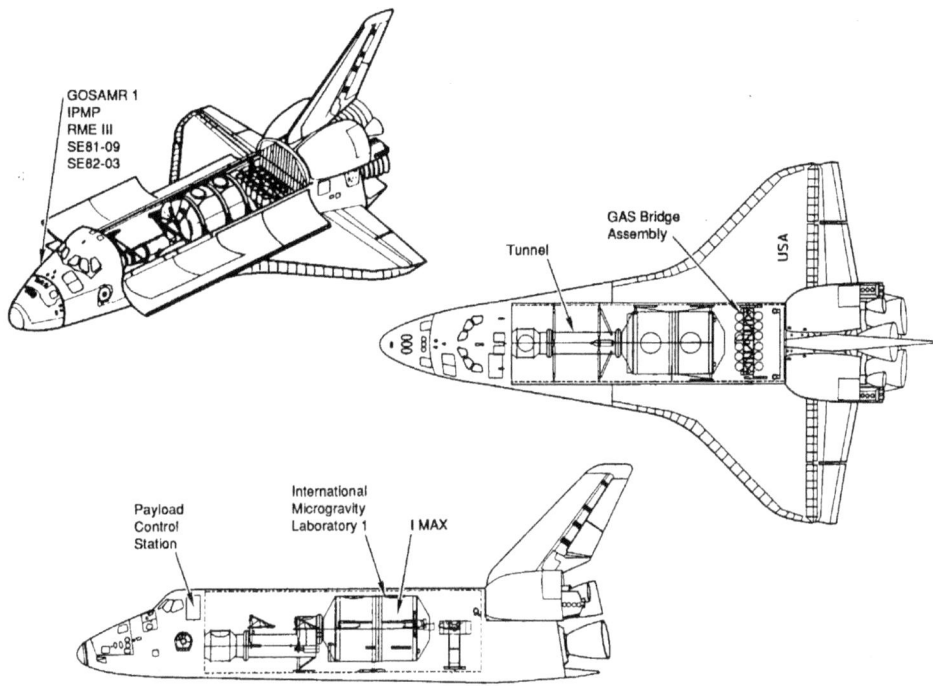

Fig. 5.10 STS-42 payload configuration

contain her own excitement as she headed with her fellow crewmates to the launch pad at Cape Kennedy on January 22, 1991. Canada's astronauts were finally returning to space, and for Bondar personally, the day finally marked the end of more than nineteen flight postponements. Still, *Discovery's* launch that morning was delayed an hour to allow for the evaluation of KSC field mill indicators and to assess a transient power surge from an orbiter fuel cell. Once everything was cleared, *Discovery* left the pad just before ten o'clock eastern standard time, achieving orbit shortly after.

Completing the IML-1 operations was the main objective of STS-42 and the lab was immediately activated following crew cabin unstow activities on the first flight day. Canada's contribution to the mission included a group of six life science experiments, collectively known as the Canadian Space Physiology Experiments (SPE), some of which continued on from research initially undertaken by Garneau during his 1984 flight. These experiments were designed to investigate human adaptation to weightlessness and other phenomena; the human vestibular and proprioceptive (sense of body position) systems; energy expenditure; cardiovascular adaptation; nystagmus (oscillating eye movement); and astronaut back pain.[30]

---

[30] Ibid., p.28.

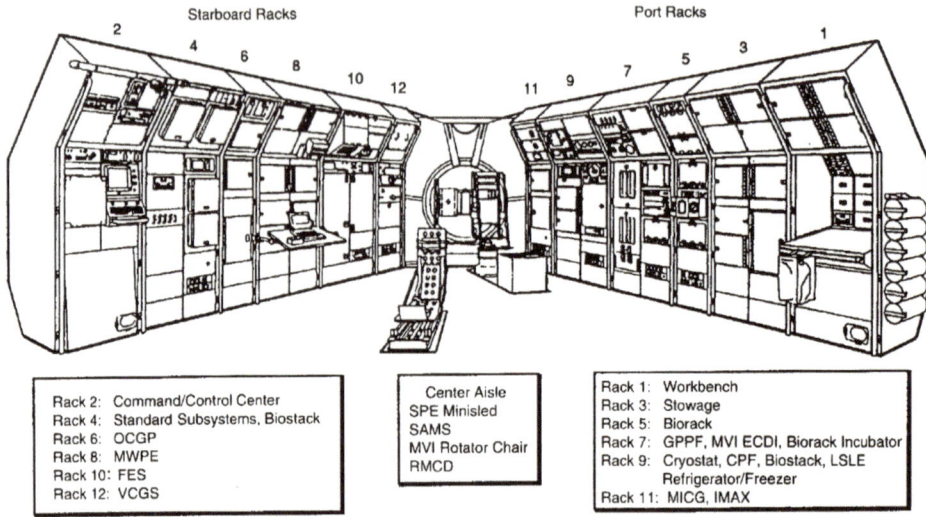

Rack 2: Command/Control Center
Rack 4: Standard Subsystems, Biostack
Rack 6: OCGP
Rack 8: MWPE
Rack 10: FES
Rack 12: VCGS

Center Aisle
SPE Minisled
SAMS
MVI Rotator Chair
RMCD

Rack 1: Workbench
Rack 3: Stowage
Rack 5: Biorack
Rack 7: GPPF, MVI ECDI, Biorack Incubator
Rack 9: Cryostat, CPF, Biostack, LSLE
            Refrigerator/Freezer
Rack 11: MICG, IMAX

Fig. 5.11 International Microgravity Laboratory-1 module layout

Fig. 5.12 Payload Specialist Dr. Roberta Bondar [CSA] and Discovery Pilot Stephen S. Oswald [NASA] conduct experiments in the IML-1

Many astronauts experienced something known as space adaptation syndrome, which often included illusions, loss of knowledge of limb position, nausea and vomiting. These symptoms may have occurred because of conflicting messages about body position and movement, which the brain receives from the eyes, the balance organs of the inner ear and gravity sensing receptors in the muscles, tendons, and joints. Seven investigations in this experiment studied the nervous system's adaptation to microgravity, hoping to learn more about motion sickness and how to treat it both in space and back on Earth. The nystagmus experiment, meanwhile, sought to investigate two types – spontaneous, where the eye oscillates at the same rate regardless of head position, and positional, where the oscillation varies according to head position. The objective was to determine whether it was possible for both types to occur simultaneously in the same crewmember, as part of a larger study to improve detection and treatment of inner ear disorders on Earth. Yet another experiment, the Assessment of Back Pain in Astronauts (BPA), sought to collect data as well as develop techniques to alleviate this condition by studying its causes. Interestingly, an astronaut's spine elongates by as much as seven centimeters in microgravity as a result of the vertebrae in the back spreading slightly apart. This elongation often caused painful tension and could even affect tactile acuity. It was hoped that, by studying the phenomenon closely, spacefaring nations might develop some countermeasures for their own astronauts.[31]

As a payload specialist on the mission it was Bondar's primary responsibility to perform these experiments on behalf of scientists from Canada and other participating countries. "Not everybody likes to be a guinea pig", Alan Mortimer, then serving as chief of space life sciences for the CSA, noted in an interview. "But we had all seven crew members participating in the Canadian experiments. They were very helpful, and it allowed extra scientific data to be obtained."[32] The IML-1 activities carried on nonstop; thus the crew split into two teams each undertaking a twelve-hour shift. Still, Bondar reflected in a later interview, "…work is inefficient because you're floating around and having to trap things". Everything she did simply took more time, which was disconcerting when she already had to try and fit ten days worth of experiments into a seven-day mission schedule. Fortunately, as it had enough surplus consumable resources remaining onboard, STS-42 was unexpectedly granted an extra day in orbit, which helped greatly in completing a number of the experiments.

Fortunately, Bondar's mission did include a bit of time for lighthearted activities. As STS-42 flew during the American Superbowl, the crew participated in a live broadcast where Bondar, clutching the official coin in her hand, was gleefully flipped by two of her crewmates for the player to call. Another time, she suddenly took a break in her work to get to the observation window of the spacelab so she could see her hometown from orbit. The opportunity to observe the Earth from space had, as it likely does for all astronauts, a profound impact on her. After her spaceflight Bondar subsequently published a book titled *Touching the Earth*.[33]

---

[31] Ibid., pp.30–31.

[32] Op. cit., Dotto, *The Astronauts*, 7p.7.

[33] Roberta Bondar, *Touching the Earth*. Toronto: Key Porter Books, 1994.

Fig. 5.13 The IML-1 logo

The STS-42 mission wrapped up on January 30, 1992 with space shuttle *Discovery* landing safely on Runway 22 at Edwards Air Force Base, California, shortly after 8am in the morning. The CSA celebrated Bondar's successful mission and safe return, but for her the end of mission was bittersweet. In the run-up to being assigned to STS-42, Bondar endured and subsequently won the demanding competition to be selected as the primary crewmember for her mission, beating out her colleague Ken Money for the coveted spot. Still, it had not been an easy road for her, partly due to her gender and partly because the selection process had essentially pitted her against her colleague on many occasions when she would have preferred to work collaboratively to achieve certain mission goals. For his part, as a result of being passed over yet again for a chance at spaceflight, Money finally decided that it was not meant to be and therefore time to move on to other things. During an STS-42 premission press conference in November 1991, he remained quiet for most of the event, and then announced towards the end that he intended to resign from the Canadian astronaut program following the successful completion of the mission. As for Bondar, she spent several months following her spaceflight assessing what role she might have in the program in the following years. Bondar was disappointed when she was not considered by the Canadian Space Agency for the NASA Mission Specialist training program that was scheduled to begin in the summer of 1992. It meant she would remain a payload specialist and thus less likely to be selected for another mission over those who had both payload and mission specialist qualifications. Not being optimistic about the possibility of another Canadian life sciences mission being scheduled anytime soon, she ultimately decided that, given her experience and research goals, she could do a lot more for the Canadian space program by leaving it than staying in it. As a result, Bondar also tendered her resignation from the astronaut corps that year, leaving the CSA in September.[34]

Though the Canadian Space Agency had effectively lost one-third of its active astronaut corps in 1992, senior administrators had already initiated steps to launch a second recruiting campaign which would hopefully yield new candidates that the agency knew would soon be needed to first supplement, and eventually succeed, the original six who were first selected in 1983. As well, the agency was already aggressively pursuing the opportunity to get Canadian astronauts placed into areas of greater responsibility, hence its pursuit of the NASA Mission Specialist training program, which it would eventually succeed in doing the following year. Otherwise, with its second space mission successfully completed, the CSA now quickly turned its attention to Steve MacLean and his backup Bjarni Tryggvason, the former being slated to fly in space in less than nine months from the end of Bondar's mission, as the sole payload specialist on STS-52. Both astronauts had relocated to the Johnson Space Centre in Houston to begin the final workup training program that took place in the last months prior to the flight, with MacLean finding the ascent profile simulation he undertook at San Antonio to be especially interesting. He also noted in various interviews that both he and Tryggvason were being integrated into general crew activities training far more than Garneau or Thirsk would have been when training for the first Canadian flight. By the early 1990s, the NASA astronaut corps' cold attitude towards payload specialists and their work aboard the shuttles had begun to thaw somewhat, with flight deck crews taking a much greater interest both in creating a single cohesive team and also in the science that was going on the middeck below. Tryggvason suggested, "I think it comes in part from the *Challenger* accident and the acceptance by NASA that they have to prepare everyone much more."[35] He also felt that NASA's future depended greatly on getting the science right. "Maybe it's sort of writing on the wall for the space station – that much more emphasis has to be placed on ensuring that good science is achieved both on the shuttle and on the space station."[36]

Mission STS-52: MacLean and CANEX-2

STS-52 was NASA's 51st space shuttle flight as well as *Columbia's* thirteenth mission. A crew of six and 11 major payloads were carried aboard, demonstrating the versatility of the space shuttle as a satellite launcher, science platform, and technology test bed. The mission commander was NASA astronaut James Wetherbee with Michael Baker serving as the shuttle's pilot. The crew's mission specialists were NASA astronauts Charles Lacy Veach, William Shepherd and Tamara Jernigan. Canadian astronaut Steve MacLean

---

[34] Op. cit., Dotto, *The Astronauts,* pp.98–99; see also Canadian Space Agency, Biography of Roberta Lynn Bondar, accessed online on Marc 30,h 2016 at URL http://www.asc-csa.gc.ca/eng/astronauts/biobondar.asp

[35] Ibid., p.124.

[36] Ibid.

was the only payload specialist on the flight, and officially the third Canadian astronaut to fly into space. *Columbia* lifted off the launch pad at KSC at 1:10pm on October 22, 1992, after a two-hour delay while unfavorable weather was allowed to pass. Nevertheless, *Columbia* achieved orbit without incident to begin a very busy ten-day mission. There were several objectives to achieve on this particular flight; the Canadian set of experiments (CANEX-2) overseen by MacLean were in large part extensions of the work carried out by Garneau on his first flight in October 1984. The CANEX-2 consisted of ten separate investigations, the results of which would have potential applications for future Canadarm and other space robotics operations. Additionally, the experiments examined applications related to the manufacturing of goods, the development of new protective coatings for spacecraft materials, improvements in materials processing, and a better understanding of Earth's stratosphere, which contains the protective ozone layer. Greater knowledge of human adaptation to microgravity was another objective of the CANEX-2 payload. Finally, MacLean conducted further experiments on back pain, body water changes, and the effect of weightlessness on the vestibular system.[37]

Fig. 5.14 Crew of STS-52 and Canadian Payload Specialist Steve MacLean

---

[37] NASA, Space Mission STS-52 Press Kit. October 1992, pp.24–25.

Fig. 5.15 Canadian payload specialist Steve MacLean

The centerpiece of MacLean's work during STS-52 was testing the SVS technology, a prototype of which was flown on this mission. Working closely with the two astronaut mission specialists responsible for Canadarm operations, Lacy Veach and Tamara Jernigan, MacLean operated the SVS while his crewmates used the Canadarm to pick up the Canadian Target Assembly from its berthing cradle in the cargo bay, execute a series of maneuvers with it, and then set it free. Then they recaptured it and returned it to its cradle. The exercise was done once without the SVS and once with it, so MacLean could evaluate the difference. Not surprisingly, perhaps, both mission specialists reported they found it much easier to operate the Canadarm precisely with the SVS. Other tests were also made, and eventually the CTA was left in space to burn up on re-entry into the atmosphere. MacLean watched as the object moved away from the orbiter and could not help but be impressed by the speed at which it also traveled, giving the crewmembers who watched it a true sense of what Mach 25 truly looks like to the observer.[38]

The STS-52 crew returned to Earth on November 1, touching down on Runway 33 at the KSC just after 9:00am. The mission had been a great success, and NASA officials were especially impressed by how well the SVS had worked in orbit. "Everywhere I went at Houston and Cape Kennedy, they were singing the praises of Bjarni and Steve", Mac

[38] Op. Cit., Dotto, *The Astronauts*, pp.129–135.

Evans later recalled.[39] He and other CSA officials were similarly pleased when NASA moved ahead immediately with plans to put the SVS on all of its shuttles in the fleet as well as on the proposed space station. It was therefore fortunate for all parties that MacLean had by then already developed an advanced version ready for testing – one of the positive outcomes of the forced hiatus in Canadian spaceflight caused by the *Challenger* accident. As a result, both he and Tryggvason spent considerable time down at the JSC right after MacLean's flight continuing to work on the advanced SVS technology and associated training. Eventually, these activities would lead to another spaceflight for one of Canada's trained payload specialists, setting the stage for Bjarni Tryggvason's mission in 1997.

Canada's Space Program: Long-Term Planning

Having successfully navigated the agency through its first crucial years, Larkin Kerwin stepped down from the post of agency director in the summer of 1992, to be succeeded by Dr. Roland Doré, an educator and former administrator at the École Polytechnique de Montréal. Kerwin had left an impressive legacy of achievement, and Doré took up his new appointment as President knowing that there remained much work to be done to put the agency on a path of long-term stability and activity.

One of Kerwin's last objectives before moving on was the initiation of a new policy framework for longer-term planning. The Canadian Space Agency's initial priorities were built on Canada's growing leadership in space science, satellite communications, Earth observation and remote sensing, and space robotics that evolved in the 1970s and 1980s. The first Long-Term Space Plan (LTSP I) produced by Industry Canada in 1985 had articulated a roadmap for what became the new space agency's preliminary objectives and budgets, and served as a primary planning guidance until the publication of the Space Policy Framework (SPF) and a new Long-Term Space Plan (LTSP II) in 1994. Both documents were considered important instruments in achieving the Government's objectives in space, and were designed to reinforce and focus the aims of Canada's space program as a whole. In addition to what might be considered traditional objectives, the SPF in particular placed increased focus on the principle of an international partnership strategy. This made sense as, arguably, by the mid-1990s almost all of Canada's space activities entailed some degree of international partnership. If the CSA was not partnered with NASA on a particular project, chances were it was partnered with the European Space Agency The SPF and the LTSP II simply formalized this ongoing strategy, ensuring that the general commercialization of national space assets gave Canada continued access to international space endeavors and the benefits that resulted from such broad collaboration.

Canada's space budget for the remainder of the 1990s was therefore largely governed by the estimates laid out in the LTSP II. After a twelve percent cut in funding in 1995, the CSA operated on an annual budget of approximately $300 million. From this, the agency managed its ongoing space station commitments, Earth observation programs, the RADARSAT program, its space science program, and all infrastructure costs associated with managing the country's space program. The implementation of these plans streamlined the budget and

---

[39] Ibid., p.140.

space activities to some degree, but still the agency could not do everything and there remained difficult choices given what it could afford. The CSA's two main high-profile projects, the shuttle/space station program and the RADASAT program, consumed the majority of the agency's operating budget, and also inflicted costly budget overruns on the agency whenever either project was delayed. This left relatively little to support other opportunities that were presented to Canada during the same period, such as the U.S.-led Mars exploration program or the European-led Galileo global navigation satellite system. [40]

With the agency's continued focus on future space shuttle and space station projects, it began a second recruiting campaign for new astronaut candidates in January 1992. Two years earlier the CSA had accepted an invitation from NASA to send two of its own astronauts to undertake Mission Specialist training at the Johnson Space Center in Houston. The demanding program involved approximately four to five years of training, culminating in a spaceflight for at least one of them and possibly both. During their training the two selected members would be largely unavailable for other CSA tasks, so when Bondar and Money resigned in the same year that the training was scheduled to begin in 1992, the agency suddenly found itself with perhaps having access to only two remaining astronauts for a large range of other tasks and duties. Thus, if it expected to be fully involved in the development of the space station, the CSA needed to begin preparing astronauts for potential future duties in support of that goal. The total number to be recruited was a matter of some debate, as were the prerequisite qualifications and experience required of the new candidates. The first group was selected to fulfill their expected roles as engineering or life sciences payload specialists. Astronauts assigned to the future space station would need strong science and engineering backgrounds of course, but as potential shuttle mission specialists and station crew they also required something more, namely a high degree of flexibility and the ability to adapt quickly. Space station operations demanded crew who could not just perform scientific experiments but also carry out all the other tasks required to run a space station.

With this in mind, the CSA set out to develop its new recruitment campaign using an independent contractor. The 1983 campaign had overwhelmed the small public affairs offices at the National Research Council and the agency was looking to avoid repeating that error. Still, its attempt to quietly set up its campaign failed when the CSA president and the government's science minister at the time, Larkin Kerwin and William Winegard respectively, let slip at a press conference prior to Bondar's 1992 spaceflight that the agency intended to recruit new astronauts sometime soon. [41] Nevertheless, the competition got under way on schedule in January 1992, and within a few months the agency had chosen 370 out of the more than 5300 applications received. Further medical and psychological tests reduced the number of suitable candidates to about 100, and then again down to fifty. This new round of candidates were being assessed against their viability as potential mission specialists and space station crewmembers so the standards were often much

---

[40] The Galileo Global Navigation Satellite System was subsequently initially led using a combined concept of the three main contributors – Germany, France, and Italy. As of 2016 its projected constellation of 30 satellites are still being deployed, 12 of which are currently operational.

[41] Op. cit., Dotto, *The Astronauts*, p.12.

higher. Many failed simply due to poor eyesight. After further tests, interviews, and elimi-
nations, on June 9 the CSA held a media briefing in Ottawa to announce the names of its
four newest astronaut candidates.

From a field of thousands the successful applicants were 28-year-old Julie Payette, a
computer engineer then working at Bell-Northern Research Ltd.; 32-year-old Chris
Hadfield, a Canadian Forces test pilot who had already logged more than 2000 hours flying and testing over fifty different types of aircraft; 37-year-old Dafydd (Dave) Williams,
an emergency services physician who was a director at Sunnybrook Health Science Centre
in Toronto; and 37-year-old Robert Stewart, an associate professor of geology and chairman of exploration geophysics at the University of Calgary. Interestingly, this was
Stewart's second attempt, having been screened out of the first competition 1983. "I got
the feeling this time that it was a little different. They were looking for a broader range of
candidates", he noted in an interview, "more science, not just MDs and engineers."[42]
Surprisingly, Stewart withdrew his candidacy soon after, choosing instead to remain in his
current position at the university. As a result, his place was offered to 29-year-old Michael
McKay, a military officer and lecturer in computer engineering and robotics at the Collège
Militaire Royale in St. Jean, Quebec. McKay was ecstatic to receive the call, having been
previously screened out of the competition.

Fig. 5.16 Canada's second astronaut group – Mike McKay, Chris Hadfield, Julie Payette, and
Dafydd Williams

---

[42] Ibid., p.30.

With the announcement made, the CSA could now move forward on filling its two allocated spots on the NASA mission specialist training program. This was a natural evolution of affairs for Canadian manned spaceflight; in the post-*Challenger* era NASA had changed its policy on who could fly and more justification would be needed to put a payload specialist on a flight instead of having one of NASA's mission specialists perform the task. Though NASA honored those agreements already in place prior to the change in policy (hence why Bondar and MacLean got to fly) any new assignment would have to meet the new criteria. Mac Evans, then serving as the CSA's vice president of operations, noted that the new criteria were first, NASA had to have a direct interest in the experiment, and second, the agency or organization wanting to fly the payload specialist had to demonstrate that none of NASA's mission specialists were capable of performing the task.[43]

Still, Canada's billion-dollar investment in the space station gave it a degree of clout in terms of securing future flights for its own astronauts. This also provided further justification for placing Canadians into the NASA mission specialist stream. On May 21, 1992 a formal agreement was signed between the CSA and NASA concerning the exchange of personnel for the course, with the agency choosing the two it would send soon after. Some of the astronauts were simply out of the running due to timing. Bondar had just returned from her space mission, and MacLean was about to launch on his spaceflight. Money had announced his departure from the astronaut corps. That left three others, but Tryggvason was already engaged as backup to MacLean, and there would be considerable risk in pulling him from his assignment or trying to double-task him. That left only Garneau and Thirsk from the original group, plus the four new candidates, to choose from.[44]

In the end, it was decided that one veteran, Marc Garneau, and one member from the second astronaut candidate group, Chris Hadfield, would be sent to become mission specialists assuming they passed another round of NASA tests. Hedging their bets, the CSA managed to arrange for Tryggvason to also undergo the mission specialist medical testing, just in case Garneau or Hadfield did not pass their own examinations. In the end the two were successful, however, and a few weeks later they received the official nod from the CSA to report to the Johnson Space Center in Houston in early August, joining others to begin the 28-week mission specialist basic training course. The core of this first phase's curriculum consisted of a thorough indoctrination into the space shuttle and its operations, as well as the theory, design, and operation of all of the shuttle's major systems. Garneau and Hadfield would become familiar with all aspects of the shuttle's central systems including its computers, electrical power, environmental systems, propulsion, navigation, and communications. They learned to work both inside and outside of the space shuttle through extensive training in NASA's many simulators and other related facilities. With the completion of the basic course, the pair then moved onto the advanced training course, designed to sustain the skills they had learned while improving their proficiencies even further. In addition, Garneau and Hadfield were assigned collateral duties in support of ongoing space shuttle operations; those these were often temporary and did not impede the learning of core competencies.

---

[43] Ibid., pp.100–101.

[44] Roberta Bondar later disputed her lack of availability to be considered for mission specialist training, and also suggested that, despite the reasons given for its decision, the agency had shown poor judgment in excluding the scientist astronauts from their choices.

Fig. 5.17 The Canadian Astronaut Program, c.1994. Back row left to right: Garneau, Hadfield, Tryggvason, and MacLean. Sitting in front left to right: McKay, Williams, Payette, and Thirsk

During their third and fourth years of training, Garneau and Hadfield were introduced to even more complex hands-on activities that would allow them to develop proficiencies in handling not just a single system but, eventually, the entire system of systems that made shuttle flight possible. During this stage the training was not only complex, but also increasingly competitive, as by the end of it those who demonstrated the greatest potential were subsequently identified for a potential assignment to a future mission. Throughout the whole endeavor officials at the CSA continued to subtly lobby for Garneau and Hadfield to get maximum exposure to space stations operations tasks. Canada's main effort in the space station project was the development of the mobile servicing system, and as such it very much wanted one of its own to be heavily involved in the deployment, testing, and integration of that system in the space station build. Mac Evans explained, "we're supposed to know our equipment better than anybody else, so we should be the ones to check it out". Further to this, as Karl Doestch, then serving as director of the Canadian space station program, admitted to author Lydia Dotto when asked about having Canadian astronauts perform EVAs to accomplish these tasks if necessary, he replied quickly, "I'm expecting it".[45]

---

[45] Op. cit., Dotto, *The Astronauts*, p.120.

The RADARSAT Program

From its formative years and through the aftermath of the *Challenger* accident, the activities of the Canadian Astronaut Program had garnered considerable attention from both the space community and the general public from its formative years, but it was not the only major Canadian program then under way. Since the early 1970s, the Canadian government had studied the ever-increasing requirements of exercising sovereignty over its vast territory. Up to that time, aircraft provided the primary means of overflight, but as these machines reached the end of their service lives, it was conceded that another means of exercising sovereignty would soon be needed. As early as 1974 the government had already directed several departments to collaborate on a study of the problem, but it was not until September 1976 that an interdepartmental task force under the oversight of the recently-created Canada Centre for Remote Sensing (CCRS) produced its pioneering study report, *Satellites and Sovereignty*.[46] In this report, particular attention was given to the use of synthetic aperture radar (SAR) due to its generally high ground resolution and its independence of visual obstacles such as weather or darkness.[47] After the country extended the limit of its territorial waters in 1977, the government conceded that the control of its geography and coastal waters would require surveillance on an unprecedented scale, and satellites appeared to be the solution. However, two questions remained for the government. First, was it technologically possible to develop such a system? Second, how much would it ultimately cost?

Though the *Satellites and Sovereignty* report focused greatly on the need for the monitoring of human activities, as the program evolved it became increasingly clear that in addition to sovereignty enforcement there was also a requirement for detailed ice mapping and forecasting throughout the Arctic in support of safe operations and navigation throughout Canada's northern waters. Additionally, the ability to do so would be required year round regardless of weather or daylight conditions. These requirements served only to further support the need for a space-based solution. The CCRS study separated the potential applications of radar remote sensing into twelve categories relating to ocean surveillance, and a further three focusing on land surveillance. Of these, about two-thirds were identified as feasible applications that could be incorporated into a space system. Further studies, however, were still needed to understand and realize the full potential of the technology. In April 1977, a new Canadian program was established with the mandate to participate in the U.S. SEASAT Program, and its project manager, Dr. S.W. McCandless, visited Canada at that time to explain the project in greater detail. The sensor payload on SEASAT included a L-band synthetic aperture radar – 23.5cm; 1.275GHz; HH polarization; 23 degree incidence angle; and a resolution range of 25m – something of great interest to Canada given its own requirements, and the government was subsequently invited to participate in an upcoming American experimental program for sensor validation. Canada accepted without hesitation, and a new domain of Canadian space technology development began.

---

[46] CCRS, *Satellites and Sovereignty: Report of the Interdepartmental Task Force on Surveillance Satellites*. Ottawa: Minister of Supply and Services, August 1977.

[47] E.J. Langham. 'RADARSAT – Canada's Program for Operational Remote Sensing', *Canadian Journal of Remote Sensing*, 8:1, July 1982, 29.

With the acceptance of the American invitation to participate in SEASAT, a complementary program was established under the oversight of the Department of Energy, Mines, and Resources, which then led Canada's effort in the development of SAR technology and applications. As part of its program of development, in September 1977 the government issued a request for proposals from approximately 250 Canadian scientists who previously had demonstrated interest in the applications of remote sensing, with an emphasis of inviting experiments that focused on ocean and sea ice surveillance. Expecting to generate about thirty returns, they in fact received in excess of a hundred proposals, about half of which dealt with land applications. The reason for this was that many scientists' research interests and affiliations were connected to the development of remote sensing techniques for the monitoring of resource management. As there was potential economic benefit in pursuing these lines of investigation, the program also supported the additional land application experiments.[48]

The American SEASAT was launched into orbit on June 27, 1978 and began operations soon after. NASA's Jet Propulsion Laboratory managed the program, and the satellite operated without incident until October 10, 1978 when a reported massive short circuit in the satellite's electrical system suddenly ended the mission.[49] Despite the sudden loss of SEASAT, Canada's program office decided that enough data had been collected – approximately 126 million square kilometers had been imaged – to support the continuation of its analysis. As a result, in March 1979 MaDonald, Dettwiler & Associates (MDA) produced the world's first digitally processed SEASAT SAR image, leading its rivals by six months with the best quality imagery then available. By the time the Canadian program office concluded its own activities in March 1980, its scientific community had gained considerable experience in SAR data applications and analysis, setting the stage for future endeavors in this field.

In its final report back to government in July 1980, the Canadian program office involved in the SEASAT trials emphasized three key recommendations. First, it recommended that Canada continue to move toward operational use of surveillance satellite data by participating in research and development of the technology and applications of surveillance satellites. Second, that the country sought cooperative programs with other nations or international agencies to share the costs of surveillance satellite activities; and third, that Canada further developed industrial capabilities in surveillance satellite

---

[48] Ibid., p.30. Another result of this initiative was Canada's success in joining the European Space Agency (ESA) as an associate member in January 1979. At that time, the ESA was conducting its own studies of space-based SAR technologies, and Canada looked to learn as much as possible for its new international partners.

[49] Some have posited that SEASAT's failure was in fact deliberately orchestrated by the U.S. government as a result of its unintended ability to detect the wakes of submerged military vessels. See Pat Norris, *Spies in the Sky: Surveillance Satellites in War and Peace*. Chichester: Praxis, 2008; pp.57–89.

hardware and software technologies to permit participation in future similar programs.[50] Cabinet concurred with the recommendations and in the same summer the Department of Energy, Mines, and Resources initiated discussions with NASA officials to explore the possibilities of a new joint program. It was agreed that studies for mission requirements were needed first, and an agreement with NASA was reached on November 26, 1980 to begin a preliminary program examining SAR technology and mission concepts. The RADARSAT program had begun.

A preliminary mission requirements study was completed by the summer of 1981 but the teams examining the problem still lacked sufficient data, particularly related to C-band SAR capabilities, to provide a definitive way forward. Again, the problem was solved by using modified airborne SAR capabilities to generate C-band images, and to change the incidence angle in flight. This provided sufficient information to produced an updated mission requirements document in February 1982.[51] From this baseline, further work was completed on improving overall image quality, developing imagery interpretation techniques, and evolving C-band SAR technologies and applications. At this point it appeared that, while an L-band SAR satellite would be technologically easier to develop, a C-band SAR satellite would provide substantially greater scientific and economic benefits. With the program's mission requirements study completed in the fall of 1982, the program office turned its attention to a series of detailed design studies.[52]

By the mid-1980s the largest portion of Canada's space program's budget was allocated to the country's remote sensing program. In 1985–1986 alone the program received CDN$73.5 million in funding to complete design proposals for an orbiting platform. Part of RADARSAT's budget allocation, however, was also earmarked to cover the design costs for Canada's space station MSC proposal. While the RADARSAT program managers protested this decision and complained that their coffers were being raided at the expense of the astronaut program, in the end they were forced to commit the funding to that endeavor. Nevertheless, the RADARSAT program definition studies continued, with a view to completing this work sometime before the end of 1986. In the end, it was decided that the RADARSAT should ultimately be capable of delivering high-resolution imagery for both commercial and scientific use in such fields as disaster management, interferometry, agriculture, cartography, hydrology, forestry, oceanography, and ice monitoring. Once these capability requirements had been approved in principle, the next stage was the actual design, construction, testing, and evaluation of the spacecraft and the ground receiving stations. If all went well and stayed relatively on schedule, it was expected to have the first RDARSAT in orbit by the mid-1990s.[53]

---

[50] CCRS, *SURSAT Final Report: Executive Summary*. Ottawa: Department of Energy, Mines, and Resources, dated September 1980.

[51] EMR, *RADARSAT Mission Requirements Document*. RADARSAT Project Office Report No. 82-7. Ottawa: Government of Canada, February 1982.

[52] Op cit., Langham, 'RADARSAT', pp.35–36.

[53] MOSST, *Interim Space Plan 1985–1986*. Ottawa: MOSST, 1985, P.5.

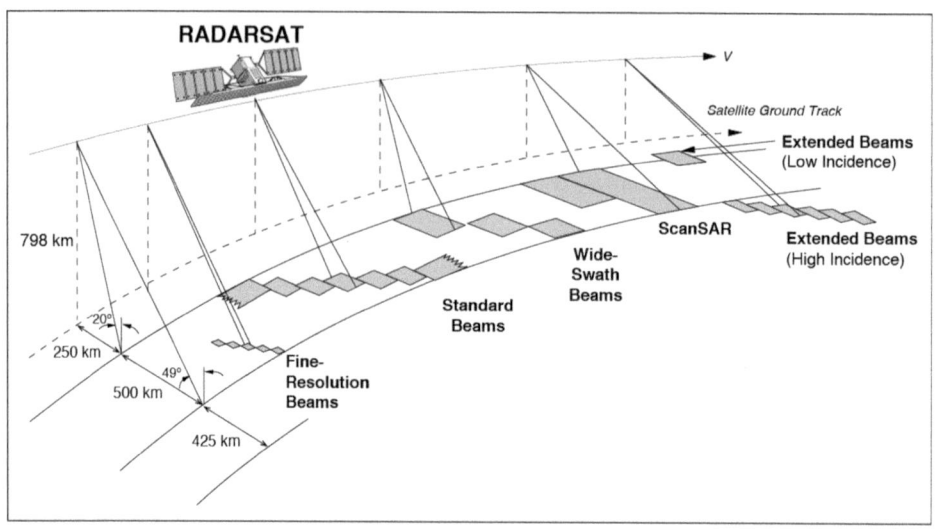

7797-D1R3

Fig. 5.18 RADARSAT beam modes

Table 5.1 RADARSAT applications

Field of application	Description
Agriculture and land cover monitoring	Measure of soil moisture, assess crop conditions, update land cover maps
Forestry (boreal and tropical)	Detect clear cuts, update forest inventories, map depletions, map forest fires, assess regeneration, monitor land use changes
Geology	Detect structural and lithologic features, assess geo-hazards (landslides), extract geomorphological information
Hydrology	Measure flood extent, improve hydrological modeling, assess flood damage
Coastal zones and oceans	Detect and track vessels, detect wind and wave spectra, detect mesoscale ocean features, monitor coastline changes
Cartography and land use	Update topologic maps, create digital elevation models, create land use and land cover maps

Table 5.2 RADARSAT beam modes

Mode	Resolution m)* (range × azimuth)	Looks**	Swath width (km)	Incidence angle (°)	No. of beams
Standard	25 × 28	4	100	20-49	7, >10% overlap
Wide (1)	48-30 × 28	4	165	20-31	3, 10% overlap
Wide (2)	32-45 × 28	4	150	31-39	
Fine resolution	11-9 × 9	1	45	37-48	5, 10% overlap
ScanSAR (N)	50 × 50	2-4	305	20-40	
ScanSAR (W)	100 × 100	4-8	510	20-49	
Extended (H)	22-19 × 28	4	75	50-60	6, 3% overlap
Extended (L)	63-28 × 28	4	170	10-23	

* Nominal: range dependent and processor dependent
** Nominal: ground range resolution varies with range

The development, assembly, and testing of RADARSAT-1 began in early 1988 under the oversight of the prime contractor, SPAR Aerospace Ltd. They in turn subcontracted a number of other companies to assist them, including Ball Aerospace, Dornier, Odetics, COM DEV, CAL Corporation, SED Systems, and MDA. In 1990, Joe McNally joined the Canadian Space Agency as Director General of the RADARSAT-1 Program. "The challenge was great and we felt the exuberance of being pioneers. With the support of countless public servants, federal and provincial governments, private industry partners and international space agencies, we rose to the challenge and delivered a satellite that would change the course of history."[54] With his team organized, assembly continued in earnest. The bus of the RADARSAT was built by Ball Aerospace and then delivered to the David Florida Laboratory in Ottawa in 1993, where it was mated to its solar panels and other components prior to testing. The final integration of the payload and bus modules, plus the SAR antennae and solar arrays, was completed by mid-1994. Final vibration, antenna, and frequency testing was completed by early 1995.

By midsummer the RADARSAT was finished. The 2750kg satellite had cost the Canadian government CDN$620 million to conceive, design, and build, and was expected to last five years, carrying 67kg of hydrazine for mission orbit adjustment, restitution, and yaw maneuver as needed. Powered by both a double solar array and batteries, it could sustain autonomous operations for 24-hour periods. The satellite contained two parallel X-Band downlinks for SAR data transmission, while its tracking, telemetry, and communications link operated in the S-band. The spacecraft was designed to provide data downlink in real time, giving scientists on the ground daily coverage of the Arctic (with the exception of the North Pole), view any part of Canada within three days, and achieve complete coverage at equatorial latitudes, using a 500km wide swath, every six days.[55]

Wrapped and packed into a special transport container, the Canadian satellite was loaded on to a flatbed at the end of the summer and moved across country to the NASA launch facilities at Vandenberg Air Force Base, California. There it was stacked atop a Delta II 7920-10 rocket and prepared for launch. On November 4, 1995 Canada's RADARSAT-1 was launched into a 798km, 98.6 degree inclination, sun-synchronous orbit and went into service soon after. The CSA headquarters in St. Hubert led the operation including the mission control center, the mission planning office, the order desk, as well as calibration and mission management.[56]

[54] M.E. McGuire, *The RADARSAT-1 Story: A Canadian Satellite*. Ottawa: ME Management Strategies, 2014, p.7.

[55] Earth Observation Resources Portal Directory. 'RADARSAT-1'. Accessed online on March 31, 2016 at https://directory.eoportal.org/web/eoportal/satellite-missions/r/radarsat-1

[56] Op. cit., M.E. McGuire, *The RADARSAT-1 Story*, p.20.

Fig. 5.19 SPAR engineers test a RADARSAT engineering model solar array fully deployed at the David Florida Laboratory, June 18, 1993

As with previous Canadian satellites, the RADARSAT-1 exceeded all design expectations. It returned its first image to Earth on November 28, 1995 shortly after beginning operations, and by 1999 the spacecraft was providing the first routine surveillance of the entire Arctic and Antarctic regions, completing maps that greatly impressed the international scientific community. The RADARSAT-1 continued running well past its five-year design life span until it suffered a major power disruption in early 2013. Over the course of its operational life, the RADARSAT-1 executed over 351,500 requests achieving a

Fig. 5.20 The RADARSAT Telemetry, Tracking, and Command Station was established at the Canadian Space Agency's RADARSAT Mission Control Facility in August 1994

95.2% average system performance. The program generated 10,000 person years of employment in Canada alone, and provided more than CDN$1 billion in benefits to Canada's public and private sectors.[57]

The RADARSAT program may have begun out of necessity to protect Canada's sovereignty, but it continued because of its technological achievement and the new opportunities it brought to Canada's scientific, engineering, public, and private sectors. Not long after the launch of RADARSAT-1 in 1995 the government, satisfied with the results of the program, approved in principle further funding for the research and development of a follow-on SAR satellite. An official project announcement soon followed with an expectation the Canada would launch a second RADARSAT sometime in the year 2001.

---

[57] Ibid., pp.120–121.

Fig. 5.21 The RADARSAT undergoes an integrated systems test at the DFL on April 28, 1995

Canada's Visit to *Mir*

As the RADARSAT program worked towards its launch date, the Canadian astronaut program likewise prepared for its next series of flights. Both Marc Garneau and Chris Hadfield had successfully worked their way through NASA's mission specialist training program, with Hadfield being assigned to the next scheduled Canadian spaceflight in 1994.

An agreement between the United States and Russia signed in July 1991 opened the door for the exchange of astronauts and cosmonauts on each other's spacecraft. A year later at their first summit in Washington D.C., the leaders of the two superpowers agreed

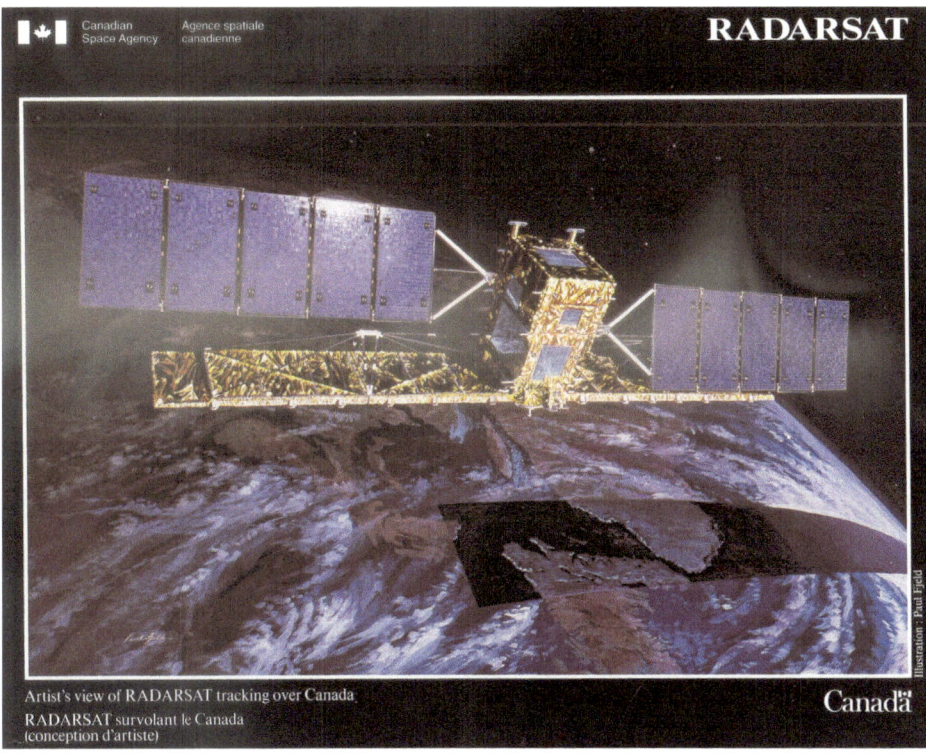

Fig. 5.22 Artist's impression of RADARSAT-1 in orbit

to give consideration to a joint mission, which in turn was followed by further agreements to incorporate Russian space technology into the evolving American-led international space station program. Among those options considered was the installation of a Russian-built docking system in the American space shuttle. This not only created the opportunity for western astronauts to visit and even serve aboard the Russian *Mir* space station, but also gave NASA the opportunity to start on the some of the space science projects that were then being planned for the international space station. Further willingness between America and Russia to collaborate in space ultimately led to the signing of the 'Human Spaceflight Cooperation' protocols on October 5, 1992, officially setting the stage for a series of joint shuttle–Mir missions.[58]

[58] David M. Harland, *The MIR Space Station: A Precursor to Space Colonization.* Chichester: Wiley Praxis, 1997, pp.219-241; see also David M. Harland and John E. Catchpole, *Creating the International Space Station.* Chichester: Springer-Praxis, 2002, pp.177–188.

Fig. 5.23 Canadian Astronaut Chris Hadfield flew as mission specialist 1 on STS-74 and was the only Canadian to visit the *Mir* space station

The Shuttle–*Mir* Program was formally announced in 1993 with the first joint U.S.–Russian space shuttle mission successfully completed in 1994. The first space shuttle flight to visit *Mir* occurred a year later, when STS-63 rendezvoused with the *Mir* space station on February 4, 1995 but did not dock. That honor fell to the crew of STS-71, which successfully docked with *Mir* on June 29, 1995, the first of nine such dockings to occur during the course of the joint program. Having successfully proven the concept that summer, the way was now clear for NASA shuttle dockings to proceed, the next of which was scheduled to occur in mid-November when STS-74 would make a planned visit the station.

NASA astronaut Kenneth Cameron commanded the STS-74 mission, which was his third and last spaceflight. James Halsell, making his second spaceflight, served as *Atlantis'* pilot. The mission specialists on STS-74 included veteran astronaut Jerry Ross, making his fifth spaceflight, William McArthur, making his second spaceflight, and the rookie Chris Hadfield. Yet despite having never flown before Hadfield had been assigned a mission-critical task. The most important cargo STS-74 was tasked to carry aloft into orbit on its mission was a Russian-built Docking Module (DM), which would be attached to *Mir's* Kristall module thereby facilitating all follow-on shuttle dockings at the station.

Fig. 5.24 The STS-74 crew

The 4.6m long, 2.2m diameter, 4.2-tonne rust-colored module resembled an extremely stretched Soyuz orbital module with androgynous ports at each end. Without the DM, the Kristall module would have had to be moved for each shuttle-docking mission to provide adequate clearance, an action that was neither practical nor desirable. The addition of the DM to *Mir* would negate this problem altogether, and once *Atlantis* rendezvoused with *Mir* it would be Hadfield's job to unstow the DM from the cargo bay using the Canadarm and install it.[59]

Years and years of training and preparation finally came to fruition for Chris Hadfield on November 12, 1995, as space shuttle *Atlantis* lifted off from Launch Pad 39A at the Kennedy Space Center and headed for orbit. Punching through a low cloud deck on the way up, the first stage was, as it often is, a bit of a rough ride. Yet once the solid rocket boosters separated and the crew passed through first staging of the vehicle, the ascent smoothed out the rest of the way up to orbit. Flying in the right-hand window seat just behind the pilot, during the ascent Hadfield was responsible for monitoring the launch

---

[59] NASA, Space Shuttle Mission STS-74 / Shuttle-Mir Mission – 2 Press Kit, November 1995, pp.24–25.

Fig. 5.25 Chris Hadfield in the Shuttle white room preparing to board *Atlantis*

progress and helping with various procedures. Had an emergency occurred during the launch, Hadfield would have had the critical task of helping the crew with backup checklists, but thankfully all went well on the ride into outer space. Approximately 43 minutes after launch *Atlantis* executed another two minute and thirteen second burn to change the shuttle's flight path into a 162 nautical mile circular orbit. Once there, the crew configured *Atlantis* for on-orbit operations and opened the payload bay doors. At about three hours into the flight the space shuttle fired its reaction control jets in a first of a series of burns that would align *Atlantis'* path to that of the *Mir* space station. After these maneuvres were completed, Hadfield activated the DM and its systems in the payload bay and prepared the precious cargo for mating to the *Mir*.

The following day Hadfield had the task of powering up the Canadarm while the crew closed the gap between the shuttle and the Russian space station. He also took time with other available crewmembers to answer some questions posed by Canadian journalists back on Earth. As only the fourth Canadian astronaut to fly into space and the first to visit a space station in orbit, Hadfield's mission attracted considerable attention back home. With the public relations task complete, the crew returned to the business at hand. On day three of the mission Hadfield's main task involved maneuvering the DM into place above the shuttle's orbiter docking system, allowing the commander and pilot to successfully mate the two objects. With this task complete, *Atlantis* was ready to arrive at *Mir* the following day.

Fig. 5.26 The Docking Module is prepared for installation into the payload bay of *Atlantis*. This cupola allowed for routine NASA space shuttle visits to *Mir* throughout the mid-1990s

Atlantis continued chasing the Russian space station on day four, closing the distance to eight nautical miles with a terminal phase initiation burn of the shuttle's engines. This brought *Atlantis* into line directly below *Mir*, and after establishing radio contact with the Russian crew the shuttle continued closing the distance between the two spacecraft until just 52 meters separated them. At this point shuttle commander Kenneth Cameron, who had been controlling *Atlantis'* final approach using the aft flight deck controls, now waited while the *Mir* space station was maneuvered into alignment to receive the docking. Once both Houston and Moscow had given the go ahead, Cameron closed the gap again to just 9.1 meters, halting momentarily one last time to make small final adjustments on the approach. At this point another piece of Canadian technology, the advanced space vision system (ASVS), first tested by Canadian astronaut Steve MacLean on STS-52, became instrumental to success. With everything looking good, he nudged the orbiter closer using the elbow camera on the *Canadarm* to assist him, until the two spacecraft touched and locked. "It happened in the blink of an eye", SPAR systems engineer Mike Hiltz later commented, "a year's worth of work, five hundred misalignments in the simulator, worrying about what happens if we don't capture, and it was all over just like that."[60] The second docking, and the first to use the new specially-designed Docking Module, was a success.

Having confirmed that a good seal existed between the two spacecraft, the crews were given the green light to open the hatches at approximately 4:02am on November 15. After the traditional handshakes and exchanges of gifts (that included among other items a new guitar and Canadian maple sugar candies), the STS-74 crew commenced the transfer of the DM and other payloads to the *Mir* 20 crew. This included over 450kg of water, spare lithium hydroxide canister for the air scrubbers, and other critical supplies. Hadfield served as the main shepherd of supplies. He spent nearly seven hours filling large containers with water produced from *Atlantis'* fuel cells, and then moved these along with food and other essential items over to *Mir*. Hadfield then transferred other items, including parts from *Mir*, back to the space shuttle for subsequent return to Earth. Although not an overly glamorous task perhaps, nevertheless it was critical to keeping the space station going.

During the *Atlantis–Mir* linkup Hadfield was also responsible for activating and running a number of experiments including the Photogrammetric Appendage Structural Dynamics Experiment (PASDE) and another Shuttle Glow Experiment (GLO-4). The aim of the PASDE was to accurately record the dynamics of one of *Mir*'s solar arrays as it reacted to the jolts from docking and from a jet firing sequence, and to the thermal stresses of moving between day and night in orbit. The GLO-4 experiment, meanwhile, recorded further data related to the strange glow phenomena visible at night as the two spacecraft moved through the thin upper layers of the atmosphere.[61] In addition to these, Hadfield was responsible for conducting a photographic and video survey of *Mir* from the space shuttle, as well as measuring the noise environment inside the Russian space station. "We're looking to gather baseline data", Hadfield noted during an in-flight interview, "at the only space station in the world – or off the world – to modify how the International Space Station is going to be built."[62]

The space shuttle's second visit to *Mir* concluded on November 18. After exchanging farewells the two crews secured their hatches, followed a short while later by the separation of the two spacecraft. The DM installed by *Atlantis* upon arrival stayed behind. Once the docking mechanism's springs had nudged *Atlantis* a meter away from the port, shuttle commander Kenneth Cameron re-engaged the steering jets and slowly eased *Atlantis* away from *Mir*. Eventually moving the shuttle to a distance of 122 meters from the space station, *Atlantis* then executed a fly around of *Mir*, circling the outpost twice while the crew performed a photographic survey of the Russian spacecraft. Once this was complete *Atlantis* fired her jets once again, departing the vicinity of *Mir* and reinserting herself into an orbit in preparation for returning home.

---

[60] Lydia Dotto,. 'STS-74: A Canadian Visits Mir'. N.d. 3-5. Accessed online on January 30, 2001 at http://schools.tdsb.on.ca/spacenet/canastroprogram/sts-74/default.htm

[61] Canadian Astronaut Office. STS-74 Mission Manoeuvres. Accessed online on December 21 2000 at www.space.gc.ca/csa_sectors/human_presence/en/canastronauts/astro/sts74/emanoe.htm

[62] Lydia Dotto, 'STS-74: A Canadian Visits *Mir*', p.6.

During re-entry Hadfield made his first journey home from space somewhat alone. The *Atlantis* had a crew of five, but as four of them were seated up on the flight deck during operations, he was the only one strapped in below on the middeck. Nevertheless, he and his crewmates arrived safely back on Earth on November 20, touching down at Runway 33 at the Kennedy Space Center, Florida. The space shuttle's 73rd mission was complete, and Canada once again took great pride in the accomplishments of its astronaut corps. Hadfield's performance had been outstanding throughout the entire mission, and he easily cleared the way for more Canadian astronauts to qualify as mission specialists for future flights.

Fig. 5.27 The STS-74 crew gives a thumbs-up after return to the KSC on November 20, 1995

Conclusion

The successful completion of Chris Hadfield's mission seemed to only whet the appetite of the Canadian space program for more. Having heavily invested in the development of the forthcoming international space station project, the country was guaranteed a number of future spaceflights that would continue to place Canada front and center as one of the main actors in one of the greatest scientific and engineering projects ever undertaken in outer space. Two more Canadian astronauts were scheduled to fly within eight months of Hadfield's flight, with another planned for about a year after that. Politically, it also sent a clear signal that Canada considered itself amongst the most technologically competent nations and strong partner in the spacefaring nations community. And though there remained much work to be done organizationally, in a little over a decade Canada's space effort had evolved from a series of separate satellite initiatives into something resembling a much more cohesive national space program. Leading the way were its two main efforts – the Canadian Astronaut Program and the RADARSAT – both of which did much to connect the Canadian population to outer space by encouraging one community to look up and another to simply look down.

6

Build Up: Canada's Space Station Era Begins, 1996–2009

Since the publication of the first five-year plan in 1980, Canada had met milestone after milestone in its evolution into a mature spacefaring nation. Though the scale of the program remained small compared to larger partners such as the United States and Russia, Canada's achievements were indeed impressive when juxtaposed to those of the European Space Agency or the Japanese Space Development Agency. The decision to focus its main effort in two fields – Earth remote sensing and human spaceflight – had returned a handsome dividend. By providing critical technologies such as the Remote Manipulator System (RMS), the government ensured that every time an American space shuttle flew, Canada flew too. Similarly, its focus on designing and building the mobile servicing center for the proposed International Space Station likewise ensured that the country could enjoy a seat at the table in the upcoming century of sustained human spaceflight.

Canada's space science program benefitted greatly as well. Not only did it leverage the astronaut program's involvement in the space shuttle, *Mir*, and the planned international space station, there was also the RADARSAT program in addition to the evolving space astronomy program. Given the success of RADARSAT-1 the government had committed to building and launching a more technologically advanced second satellite by 2001, whilst at the same time contributing to other international satellite projects that explored the solar system and the universe beyond. The agency collaborated with both France and the U.S. on the Far Ultra-Violet Spectroscopic Explore (FUSE) satellite that launched in 1999, and also developed the MOPITT (Measurements of Pollution in the Troposphere) package for the NASA *Terra* satellite, the flagship spacecraft in its Earth Observing System (EOS) launched in December 1999.

A Busy Year in Space: STS-77 and STS-78

Though Canada's unmanned space program matured over the course of the 1990s and could note a number of its own successes, the public's attention remained largely fixed on the trials and accomplishments of the Canadian astronauts. The year 1996 was a

© Springer International Publishing AG 2017
A.B. Godefroy, *The Canadian Space Program*, Springer Praxis Books,
DOI 10.1007/978-3-319-40105-8_6

remarkable one for the Canadian space program. Still riding high on the success of Chris Hadfield's central role in the delivery of the Docking Module to the *Mir* space station on STS-74, Canada enjoyed further achievement in its space program with back-to-back performances by two of the original astronaut corps. The first of these missions was remarkable for its own "firsts" for Canada – Marc Garneau's trip on STS-77 marked the first return of a Canadian astronaut to space, and it also marked the second occasion when a Canadian would serve as a mission specialist on a space shuttle flight. Robert Thirsk's mission on STS-78 marked the first flight for a Canadian backup astronaut, and the first long duration flight for a Canadian – STS-78 would spend 17 days in space, much longer than any other space shuttle mission up to that point.

Marc Garneau returned to space on NASA's fourth space shuttle mission of the year, a flight specifically aimed at fostering the continued development of the commercial space frontier. Its Office of Space Access and Technology sponsored over ninety percent of the payloads aboard STS-77 in an effort to engage industry in the advancement of space-based science and technology. Space shuttle *Endeavour* carried as its main payload the commercially-owned and operated SPACEHAB-4 Module, stuffed with nearly 1361kg of support equipment, experiments, electronic materials, polymers, and agricultural items. In addition, it also carried aboard the free-flying Spartan-207 spacecraft, which itself carried the Inflatable Antenna Experiment (IAE). *Endeavour*'s other main payload was a suite of four technology experiments collectively known as TEAMS (Technology Experiments for Advancing Missions in Space). Canada also flew four experiments of its own, including the Commercial Float Zone Furnace (CFZF), the Aquatic Research Facility (ARF), the Nano crystal Getaway Special (NANO-GAS) experiment, and the Atlantic Canada Thin Organic Semiconductors (ACTORS) experiment.[1]

Space Shuttle *Endeavour* lifted off from launch pad 39B at KSC on May 19, 1996, carrying its crew of six astronauts 153 nautical miles into orbit without incident and prepared to execute a ten-day mission devoted to commercial science and exploration. Veteran astronaut John Casper served as STS-77 commander with Curtis Brown as his pilot. The crew's four mission specialists were Andrew Thomas, Dan Bursch, Mario Runco, and Marc Garneau. Once satisfactorily on station, *Endeavour* opened her payload bay doors, checked out the RMS, and then used it to conduct a survey of the payload bay. With everything looking as it should, the crew prepared for the deployment of the Spartan-207 spacecraft on flight day two.

The Spartan-207 was a versatile spacecraft, designed to provide short duration free flight opportunities for a variety of scientific studies. For this particular mission it carried the IAE experiment, which it would jettison once the astronaut's investigation was complete. Mission specialist Mario Runco used the Canadarm to release the Spartan, after which shuttle commander John Casper backed *Endeavour* approximately 122m away from the satellite. After completing a partial fly around the satellite, Casper moved the space shuttle back to within communication distance so that the IAE experiment could begin. The IAE itself was a 14m-diameter inflatable antenna reflector structure mounted on three 28m struts. After the Spartan was released from *Endeavour* and the spacecraft had

---

[1] NASA. Space Shuttle Mission STS-77 Press Kit, May 1996, pp.1–3.

Fig. 6.1 The STS-77 Crew. Standing left to right – Burtch, Runco, Garneau, and Thomas. Seated left to right – Brown and Casper

moved to a safe distance from the orbiter, the IAE inflation began, resulting in the deployment of an antenna approximately the size of a tennis court.[2] Despite some unexpected deployment dynamics – the arms of the circular antenna appeared to "flap in the wind" as they deployed, briefly making the whole object resemble something like a giant jellyfish swimming in space – the IAE inflated as designed and maintained its rigid shape for a complete orbit whilst rotating at approximately 180 degrees per minute. Overall, the test was considered a success, and after the IAE was jettisoned from the Spartan, *Endeavour* was moved into position where Garneau, operating the Canadarm, retrieved the Spartan spacecraft for its eventual return to Earth.

The other main payload of Canadian interest was the Aquatic Research Facility (ARF), a space lab jointly sponsored by NASA and designed and built in Canada that allowed researchers to use the microgravity environment to study and better understand early birth defects, bone calcium loss, and ocean ecology. The research provided important information on the earliest stages of marine life development, and on the feeding patterns and

---

[2] NASA, 'STS-77 Space Shuttle Mission Report'. Houston, Lyndon B. Johnson Space Center, June 1996, p.9.

Fig. 6.2 Marc Garneau in the ready room suits up for his second spaceflight on May 19, 1996

distribution of specific marine species in Canada's coastal waters. The space shuttle crew reported two anomalies with the ARF equipment during the mission. First, it was discovered that the needle in the zero-G fertilization unit was bent, requiring it to be bagged and taped for crew safety reasons. Second, one of the samples resisted the 1-G needle fertilization attempt, so the crew simply bypassed this unit. Otherwise, all went well with the ARF and the experiment was considered a success. In addition to the ARF, Canada also flew two Get Away Special (GAS) experiments on STS-77, one examining high-quality nanocrystal growth and the other – ACTORS – to produce enhanced quality organic thin films that could find applications as gas detectors, in computers, in lasers, and other related high-performance electronic equipment.[3] Garneau was responsible for overseeing these two

---

[3] These GAS experiments were designated as G-564 and G-565. For details, see NASA, Space Shuttle Mission STS-77 Press Kit, May 1996, p.27; see also, NASA, STS-77 Space Shuttle Mission

Fig. 6.3 Space Shuttle *Endeavour* on pad 39B at Kennedy Space Center ready for mission STS-77, May 1996

experiments, though as they were self contained and autonomously run, he was not required to physically do anything with them.

Garneau also enjoyed the sound of a familiar voice on CAPCOM throughout his flight. Chris Hadfield, who had flown on STS-74 just a few months before, served as lead Capsule Communicator (CAPCOM) for the mission, providing the only direct voice-link between Mission Control in Houston and the orbiter. This again was indicative of the increasing trust NASA placed in Canada's astronaut corps to integrate fully into the American way of doing things. For astronauts like Hadfield and Garneau, both of whom came from military backgrounds and had previous experience working with the American military, being interoperable with their American partners at NASA was largely second nature.

Report, June 1996, p.14.

Fig. 6.4 Space Shuttle Endeavour lifts off from Pad 39B at KSC on May 19, 1996, carrying Canadian astronuat Marc Garneau on his second mission into space

Fig. 6.5 The Spartan-207 spacecraft with the Inflatable Antenna Experiment was deployed on May 20, 1996

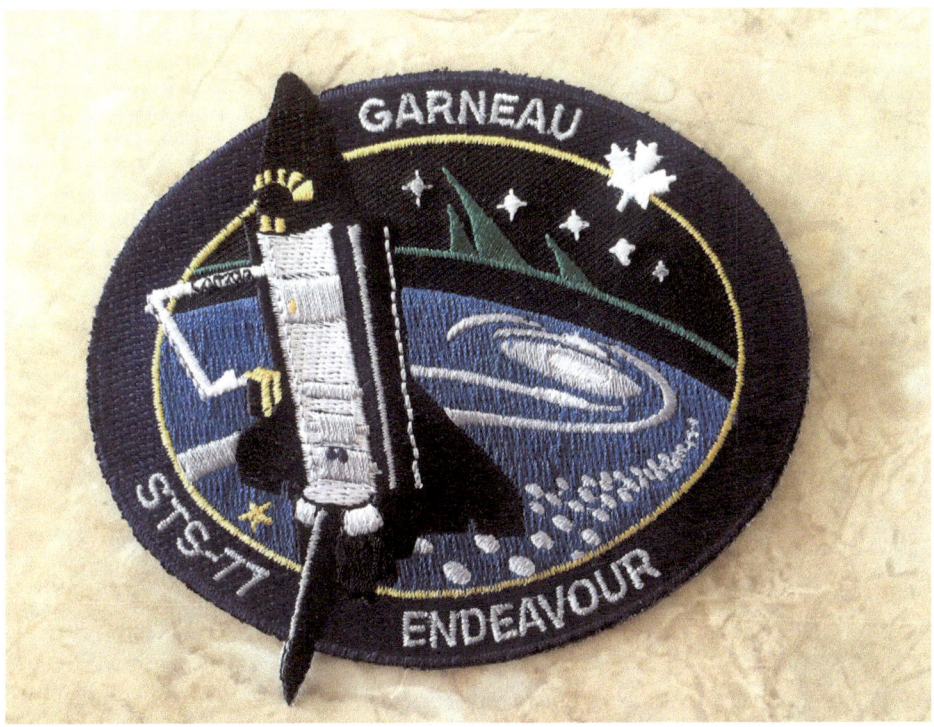

Fig. 6.6 A tradition continues. Marc Garneau's personal STS-77 space patch

Space shuttle *Endeavour* successfully concluded its ten-day mission on May 29, and safely returned the crew of STS-77 to the KSC in what was once again becoming something of a routine for shuttle operations. The next mission, STS-78, was scheduled to launch in less than a month's time and would send *Columbia* on a long-duration flight (nearly 17 days) as part of the preparation for the International Space Station Project. As such, mission STS-78 had as its main objectives research into the effects of long duration spaceflight on human physiology in preparation for flights on the ISS; over twenty life science and microgravity experiments using a pressurized Life and Microgravity Spacelab Module (LM2); and tests into the use of the space shuttle's reaction control system (RCS) to raise the altitude of orbiting satellites.

The main payload on STS-78 was a pressurized Life and Microgravity Sciences (LMS) spacelab module specially fitted with equipment to pursue two main areas of investigation. The LMS life science studies program probed the responses of living organisms to the low-gravity environment and also highlighted studies associated with musculoskeletal physiology. The LMS microgravity experiments focused on understanding the subtle influences at work during the processing of various samples, such as alloy materials, when gravity's effects are greatly reduced. This was especially important to understanding how gravity distorts scientific results on Earth.

Fig. 6.7 The crew of STS-78. Canadian astronaut Robert Thirsk is standing at the far right

Fig. 6.8 Launch day! Thirsk suits up for his first spaceflight, June 20, 1996

Terence Hendricks, a veteran astronaut making his fourth space shuttle flight, commanded the STS-78 mission. Kevin Kregel served as the pilot for space shuttle *Columbia*, with Susan Helms, Richard Linnehan, and Charles Brady serving as the shuttle's mission specialists. Two foreign astronauts, Jean-Jacques Favier from CNES and Robert Thirsk of the CSA, flew on STS-78 as payload specialists. As a medical doctor, Thirsk was the ideal astronaut to fly this long-duration mission. Having served as backup for the very first Canadian spaceflight in 1984, Thirsk had devoted his time whilst waiting for his own flight to further NASA's knowledge of the effects of spaceflight on the human body. He had been the principle investigator on a number of space shuttle experiments, and also even developed an experimental zero-G flight suit designed to protect shuttle astronauts from dizziness or blackout when returning from orbit to the effects of Earth's gravity.

Fig. 6.9 STS-78 lifts off from Launch Pad 39B on June 20, 1996, carrying Canada's fifth astronaut into orbit

Space shuttle *Columbia* lifted off from KSC Launch Pad 39B on June 20, 1996, reaching orbit without incident. Once on station, the crew immediately got to work in the LMS spacelab module. Dr. Thirsk was directly involved in a number of the experiments carried out, and often served both as researcher and subject for various tests. He had a major role in McGill University's Torso Rotation Experiment (TRE) and he was also involved in four other muscle physiology experiments. Thirsk was particularly interested in the lung function experiment, the goal of which was to explain the large differences in the ventilation and the perfusion to the top and bottom of the lung.[4]

Fig. 6.10 Thirsk sports his Team Canada hockey jersey while collecting data during the Torso Rotation Experiment on STS-78

In addition to his other duties, Thirsk performed two separate sets of experiments on behalf of the Young Space Scientists Program (YSSP), some of which were designed by young Canadian research teams. These experiments were aimed at both elementary and high school audiences, and included the production of an educational video for high school students demonstrating the principles of Newtonian physics. Beyond these extracurricular activities, Thirsk used his off-duty time during STS-78 to observe the Earth and record his

---

[4]Canadian Astronaut Office, 'STS-78. Bob Thirsk: Mission'. Accessed online on September 19, 2000 at http://www.space.gc.ca/csa_sectors/astronauts/astro/sts78/s78btpae.htm

Fig. 6.11 Thirsk looking like he is enjoying himself during the Torso Rotation Experiment on STS-78

thoughts in various letters that were subsequently published on the Canadian Space Agency website. "What an adventure it has been … I feel fortunate to be doing this work. I have spent the 15 preceding months training for this mission and now I must pinch myself that I am actually here living and working aboard a spacecraft that is orbiting Earth," Thirsk reflected in a communication from orbit on flight day seven (June 27), "the research work that we are performing as part of the Life and Microgravity Spacelab mission has been going very well. The shuttle systems and payloads are operating well with only a handful of small problems to deal with. The mood amongst my crewmates and I has been understandably upbeat."[5] Another letter, composed on flight day 15, noted "While I feel proud of the work that we have accomplished during the mission and long to

[5] Robert Thirsk, 'Letter From Space', (June 27, 1996), 1. Accessed online on December 21, 2000 at hrttp://www.space.gc.ca/csa_sectors/human_presence/en/canastronauts/astro/sts78/s78jlee.htm

be with my family again, I feel some sadness as we approach the end of our mission. There are many things about this flight that I will miss." His final line observed, "I am sure that I will be asked what it felt like to be a crewmember on the longest shuttle mission to date. My response will be that 17 days wasn't long enough!"[6] Fortunately for Thirsk, many years later he would have the opportunity to break his own record.

Space Shuttle *Columbia* returned safely to Earth on July 7, 1996, successfully concluding the orbiter's twentieth mission into space. Thirsk's return also signaled the end of Canadian astronaut flights for the moment, though it would not be long before another Canadian headed into orbit. The only remaining member of the first Canadian astronaut yet to fly was Bjarni Tryggvason, but he would soon have his chance after being assigned to the crew of STS-85, scheduled to fly in the late summer of 1997.

Mastering Microgravity: Tryggvason and STS-85

As one of the original Canadian astronauts chosen in 1983 Tryggvason continued to wait patiently for his own chance to fly, but he was not idle during the long years between his initial selection by the NRC and his final assignment to a space mission. In the aftermath of the 1986 space shuttle *Challenger* accident Tryggvason had returned to teaching and research while awaiting the fate of his space career. His continued focus on physics and fluid dynamics subsequently led him to the research of Dr. Tim Salcudean, which investigated the possibility of isolating microgravity vibrations that might interfere with space-based scientific experiments and other research. Intrigued by the idea of eliminating these vibrations on spacecraft Tryggvason, along with other scientists, pursued the technological development of a new device that could assist astronauts when conducting sensitive experiments on future space missions. Together, they built two machines, a Large Motion Isolation Mount (LMIM) for parabolic flight research and a smaller Microgravity vibration Isolation Mount (MIM) for orbital flight research.

As work proceeded on the development of the LMIM and the MIM, Tryggvason served as the Canadian Space Agency representative on the NASA Microgravity Measurement Working Group and, later, the ISS Microgravity Analysis and Integration Team. As a principle investigator in the development of the LMIM, he flew with it numerous times on NASA's KC-135 and DC-9 aircraft. The data from these flights served to inform the design of the MIM, which Tryggvason was also involved with as the principle investigator. As researcher Niall Parker and his associates noted on the University of British Columbia's website for the Department of Electrical and Computer Engineering:

> "The approach taken for MIM was to develop a mechanism having two parts: a stator attached to the structure and a payload-carrying flotor, with the only coupling between the two components being a flexible umbilical carrying signals and power to the flotor. The flotor is actively magnetically levitated by a set of wide-gap voice coil actuators. Its position relative to the stator is sensed by an optical position sensor while its absolute acceleration is sensed by an inertial accelerometer system.

---

[6] Ibid., p.4.

Fig. 6.12 CSA astronaut Bjarni Tryggvason was selected as payload specialist for STS-85

A digital controller uses the sensed information to compute actuator currents based on a control law that regulates acceleration and steady-state position to zero. The weak coupling between stator and flotor and the insensitivity of actuator force with position makes such a system extremely effective for vibration isolation."[7]

MPB Technologies Inc. of Dorval, Quebec, built a prototype of the design, which weighed about 30kg in total, under the sponsorship of the CSA's Microgravity Sciences Program. Then Tryggvason and three other Canadian scientists took the MIM to Russia and Star City, where they trained American astronauts and Russian cosmonauts in how to operate it. Ultimately, the MIM was destined for the *Mir* space station. Like any great machine, *Mir* vibrated. "If you put your hand against the side of a car as it is running, a vibration is always there," Tryggvason once told western reporters during a brief tour of Star City in the summer of 1996, "you can't feel it, but the space station also vibrates about one tenth of a

[7] Niall Parker et al., 'Design of Magnetically Levitated Vibration Isolation Platform', Department of Electrical and Computer Engineering, University of British Columbia. N.d. Accessed online on March 31, 2016 at http://www.ece.ubc.ca/~tims/vib_design.html

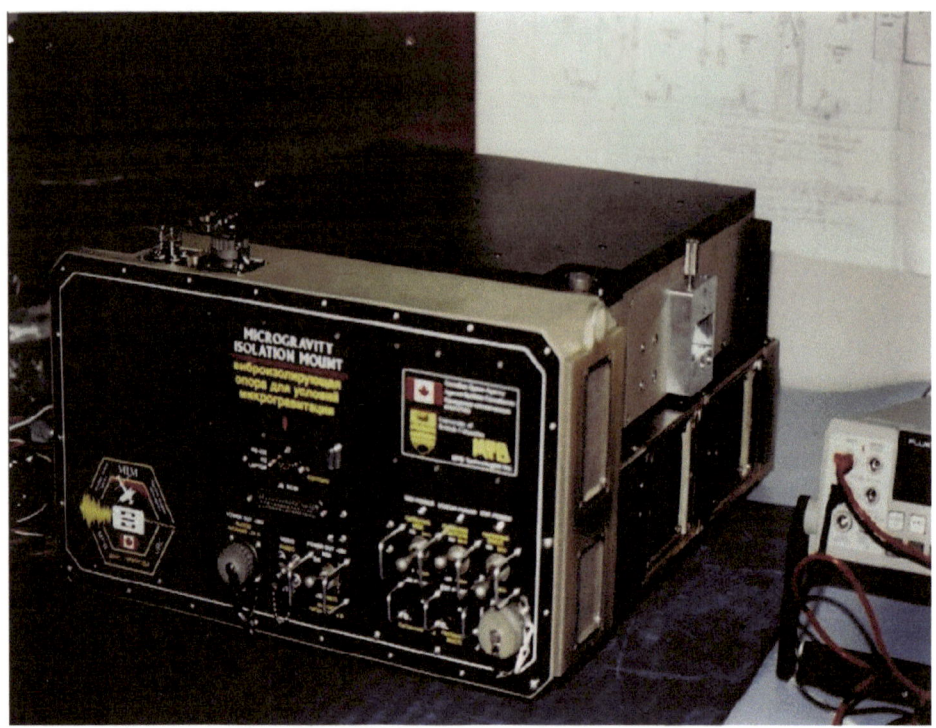

Fig. 6.13 The MIM carried into space to *Mir* and aboard STS-85

Fig. 6.14 The MIM-2 machine carried into space to *Mir* and aboard STS-85

Fig. 6.15 CSA astronaut Bjarni Tryggvason at KSC prior to his flight

G (gravity force)." With the application of new technologies like the MIM Tryggvason explained, "We're trying to get it down to a micro G, or one thousandth of a G."[8]

On May 23, 1996 the MIM was launched into space aboard the Russian Priroda module via a Proton rocket, the seventh and final installment of the *Mir* space station. Problems with the spacecraft's power system, however, seemed to put the MIM's future in jeopardy for a while, but the ground controllers were able to overcome the challenges posed by the faulty module and the Priroda successfully docked with *Mir* on 26 April. The MIM was officially activated by NASA astronaut Shannon Lucid after she arrived onboard *Mir* via STS-76 on March 24. The CSA's Microgravity Sciences Program sponsored the first experiments using the MIM, including the Queen's University Experiments in Liquid

---

[8] Matthew Fisher, 'Good Vibrations: Canadian astronaut puts cosmonauts to the test', *Toronto Sun*, July 21, 1996; 2. Accessed online on Marc 31,h 2016 at http://www.canoe.ca/SpaceArchive/960721_bjarni-sun.html

Diffusion (QUELD) II furnace to develop new alloys and semi-conductor materials. Other experiments followed, and before long the MIM became an essential tool for space station research and experimentation.[9] The success of the MIM on *Mir* led to the decision to develop a second device to be carried aboard a space shuttle flight in order to thoroughly characterize the technology using advanced control strategies. Data collected from both the *Mir*'s operating unit as well as the one carried on the space shuttle would assist NASA engineers in refining the requirements for isolation systems on the International Space Station. As the principle investigator for the MIM, it made sense that Tryggvason be assigned to the flight that would carry the MIM-2 into orbit. After waiting almost fourteen years, Tryggvason finally got his opportunity to fly in space.

NASA's sixth shuttle flight of 1997, lasting from 7 to 19 August, focused on Earth observation and the evaluation of future space station technologies. The main payload for this flight was a package known as the Cryogenic Infrared Spectrometers and Telescopes for the Atmosphere-Shuttle Pallet Satellite-2 (CRISTA-SPAS-2), a joint German Space Agency (DARA) and NASA venture to study the Earth's middle atmosphere. The free-floating satellite was deployed using the Canadarm and then allowed to free fly for over two hundred hours as it executed its task. Additionally, *Discovery* carried the Manipulator Flight Demonstration (MFD) experiment, sponsored by the Japanese Space Agency (NASDA). This investigation sought to demonstrate applications of a mechanical arm for possible use on a Japanese Experiment Module destined for the future International Space Station. Other payloads on this mission included Technology Applications and Science-01 (TAS-01) and the International Extreme Ultraviolet Hitchhiker-02 (IEH-02). The TAS held seven separate experiments providing data on the Earth's topography and atmosphere, the sun's energy output, and the testing of new thermal control devices. The IEH, meanwhile, carried four experiments to study the ultraviolet radiation from the stars, the sun, and other sources in the solar system. Lastly, the MIM-2 series of tests would be Tryggvason's main focus throughout the mission. Officially identified as a Risk Mitigation Experiment (RME 1328), he was responsible for operating the device using advanced control techniques to determine what level of microgravity quality was obtainable with the MIM-2.

Mission STS-85 lifted off as planned from Pad 39A at the KSC at 10:41am on August 7, with *Discovery* ferrying her crew safely to an orbit of 160 nautical miles shortly after. STS-85 was commanded by Curtis L. Brown, with Kent Rominger serving as his pilot. The three mission specialists serving on the flight were Robert Curbeam, Stephen Robinson, and Jan Davis. Bjarni Tryggvason served as the sole payload specialist on the flight. Once *Discovery*'s cargo bay doors were opened, the crew activated key systems and experiment hardware. A little less than eight hours into the flight, mission specialist Jan Davis employed the Canadarm to deploy the CRISTA-SPAS 2 payload, which would fly free for the next 200 hours using three telescopes and four spectrometers to measure infrared radiation emitted by the Earth's middle atmosphere. Tryggvason activated the MIM-2 on the first flight day as well, and scientists later reported a 'fantastic' first day of data return.[10]

[9] Canadian Astronaut Office, 'MIM (Microgravity Vibration Isolation Unit)'. Accessed online on September 19, 2000 at http://www.space.gc.ca/csa_sectors/human_presence/en/canastronauts/astro/sts85/mim.htm

[10] KSC. STS-85 Day 2 Highlights. Accessed online on March 31 2016 at http://science.ksc.nasa.gov/shuttle/missions/sts-85/sts-85-day-02-highlights.html

Fig. 6.16 STS-85 payload specialist Bjarni V. Tryggvason and mission specialist Stephen K. Robinson go through countdown procedures aboard the Space Shuttle orbiter *Discovery* during Terminal Countdown Demonstration Test (TCDT) activities for that mission

Fig. 6.17 August 7, 1997. Launch of the STS-85

Fig. 6.18 (**a**) Bjarni Tryggvason shows off the Microgravity Vibration Isolation Mount (MIM) fluid disk. One of five fluid loop experiments (FLEX), this one deals with the growth of resonance patterns (GORP) in gaseous liquid systems. (**b**) Bjarni Tryggvasoninputs data into a computer regarding the Microgravity Vibration Isolation Mount (MIM) experiment on the mid-deck of the Space Shuttle *Discovery*

Tryggvason continued to operate the MIM-2 throughout the flight, continuously collecting data as the rest of the crew went about their assigned tasks. He took a short break from his work on spaceflight day 6 to speak with the Canadian Prime Minister, Jean Chretien. Further tribute to Tryggvason's work appeared with the crew's wake-up music on flight day 7, with the Beach Boys song 'Good vibrations' being piped through the system. He concluded his data collection on flight day 11, and then stowed the MIM-2 in preparation for return to Earth the following day, but was pleasantly surprised when the crew was informed that, due to poor weather at the landing site, *Discovery* would remain in orbit one more day. STS-85 finally returned safely home in the early morning hours of 19 August 1997. Tryggvason, the sixth Canadian astronaut to fly in space and the last of the originals to do so, had spent nearly twelve days in orbit travelling 4.7 million miles. As fate would have it STS-85 would be Tryggvason's only spaceflight, but nevertheless he could take pride in his accomplishment and the invaluable work he had done in advancing the MIM technology in preparation for future ISS operations.

Neurolab: Dafydd Williams and STS-90

Canada's astronaut program continued at a steady pace, with its next flight scheduled just eight months after Tryggvason's mission. In the fall of 1997 it was announced that Dafydd (Dave) Williams, a member of Canada's second class of astronauts who had recently successfully completed the NASA mission specialist training program, was assigned to the STS-90 crew as Mission Specialist-1. "I feel very fortunate", Williams wrote at the time, "to have been chosen to represent Canada as one of the mission specialists at [the] Johnson Space Centre. Next April, I will be flying aboard the Space Shuttle *Columbia*, with six of my colleagues, on a sixteen-day life science mission."[11] This latest mission was devoted to the study of life sciences in microgravity, with a particular focus on the human nervous system. Space shuttle *Columbia*, undertaking her 25th flight, would carry a specially fitted European Spacelab module known as the Neurolab on a 16-day mission in orbit.[12] The results of the experiments undertaken were expected to contribute key answers to a number of questions associated with the effects of microgravity on the human body, which was essential to clarifying the requirements for future residency during long duration flights on the International Space Station.

Williams, a trained physician, was ideally suited for this mission. "The pursuit of answers to questions about how the body functions in space is like a search for pieces to a grand puzzle," he wrote at the time. "Many pieces are easily identified and slip readily into place. Others remain elusive." He observed his assignment to STS-90 as "a very exciting quest, seeking to explore the two remaining frontiers of this century, outer space and inner

---

[11] Dafydd Williams, 'Letter to the Science Teachers Association of Ontario'. Accessed online on December 21, 2000 at http://www.space.gc.ca/csa_sectors/human_presence/en/canastronauts/cruciblenov.htm

[12] ST-90 also marked the last flight of the ESA Spacelab Module, first flown in space on STS-9 (Columbia) in 1983.

Fig. 6.19 CSA astronaut Dafydd Williams first flew on 17 April 1998 aboard STS-90

space – the function of the human nervous system."[13] Williams would be responsible for overseeing the Canadian Space Agency-sponsored experiments on Neurolab, whilst at the same time serving as a subject for other experiments also being conducted in the Neurolab while in orbit.

Veteran NASA astronaut Richard Searfoss, making his third spaceflight, commanded the mission. His pilot was NASA rookie Scott Altman who, interestingly, already had a claim to fame for having performed many of the aerial stunts in the Hollywood film *Top Gun*. Dave Williams, Richard Linnehan, and Kathryn Hire served as the mission specialists while Jay Buckey and James Pawelczyk served as payload specialists. Space Shuttle *Columbia* lifted off from launch pad 39B at the KSC at 2:19pm on April 17, carrying the STS-90 crew 150 nautical miles into orbit without incident. Once on station and given the go-ahead for on orbit operations, the Neurolab mission got under way. For over two weeks the crew performed studies focused on better understanding the human nervous system. The results of these studies lent greatly to an improved understanding of the human in space, and also advanced a number of areas of investigation in the domain of space physiology. The Neurolab mission was very much a collaborative effort, with several international space agencies represented on the flight. The Canadian, French, German, and Japanese space agencies all partnered with NASA for the mission, providing both flight and ground hardware for various research experiments as well as expertise.

---

[13] Ibid.

Fig. 6.20 The Neurolab payload for STS-90, scheduled to launch aboard the Shuttle *Columbia* from Kennedy Space Center (KSC) on April 2, 1998, is moved to its workstand in the Operations and Checkout Building at KSC. It eventually launched on 17 April. Investigations during the Neurolab mission focused on the effects of microgravity on the nervous system

The Neurolab mission concluded with the safe return of *Columbia* on May3, 1998. For Dafydd Williams, STS-90 had proven to be an exceptional and rewarding experience. NASA was also impressed with his work. Following the mission he was offered and accepted the position of Director of the Space and Life Sciences Directorate at the Johnson Space Center, an appointment that was (erroneously) identified widely in subsequent public affairs releases as being the first non-American senior management appointment within NASA. Canadians had in fact been an integral part of NASA since its genesis in the late 1950s and some, such as Owen Maynard for example, held key NASA offices during the

Fig. 6.21 The STS-90 Neurolab payload is prepared to be positioned into the cargo bay of Space Shuttle *Columbia* in Orbiter Processing Facility bay 3

Fig. 6.22 STS-90 mission specialist Dafydd (Dave) Williams is assisted by NASA and USA closeout crewmembers immediately preceding launch for the nearly 17-day Neurolab mission

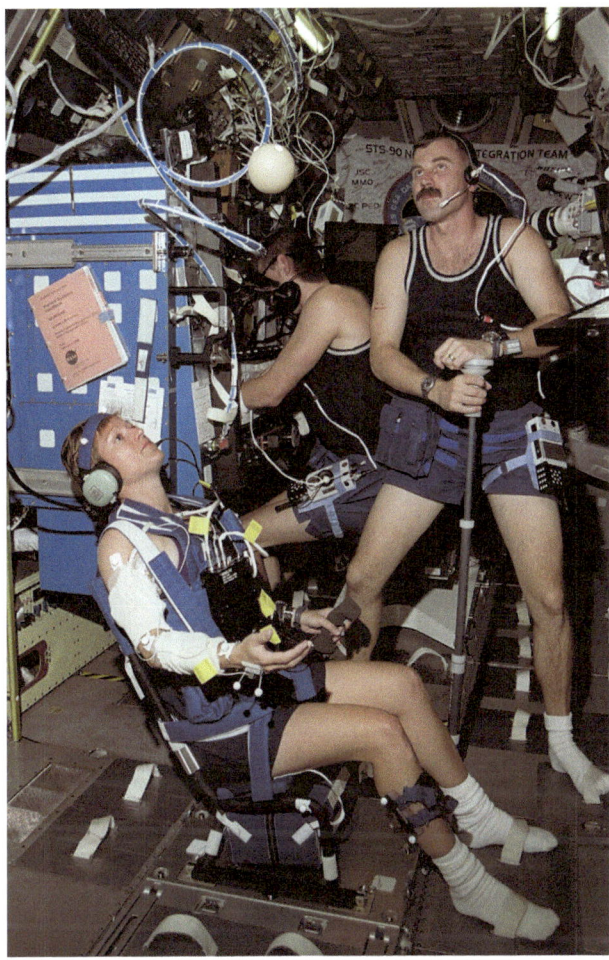

Fig. 6.23 Mission specialist Kathryn P. Hire participates in a sensory motor and performance test by using a simple, but effective, ball-catching experiment that evaluated the ability of the central nervous system to accept and interpret new stimuli in space. Canadian astronaut Dafydd Williams monitors the experiment. The experiment uses the kinelite system, an apparatus developed by the Centre National d'Etudes Spatiales

Apollo Program.[14] Regardless of its historical inaccuracy, however, the Canadian Space Agency took Williams' appointment as yet another indicator of the maturity and success of its own astronaut program.

---

[14] For a detailed collective biography and history of several Canadians who joined NASA at the height of the space race, see Chris Gainor, *Arrows to the Moon: AVRO's Engineers and the Space Race.* Burlington: Apogee Books, 2001.

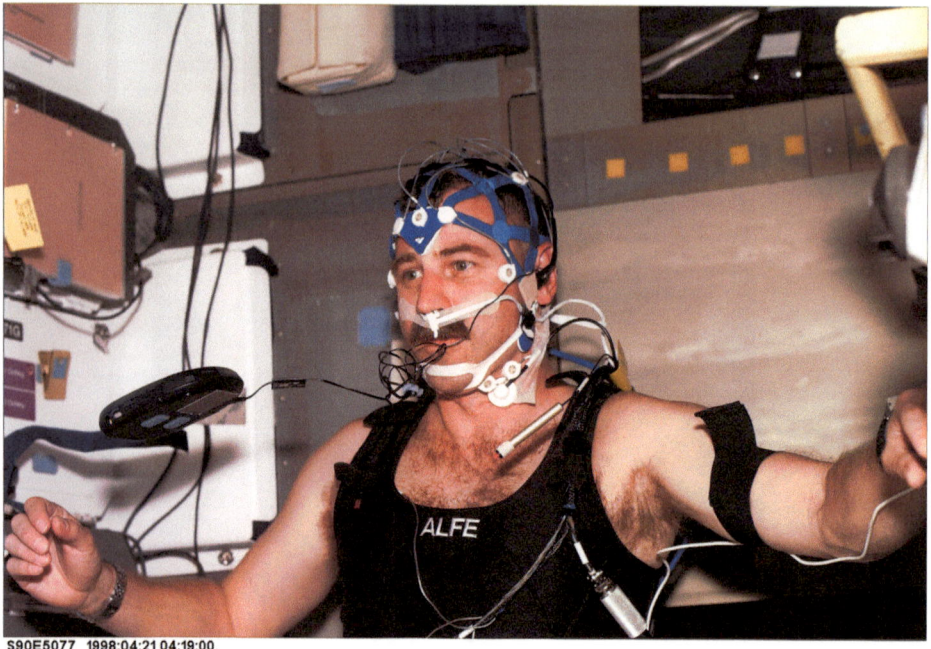

S90E5077 1998:04:21 04:19:00

Fig. 6.24 Mission specialist Dafydd (Dave) R. Williams is a test subject for the Sleep Studies experiment in the Neurolab. The sleep cap monitors and measures electrical impulses from the brain, muscles, eyes, and heart

The New Canadian Space Program

After nearly a decade of operations, it was time for the Canadian Space Agency to pause and assess all it had done. Having completed many of the original goals set out in the Canadian Space Agency Act of 1990, the CSA took the opportunity to review its accomplishments to date with a view to producing an updated strategy to see it through the next stage of international space exploration and exploitation. Led by CSA President William Macdonald 'Mac' Evans, the agency launched a new Canadian Space Program (CSP) policy supported by a Space Program Management Framework (SPMF). These strategy and planning guides directed Canada's space activities through the beginning of the 21st century. Focused on five key areas – Earth and environment, satellite communications, space science, generic space technologies, and human presence in space – the revised policy emphasized research and development, science and technology, economic and industrial development, export expansion and employment, and improved efficiency and effectiveness of government operations. Similarly, the new policy stated that its key instruments of success were its niche commercialization, and partnership strategies.[15]

---

[15] CSA, *The Canadian Space Program: A New Era for Canada in Space.* Ottawa: Department of Industry, 1999.

Fig. 6.25 Canadian Space Agency President, William 'Mac' Evans served from 1994 to 2001. He was instrumental in sheparding the agency through its first decsde of operations

No less important to its future success was a revitalization of the government's space sector decision-making process. Canada had not enjoyed a legacy of sound space program policy frameworks. Both its predecessors, the ICS and the MOSST, had suffered from a lack of structure as well as a lack of political and budgetary authority. The new policy promised to avoid this mistake, instead "ensuring ongoing consultation and collaboration … a new management framework which formalizes the structure and process by which the CSA and key stakeholders interact".[16] Along with the new framework came a new CSA Board of Advisors, appointed by the Minister of Industry. It was hoped that this new oversight body would improve relationships between the CSA and other government departments, some of which continued to direct various aspects of Canada's space program.

Given this evolution it was perhaps most appropriate that the CSA titled its 1999 policy update, *The Canadian Space Program: A New Era for Canada in Space*. Up to then, the Long-Term Space Plan II and the Space Policy Framework, approved in 1994, had served as the primary policy guidance for Canada's space program. By the end of the decade, however, there was an acknowledgement that the country's space sector was in the process of transforming. For the first time, Canada's export-oriented industry and public sector agencies generated nearly half of the space program's $1.2 billion in annual revenues, the highest percentage in the world that year. As well, by sharing the costs and benefits of the country's space exploration program, Canada was able to maximize its spin-offs for the Canadian space industry and economy. Even more important, perhaps, cost sharing between government and industry allowed the CSA to achieve several goals in space that would have otherwise been unattainable.

---

[16] Ibid., p.10.

Fig. 6.26 *The Canadian Space Program: A New Era for Canada in Space* was released in 1999

Chief among these goals was the continued focus on Canada's human presence in space. Dafydd William's flight on STS-90 marked the end of the first act in Canada's astronaut program. The eight missions flown between 1984 and 1998 focused largely on life sciences research and engineering technology demonstration, during which Canadian astronauts served as first payload specialists and then later mission specialists in support of those flights. As these missions played out NASA, along with its main partners, continued to push forward with the International Space Station project, and after many challenges and delays, construction finally began by the end of 1998 with the launch of the first two core modules, the Russian-built *Zarya* and the American-built *Unity*. Once these essential pieces were on orbit, the focus of future space shuttle flights shifted greatly towards building the ISS, which at the time NASA estimated might take at least forty missions to complete. For Canada's astronaut program this meant that most, if not all, future flights would result in direct participation in ISS construction. As such, the first act of Canadian spaceflight was complete and the second act was set to begin.

Canadians Building the ISS: STS-96 and STS-97

As with any great engineering project, missing some deadlines was to be expected. Though the first two core modules of the ISS were delivered to orbit in 1998, the next essential piece, a Russian-built service module named *Zvezda*, faced serious delays. This had an inevitable impact on the American space shuttle program, whose schedule was now tied to that of the Russian space program. When the module did not launch as planned NASA was forced to anticipate that its next mission, STS-96, might be the only flight of 1999 and that the first ISS expedition crew might not go into space until the year 2000. Having scheduled flight STS-96 as a cargo-hauling mission in advance of *Zvezda*'s and the Expedition I crew's arrival, NASA now faced a serious complication. The Russian module carried the space station's new sophisticated attitude control system, an essential technology required for space shuttle dockings. Without it, STS-96 would be required to adjust its flight path, approaching the ISS from above as opposed to the typical ascent from below which ensured that it did not block the antenna on *Zarya*'s exterior that received the vital command to deactivate other station thrusters immediately upon docking with the space shuttle. If this failed, any number of mission-critical errors might occur, including seriously damaging the ISS. As a result, NASA worked hard to find a way through.[17]

Agreeing that STS-96 would launch regardless on the assumption that the *Zvezda* module would be delayed, over the month of December 1998 ground controllers in Houston and Korolev continued to monitor the existing ISS's systems by telemetry. A series of programmed maneuvers adjusted the orbit of the space station while other commands adjusted its orientation so that its batteries could be deep cycled and its other power generation performance demonstrated. As with any new machine there were small challenges to overcome. Power generation and sustainment was a constant concern, as was ensuring that the space station maintained a passive thermal roll so as not to overheat. Other

---

[17] David M. Harland and John E. Catchpole, *Creating the International Space Station*. Chichester: Springer-Praxis, 2002, p.223.

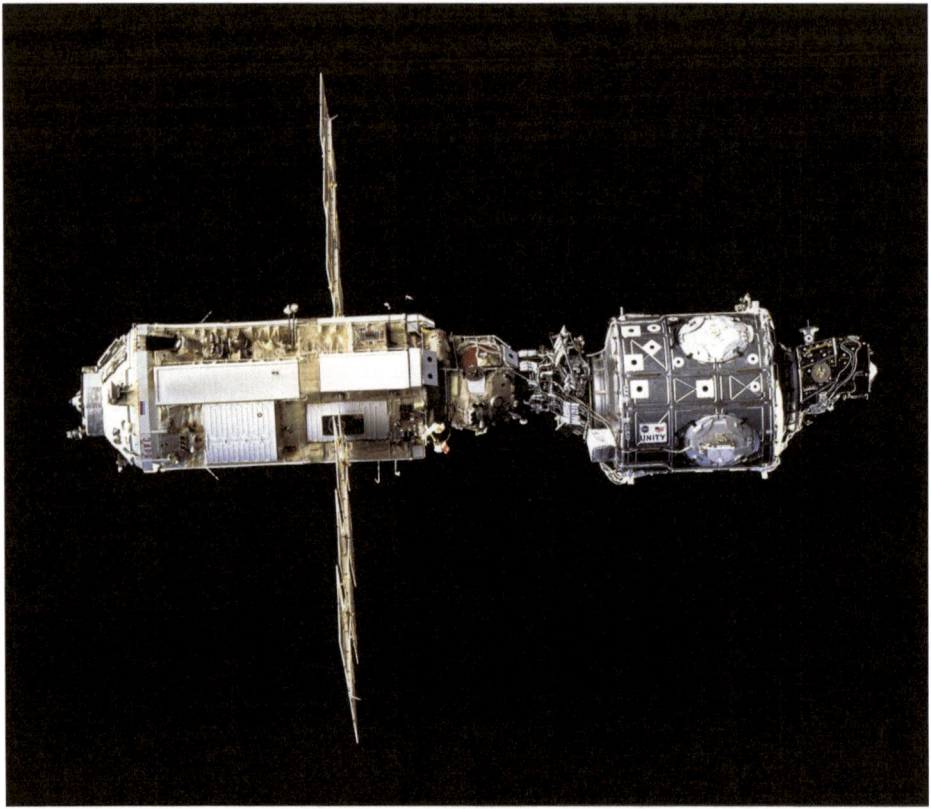

Fig. 6.27 The new space station era begins. *Zvezda* (*right*) and *Unity* (*left*) formed the core of what would eventually become the greatest engineering endeavor ever undertaken in spaceflight

troubleshooting involved repairing various communications antennae, correcting an accidental command to the ISS to power up one of the *Zarya* module's solar power retraction motors, and dealing with various space station electrical issues. On May 12, 1999 ground controllers issued commands to adjust the orientation of the ISS in preparation for the rendezvous and docking of STS-96. On May 24, just a few days before the mission began, ground controllers activated the heaters on the ISS in preparation for the arrival of its first human visitors.[18]

The first logistics and resupply mission to the International Space Station got under way in the early morning hours of May 27, 1999, as STS-96 lifted off from launch pad 39B at the KSC at 6:49am and ascended into the hazy sky over Florida's east coast. Commanded by Kent Rominger and piloted by Rick Husband, space shuttle *Discovery* reached orbit

---

[18] Ibid.

Fig. 6.28 Mission specialist Julie Payette poses next to the Canadarm in the payload bay of the orbiter *Discovery* during a STS-96 Crew Equipment Interface Test

without incident approximately nine minutes later and began its chase around the world after the ISS. By the beginning of flight day 2, *Discovery* was only trailing the space station by 1435km, but the space shuttle closed this gap by 111km with each orbit. Quickly catching up with its quarry, STS-96 reached the ISS the following day.

The space shuttle spent six days linked to the new space station as its crew transferred critical equipment and supplies that could not be previously ferried up to orbit aboard *Unity* or *Zarya*. The five mission specialists – NASA astronauts Tamara Jernigan, Ellen Ochoa, Daniel Barry, Russian air force cosmonaut Valery Tokarev, and CSA astronaut Julie Payette – carried out much of this work. A double Spacehab module in the payload bay contained 1633kg of supplies such as food, clothes, laptop computers, printers, and cameras, all of which had to be transferred over to the ISS and safely stowed. Additional equipment, including two small cranes and other spacewalking gear, was also transferred from an integrated cargo carrier held in *Discovery*'s payload bay during the mission's sole EVA on flight day 4.

Fig. 6.29 Mission Specialist Julie Payette speaks to reporters prior to her flight on STS-96. Her personal astronaut patch is clearly visible on her right arm below the Canada flag

As the designated intravehicular operator for EVA support, Payette was responsible for helping the mission's spacewalkers – mission specialists Jernigan and Barry – suit up and conduct their tasks. She operated the hatch both on their exit from *Discovery* and when they returned. In a later interview, when asked what was the most uprising thing about her first spaceflight, she reflected on a distinct memory of that experience, "… when I opened the hatch from inside, well this cold smell, this smell of nothing really because there's nothing left in there, was the smell of space, and that I did not expect".[19]

---

[19] NASA, 'In Their Own Words: Canadian Astronaut Julie Payette'. Public Affairs and Outreach Multimedia Video. Accessed online on March 1, 2016 at https://www.youtube.com/watch?v=Roi6a9VMdrQ

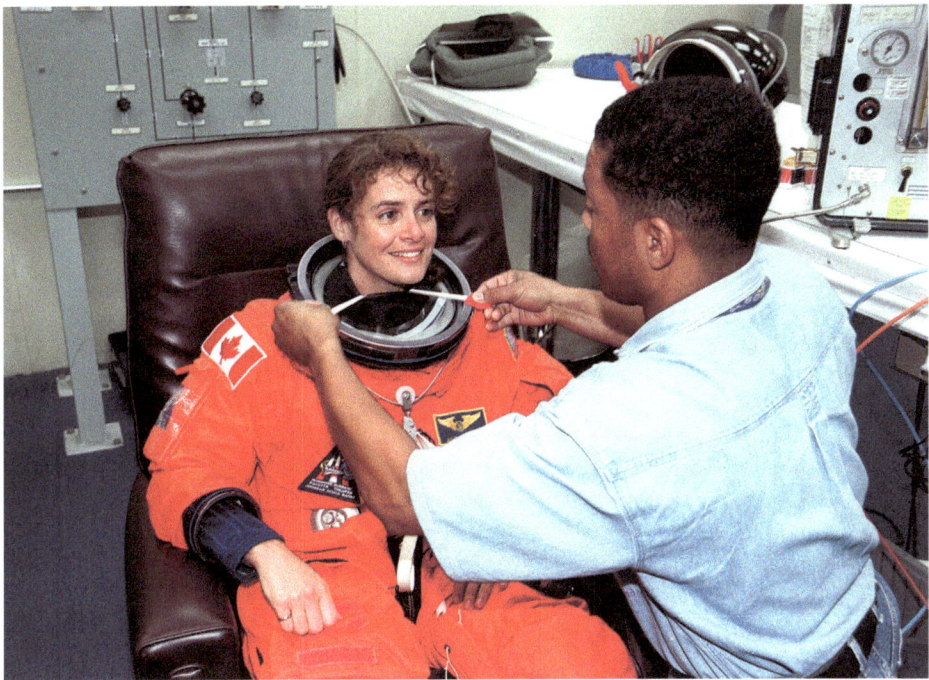

Fig. 6.30 MissionSpecialist Julie Payette suits up in preparation for launch on STS-96

In addition to EVA support and choreography, Payette had numerous other responsibilities during the mission. She was assigned the task of looking after space station contingencies and in-flight maintenance and, indeed, a problem did occur during the flight that required her and fellow crewmate Valery Tokarev to swap out all 18 integration units, part of the solar battery system that powered the Russian *Zarya* module. The small, smartphone-sized units had malfunctioned, resulting in faulty charging of the *Zarya* module batteries. Payette also operated the Canadarm on three different occasions during the mission. On flight day 2 she performed two tasks, first checking out the Canadarm in preparation for other operations, and second, using the arm to conduct a test of a robotic tool designed to assist RMS operations. After the EVA on flight day 4, Payette operated the Canadarm once more to perform a close inspection of the targets of the SVS installed on the exterior surface of the ISS. Responsible for all camera equipment onboard *Discovery*, Payette was tasked to collect a record of the less well-photographed regions of the Earth as well as photo document the ISS as it existed at the time. She oversaw and executed the deployment of the STARSHINE satellite, an educational payload that could be observed from Earth for various projects. Finally, at the conclusion of the orbital mission Payette was responsible for carefully aligning and closing *Discovery*'s payload bay doors in preparation for the crew's return to Earth.

STS-96 undocked from the ISS on June 3 and returned safely to Earth three days later. Payette was the eighth Canadian astronaut to fly a mission and the second female Canadian

Fig. 6.31 Space Shuttle *Discovery* carrying the STS-96 crew lifts off from the KSC on May 27, 1996

astronaut to reach orbit. Yet, most importantly, she was the first Canadian to visit the International Space Station. As with her predecessors, Payette's performance as a mission specialist on STS-96 had been excellent, and she brought great credit to both NASA and the CSA for her work to get the ISS up and running. At the time of the conclusion of Payette's mission, her colleague Marc Garneau was already well into the final stages of training for the

Fig. 6.32 The crew of STS-96. Mission Specilaist Julie Payette is standing at back second from the left

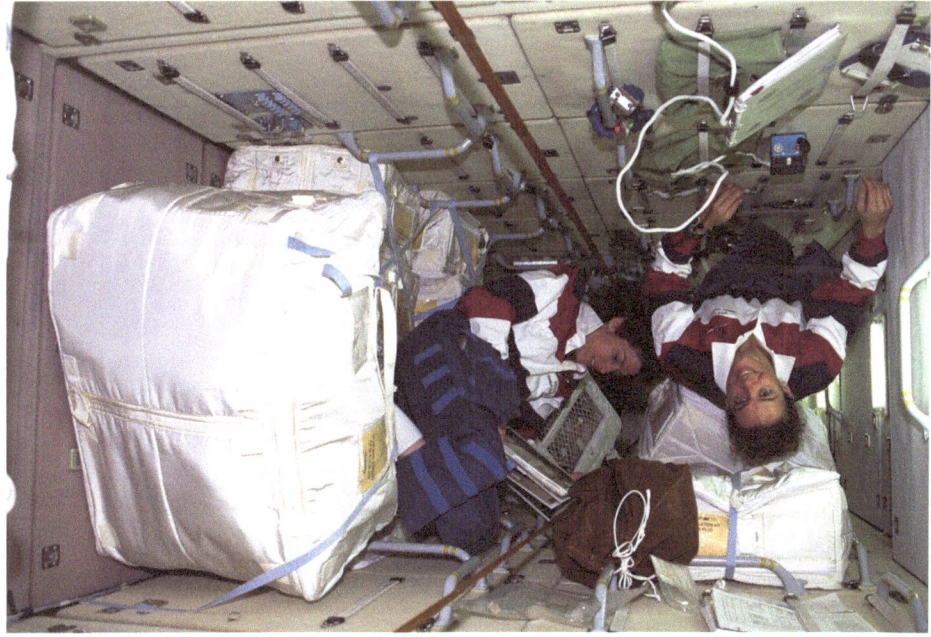

Fig. 6.33 Onboard the Russian-built *Zarya* ISS module, astronauts Julie Payette (*upside down*) and Ellen Ochoa handle a portion of the supplies which have been moved over from the docked space shuttle *Discovery*

next Canadian spaceflight, scheduled to occur the following year. Once again, there would be only limited time to reflect on Canadian accomplishments before the next mission into space.

There would be four other missions to the ISS before Marc Garneau made his third and last spaceflight as part of the STS-97 crew. NASA flights STS-101 and STS-106 brought more supplies and boosted the space station's orbit, while STS-92 delivered the next critical piece of the ISS, the Z1 Truss assembly. Just a week after STS-92 returned home, Russia's Soyuz TM-31 delivered the first ISS expedition crew to their new home. Garneau's mission was tasked to deliver the P6 Integrated Truss Segment (ITS) including giant solar arrays, the first to be added to the space station in order to power it. "Up until now," Garneau told one interviewer, "the station has been – if you look at it – a series of modules. When we bring up the first pair of solar panels, you're going to see a remarkable transformation in the appearance of the station."[20] He was later proven correct. After considerable scheduling delays, STS-97 finally lifted off from launch pad 39B at the KSC on December 1, 2000 to begin a ten-day mission to install the P6 ITS on the ISS. NASA astronaut Brett Jett commanded the mission with Mike Bloomfield as his pilot. Marc Garneau, Joe Tanner, and Carlos Noriega served as mission specialists, with Garneau acting as flight engineer for this mission.

Fig. 6.34 The crew of STS-97. Canadian astronaut Marc Garneau, participating in his third spaceflight, is seated at center

[20] NASA. Preflight Interview: Marc Garneau. Human Spaceflight – The Shuttle. Accessed online on 2 May 2001 at URL http://spaceflight.nasa.gov/shuttle/archives/sts-97/crew/intgarneau.html

Fig. 6.35 Space Shuttle *Endeavour* on Pad-39B prior to the flight of STS-97

After reaching orbit without incident, STS-97 began its chase of the ISS, catching up to and docking with the space station on flight day 3. Having tested the Canadarm and also having conducted a survey of the payload bay cargo the day before, Garneau spent the third day of the flight unpacking *Endeavour*'s precious cargo. He used the RMS to reach into payload bay and pick up the P6 ITS, moving the 14m-long 16,000kg piece of hardware over to within a few centimeters of the ISS's Z1 Truss. This section contained a claw-like latch, which was used to grasp the P6 ITS and hold it in place. Once held, Garneau left the P6 in this overnight 'parking' position in preparation for its full attachment during an EVA on flight day 4.

The next day, mission specialists Joe Tanner and Carlos Noriega conducted the first of two scheduled spacewalks to fully close the Z1 latch and then bolt down the four corners of the P6, holding it firmly in place. A second space walk, conducted on flight day 6, saw the completion of the task, during which Garneau served as the designated intravehicular activity crewmember. Though it was not common at the time to train the IVA astronaut as a backup for EVA, given the importance of successfully completing the mission nothing was left to chance. Had something gone wrong that prevented either Tanner or Noriega from conducting their spacewalk, Garneau would have stepped in – or as one report put it, stepped out – to see the task through. He might have become the first Canadian to walk in space, but that honor would ultimately fall to another.

S97E5038 2000/12/06 17:39:18

Fig. 6.36 Marc Garneau, mission specialist, looks over a checklist on the aft flight deck of the Space Shuttle *Endeavour* during flight day 6

S97E5107 2000/12/09 19:50:44

Fig. 6.37 This high-angle view of the International Space Station clearly shows the new P6 Truss with its solar array fully deployed

With the P6 Integrated Truss Segment successfully installed and appearing to function normally, the crew of STS-97 and Expedition One opened the hatches between the spacecraft. After the traditional exchange of greetings and gifts, *Endeavour*'s crew began the transfer of supplies and equipment. This gave Garneau his first opportunity to visit the International Space Station, becoming the second Canadian to do so in as many years. Looking out at the massive solar arrays he had just helped install, Garneau later remarked that the ISS came to resemble "a giant bird". With the solar arrays now fully deployed, the ISS indeed did finally have wings.

Space shuttle *Endeavour*'s visit to the ISS was a brief one, with STS-97 undocking from the space station the following day. Bloomfield eased his spacecraft away from the station and then piloted it one full circle while the crew completed a visual and photo inspection of the newly-expanded station. With everything looking as it should and operating normally, *Endeavour* bid their fellow astronauts goodbye and prepared for their return to Earth. STS-97 landed safely two days later on December 11, 2000, successfully concluding their mission. For Garneau the return was bittersweet. He knew he was not likely to fly in space again any time soon, and as events played out, within a year of his last space mission he had left the astronaut corps for good to lead the Canadian Space Agency itself.

Fig. 6.38 A veteran of three spaceflights, Marc Garneau retired from flight duty after STS-97 to become the president of the Canadian Space Agency

A Defining Moment: STS-100/A6 and Canadarm 2

Garneau's mission to the ISS had impressed the Canadian public which had watched it literally unfold in space. STS-97 also did much to revive the character of the space station by giving both its builders and observers the very first sense that the orbiting spacecraft would indeed one day look like the many conceptual drawings that had been made of it. Still, not all original concepts and designs had survived the constant redesigning of the station. Canada's original ISTF design for space station *Freedom* had been scrapped and re-emerged as a new Mobile Servicing System for the International Space Station. Whereas the ISTF was essentially a retrieval system and garage for servicing satellites, the final MSS design allowed the station astronauts to perform any number of different tasks on location throughout the space station.

Its redesign was considered simple and elegant, but as James Middleton, the Vice-President of business development for MD Robotics, later commented, transforming the redesign into working hardware presented quite a challenge. Middleton, who led the engineering team for the CSA that designed and built the SSRMS, now being referred to more commonly as the Canadarm2, said that "the space station was very large and it was clear that we had to be able to provide mobility—the ability to relocate the arm from spot to spot."[21] Another difficulty Middleton's team faced was designing the SSRMS to withstand the forces of liftoff. Much larger and heavier that the original Canadarm – the space shuttle's arm weighed about 410kg whereas the Canadarm2 weighed approximately 1800kg – it had to be able to sustain greater loads during the eight-and-a-half minute ride into orbit. Lastly, yet another issue was its length. Much longer than the original Canadarm, the SSRMS was folded into a U-shaped pallet that fit within the width the space shuttle payload bay. "We had to bend it in pieces," Middleton explained. "How to fasten it down was a very complex design process. To be able to hold it in a manner that the loads were adequately relieved during the shuttle launch was a tough challenge for the engineering team."[22]

The Canadian Mobile Servicing System represented approximately two decades of research, design, and testing, and was perhaps once of the most sophisticated robotic systems ever deployed in support of space operations. The MSS itself consisted of eight distinct components, not including the several smaller attachments and tools that were designed for use alongside it. The main parts of the MSS were as follows:

- MT: The Mobile Transporter allowed the entire Mobile Servicing System to relocate anywhere along the central truss of the space station. Constructed using a high-strength aluminum, the MT could support payloads up to 20,911kg.
- MBS. Constructed from the same materials as the MT, the Mobile Base System was mounted on the MT. Its main function was as a storage location and work platform for astronauts working outside the station. The MBS also provided a number of other capabilities, including a Common Attach System to accommodate transportation and servicing, a structural and electrical interface for users, Power Data Grapple Fixtures (PDGF) and a Latching End Effector (LEE) to hold payloads or accommodate the Special Purpose Dexterous Manipulator (SPDM).

---

[21] NASA, Press Kit STS-100/A6: Extending the Reach of the Space Station. 9 April 2001, p.22.
[22] Ibid., pp.22–23.

- SSRMS: The Space Station Remote Manipulator System (SSRMS) was the first component of the MSS deployed on the station. Weighing approximately 1641kg and measuring 17.6m in length, the SSRMS provided seven degrees of freedom, making it slightly more maneuverable than the human arm. Normally connected to the MBS, the SSRMS was, however, by no means constrained to moving only along the central truss of the space station. Instead, using its LEE the arm was able to lock on to any PDGF, detach its other end, and quite literally inchworm its way around the exterior of the space station. Only the number of PDGFs could limit the range of accessibility of the SSRMS around the station. The SSRMS was also equipped with four TV cameras feeding views to the operator at the robotic workstation inside the ISS, as well as an advanced space vision system that had the ability to track payloads. Given its size and maneuverability, the SSRMS was also given a collision avoidance capability and a stopping distance under maximum load of just 0.6m.
- SPDM: The Special Purpose Dexterous Manipulator (SPDM) resembled a two-handed robot torso that was capable of manipulating small payloads and operating robotic tools. Sometimes referred to as a robotic "stunt double" by its designers and operators, the SPDM could perform many of the extravehicular activities that once mandated a human space walk to accomplish.
- OTCM: The Orbital Replacement Unit/ Tool Change Out Mechanism (OTCM) was the "hand" located at the end of the each SPDM arm. The OTCM provided a number of tools, including a set of keyed parallel jaws, a retractable nut drive unit, an offset camera and a light. Finally, a motorized socket wrench was built into the palm of the OTCM. The OTCM had been designed for use with a micro-fixture that provided a standard mechanical interface between SPDMs OTCM and small ORUs, which included many of the electronics units that were mounted on the MBS and the SSRMS.
- LEE: The Latching End Effector was designed to provide power and data signals at both ends of the arm by acting as the coupling mechanism between the SSRMS, SPDM, and the PDGF. The LEE provided a significant increase to the mass handling capability and mobility of the SSRMS.
- PDGF: The Power Data Grapple Fixtures (PDGF) were designed for mechanical actuation for the SSRMS and SPDM. The PDGF also allowed for the transfer of power and data through the SSRMS and any payload equipped with a PDGF.
- RWS: The whole operation of the MSS was controlled from one of two Robotic Work Stations (RWS) located within the International Space Station. The first RWS was situated within the U.S. Pressurized Lab Module and the second was originally located in the Cupola. Each RWS could be commanded by space station management and control to operate either as the primary or the secondary workstation. Only one RWS, however, could be designated as the primary. The Canadian robotics software to operate the SSRMS was delivered to the station by the crew of STS-98 in February 2001, with the RWS terminals arriving on the following mission.[23]

---

[23] CSA, "One Step Closer to Canada's Historic Mission in April – Canadian Software Aboard the Shuttle Atlantis", *CSA Electronic News Release*, St. Hubert: CSA, February 7, 2001.

Table 6.1 Comparison of Canadarm and Canadarm2

	Canadarm	Canadarm2
Range of motion	Returns to Earth after every shuttle mission. Reach limited to length of arm	Permanently in space. Moves end-over-end to reach many parts of International Space Station in an inchworm-like movement; limited only by number of Power Date Grapple Fixtures (PDGFs) on the station. PDGFs located around the station provide power, data and video to the arm through its Latching End Effectors (LEEs). The arm can also travel the entire length of the space station on the Mobile Base System
Fixed joint	Fixed to the shuttle by one end	No fixed end. Equipped with LEEs at each end to provide power, data and video signals to arm
Degrees of freedom	6 degrees of freedom. Similar to a human arm: shoulder (2 joints), elbow (1 joint) and wrists (3 joints)	7 degrees of freedom. Much like a human arm: shoulder (3 joints), elbow (1 joint) and wrists (3 joints). However, Canadarm2 can change configuration without moving its hands
Joint rotation	Limited elbow rotation (limited to 160 degrees)	Full joint rotation. Joints (7) rotate 540 degrees. Larger range of motion than a human arm
Senses	No sense of touch	Force moment sensors provide a sense of touch. Automatic vision feature for capturing free-flying payloads. Automatic collision avoidance
Length	50 feet, 3 inches 15m	57 feet, 9 inches 17.6m
Weight	905 lbs. 410kg	3,960 lbs. 1800kg
Diameter (ext. diameter of composite boom)	13 inches 33cm	14 inches 35cm
Mass handling capacity	66,000 lbs. 29,937kg design case handling payload	255,700 lbs. 116,000kg design case handling payload
Speed of operations	Unloaded: 2 feet per second 60cm/sec. Loaded: 2 inches per second 6cm/sec	Unloaded: 15 inches per second 37cm/sec. Loaded: For station assembly, less than 1 inch per second 2cm/sec. For EVA support, 6 inches per second 15cm/sec. Less than half an inch per second with 100-ton load 1.2cm/sec. Stopping distance under maximum load: 2 feet 0.6m
Composition	16 plies of high modulus carbon fiber–epoxy.	19 plies of high strength carbon fiber–thermoplastic.
Repairs	Repaired on Earth	Designed to be repaired in space by replacing ORUs (Orbital Replacement Units). Built-in redundancy
Control	Autonomous operation or astronaut control.	Autonomous operation or astronaut control.
Cameras	2 (one on the elbow and one on the wrist).	4 color cameras (one at each side of the elbow, the other two on the LEEs).

"The software is critical to the operation of the station arm", Mac Evans, then serving as President of the Canadian Space Agency, related in an interview, "because it will give astronauts and cosmonauts the freedom to operate the SSRMS in complete safety." The software, along with another Canadian electronic device known as the Video Signal Converter (VSC), was installed during the STS-98 mission. While this was being carried out, down on Earth the entire MSS underwent rigoros engineering testing, finally being delivered to the KSC for inclusion in the payload of STS-100 in May 1999. In August 2000, it was folded into its final launch configuration and attached to its payload bay pallet, ready to be placed into the hold of space shuttle *Endeavour* a few weeks before its launch date.[24]

While the Canadarm2 was prepared for launch, the crew of the STS-100/A6 mission likewise prepared for their mission into space. Kent Rominger, making his fifth and last spaceflight, was assigned as commander with Jeffrey Ashby serving as his pilot. Interestingly, Rominger had commanded two previous missions containing Canadian astronauts – STS-85 and STS-96. In addition to Rominger and Ashby, the crew of STS-100 included five mission specialists – Chris Hadfield, John Phillips, Scott Parazynski, Umberto Guidoni and Yuri Lonchakov. The Italian Guidoni was flying for the European Space Agency, and Lonchakov represented the Roscosmos State Corporation for Space Activities. In preparing for his second spaceflight Hadfield noted, "Most thrill rides last three minutes and then your five bucks are used up, but the shuttle far surpasses any of those rides and it lasts for days and days or weeks or, if you're on station, for months. It is an amazing, intoxicating, and habit-forming experience."[25] Even more exciting for the veteran test pilot was the fact that he had been designated to conduct the mission's EVAs along with Mission Specialist Scott Parazynski. "I was never more excited in my life than when I was told that I was going to get a chance to walk in space," Hadfield said in an interview prior to his flight. "It is that 9-year-old boy's dream absolutely coming true."[26]

STS-100/A6 began its mission on the afternoon of April 19, 2001, lifting off from launch pad 39A at the KSC and reaching orbit approximately nine minutes later without incident. At the time of launch the ISS was south of India, so once *Endeavour* received the go-ahead for on-orbit operations it began its chase of the space station in order to dock on flight day 3. It had also been a long day for the crew, so they went to sleep about four hours after reaching space, resting up for the challenging days ahead of them.

On flight day 2 the crew spent a good portion of the day checking out the equipment that would be used for the mission's main activities. Hadfield and Parazynski completed their checkout of their spacesuits and spacewalking gear. Other crewmembers tested the Canadarm and also the tools used for final rendezvous and docking. The following day *Endeavour* closed the gap between itself and the ISS and successfully docked at the station, though the two crews would not greet each other in person until flight day 5. This was because the space shuttle maintained a lower cabin pressure as part of crew preparations for spacewalks. Meanwhile, the crew of STS-100 worked alongside the ISS Expedition 2 crew – commander Yury Usachev and flight engineers Jim Voss and Susan Helms – to get ready for the delivery of the space shuttle's critical payloads, both of which would add considerable capacity to ISS operations.

---

[24] Ibid., p.1.

[25] Andrew Adamson, 'Chris Hadfield: Canada's Top Astronaut'. *EXN.CA*, 14 May 1999. Accessed online on January 30, 2001 at http://exn.ca/Stories/1999/05/14/54.cfm

[26] Ibid.

Fig. 6.39 The crew of STS-100

Fig. 6.40 Canada's first spacewalker – astronaut Chris Hadfield

Fig. 6.41 Space shuttle *Endeavour* lifts off from KCS Florida, April 22, 2001

Sunday April 22, 2001 marked a tremendous milestone in the history of Canada's space program. Woken by the late Canadian musician Stan Rogers' song 'Take it From Day to Day', Hadfield prepared for the historic spacewalk he would take that day, accompanied by fellow crewmate Scott Parazynski. Hadfield's spacesuit was distinguished by red stripes on the legs, but even more prominently by the large Canadian flag on his suit's left shoulder. As the spacewalkers prepared to exit *Endeavour*, mission specialist John Phillips served as the designated IVA member for them, while Jeff Ashby and Umberto Guidoni operated the shuttle's RMS to install the new Canadarm2 on the outside of the ISS *Destiny* module.

At 6:45am EDT the 41-year-old Chris Hadfield floated out of the *Endeavour*'s airlock, becoming the first Canadian in history to float free outside a spacecraft. He described in a later interview that, "Entering the vacuum of space for the first time is a form of rebirth. You push your head out of the space shuttle's airlock, and it's like coming out of a womb into the world. After years of training and simulation, suddenly you are there. You're outside, and the vista is just overwhelming. I am absolutely slapped in the face with an incredibly vivid view of the International Space Station and the world below from horizon to horizon. It's as stunning a juxtaposition as you can imagine."[27] A personal dream, and goal that he set for himself at age 9, was finally achieved.

---

[27] Chris Hadfield, 'Out of This World'. *TIME Magazine*, Canadian Edition, May 21, 2001, p.45.

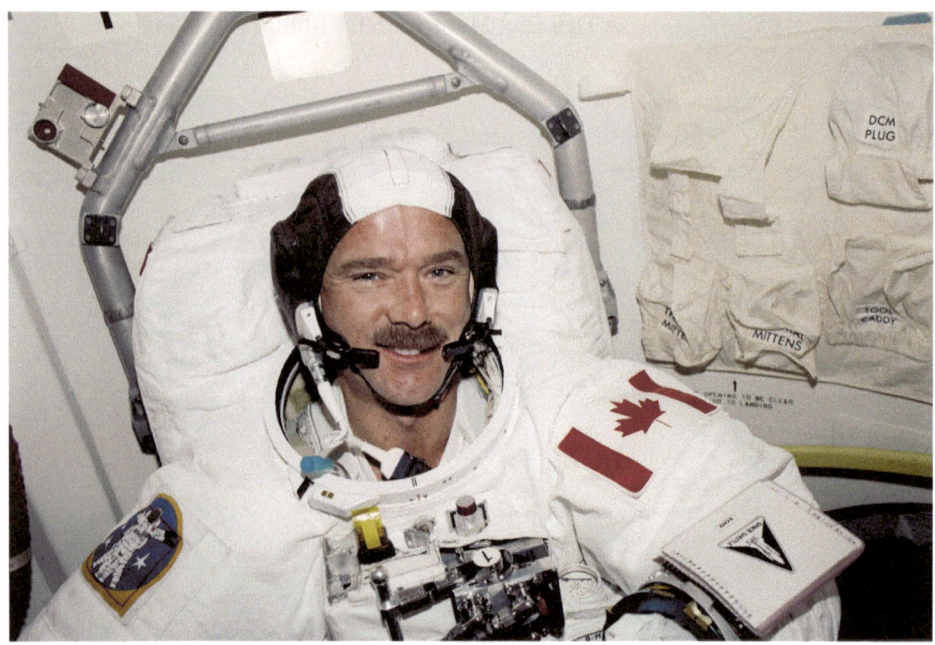

Fig. 6.42 Chris Hadfield is suited up and ready to make history – the first Canadian to walk in space, April 22, 2001

Fig. 6.43 Astronaut Chris A. Hadfield, Canada's first space walker, stands on the portable foot restraint (PFR) connected to the *Endeavour*'s remote manipulator system (RMS) robotic arm, backdropped against the blue and white Earth

Fig. 6.44 Mission specialist Scott E. Parazynski works with cables associated with the Space Station Remote Manipulator System (SSRMS), or Canadarm2, during one of two days of extravehicular activity

S100E5217 2001/04/22 10:02:17

Fig. 6.45 Cosmonaut Yuri V. Lonchakov (*right*), assists astronaut Chris Hadfieldas he dons his Extravehicular Mobility Unit (EMU) spacesuit onboard the *Endeavour*

Historic moments aside, Hadfield and Parazynski had a long spacewalk to get through. Spending a total of seven hours ten minutes working outside the ISS, they first installed a new Ultra High Frequency (UHF) antenna on the *Destiny* lab module that enabled the station to conduct future spacewalk communications as well as improve future shuttle-station communications. "Even tethered, you're basically a one-person spaceship, absolutely separate and isolated from everything else on Earth," Hadfield later wrote, "You're moving at about 8km per second or about 28,000km per hour, orbiting the Earth every 92 minutes."[28] Working together, the two spacewalkers took about two hours to install the antenna, after which they turned their attention to the Canadarm2.

First, Hadfield and Parazynski removed the insulating blankets from the arm and pallet and connected a series of cables, ensuring that station electrical power and computer commands were being supplied to the arm before releasing the bolts that held it in place during launch. Next, a series of smaller jackbolts were sequentially released, followed by eight large, 1.2m-long "super bolts" that held the arm in place on the pallet. With these removed, Ashby maneuvered Hadfield, standing on a footpad at the end of the shuttle RMS, to a point where he could begin to unfold the new SSRMS. As each section was unfolded, Hadfield and Parazynski tightened the Expandable Diameter Fasteners, a specialized tightening bolt that made the booms of the new Canadarm2 rigid. During these operations the spacewalkers encountered some problems ensuring that their pistol grip tool applied proper torque to each bolt. It was decided to simply switch the tool from automatic to manual mode, after which the pair successfully completed their task. For Hadfield there were also other challenges on this historic spacewalk. He experienced severe eye irritation due to the anti-fog solution used to polish his spacesuit visor. At one point the irritation momentarily blinded him, forcing him to vent oxygen into space to rectify the problem.[29] In spite of these difficulties Hadfield pushed on, and the two spacewalkers completed all their assigned tasks, returning to *Endeavour* at 1:55pm. As Hadfield wrapped up his tasks and prepared to return inside, Mission Control in Houston sent up a congratulatory message from fellow Canadian astronaut Steve MacLean and piped through the Canadian national anthem as an acknowledgement of his achievements. Chris Hadfield, Canada's first spacewalker, undoubtedly felt a sense of pride in a job well done.[30]

As the crew of STS-100/A6 conducted its post-EVA activities, the Expedition 2 crew commanded the first motion of the Canadarm2 and everything responded as it should. The following day, ISS crewmembers Helms and Voss "walked" the arm off the pallet and attached it to a grapple fixture on the *Destiny* module. The day after that they used Candarm2 to pass the pallet back to the space shuttle RMS, which in turn returned it to *Endeavour*'s payload bay. With the new Caandarm2 fully secured to the ISS, the next

---

[28] Ibid., p.48.

[29] Chris Hadfield, *An Astronaut's Guide to Life on Earth: What Going to Space Taught Me About Ingenuity, Determination, and Being Prepared for Anything*. New York: Little, Brown and Company, 2013, pp.86–96.

[30] JSC, STS-100 Mission Control Center Status Report #07. Sunday April 22, 2001 – 3:30pm CDT. Accessed online on March 31, 2016 at http://spaceflight.nasa.gov/spacenews/reports/sts100/STS-100-07.html

S100E5262 2001/04/22 18:14:19

Fig. 6.46 Chris Hadfield peers into the crew cabin of the *Endeavour* during a lengthy space walk to perform important work on ISS. The Pressurized Mating Adapter (PMA-2), which temporarily anchors the orbital outpost to the shuttle, can be seen behind him

delivery from STS-100 was the *Raffaello* Multi-Purpose Logistics Module (MPLM), which pilot Jeffrey Ashby plucked from *Endeavour*'s payload bay and latched to the space station's *Unity* module. Once docked, supplies and equipment were transferred from the module to the ISS over several days before being undocked and re-berthed in the space shuttle payload bay. During these operations, Hadfield and Parazynski conducted a second spacewalk lasting seven hours and forty minutes to install PDGF circuits for the new Canadarm2 on the *Destiny* module, as well as remove an old antenna no longer needed and transfer a spare Direct Current Switching Unit from the shuttle's payload bay to an equipment storage rack on the outside of the *Destiny* module.[31]

On flight day 11 the *Endeavour* bid farewell to the ISS and Expedition 2, nudged away slowly from the space station and then conducted the usual fly around while the crew completed a photographic survey. With this last task complete, STS-100 fired a separation burn and the space shuttle departed from the orbiting ISS to make its way home. With her mission complete, the *Endeavour* landed safely on Runway 22 at Edwards Air Force Base California. Canada's first spacewalker was safely home.

---

[31] JSC, STS-100 Mission Control Center Status Report #10. Tuesday April 24, 2001 – 4:00am CDT. Accessed online on March 31, 2016 at http://spaceflight.nasa.gov/spacenews/reports/sts100/ STS-100-10.html

S100E5958 2001/04/29 17:51:35

Fig. 6.47 Backdropped against the blue and white Earth and sporting a readily visible new addition in the form of the Canadarm2, the ISS was photographed following separation from the *Endeavour*

MSS Missions, Operations, and Support

The successful installation of the Canadarm2 on the ISS during STS-100 was not the end of Canada's task, but rather just the beginning. The complexity of the Mobile Servicing System demanded additional support from the ground, and as part of its MSS contribution to the ISS, the Canadian Space Agency opened a new facility in 2000 dedicated to its operation. Known as the Space Systems MSS Operations Complex, this new organization was located at its St. Hubert headquarters to manage the many new tasks that involved the MSS. In addition to real-time mission support, the complex was designed to conduct analysis of all MSS operations as well as train both astronauts and mission controllers in Mobile Servicing System operations. The Space Systems Operations Complex itself was initially divided into several smaller units, including the Space Operations Support Center (SOSC), the Mission Operations and Training Simulator (MOTS), the SPDM Task Verification Facility (STVF), the Virtual Operations Training Environment (VOTE), and the Multimedia Learning Center. Each facility was detailed to a particular aspect of the MSS though all units were considered essential to the overall success of ISS operations. The SOSC acted as a "back room" for MSS operations and was tied directly into NASA's Mission Control Center in Houston. Whenever ISS robotic tasks were scheduled, the SOSC assisted Houston in the analysis of SSRMS telemetry and provided quick (real-time) solutions to any technical problems that might have occurred with the arm. The Engineering Support Center's (ESC) robotic engineers were also on station if the problem required more technical troubleshooting. When not on call, the ESC's robotic engineers played a regular role in planning, simulating, and verifying SSRMS procedures and tasks. [32]

The Mission Operations and Training Simulator augmented the ESC with its own completely integrated space robotics training facility. The MOTS was used to train Canadian, American, and Russian astronauts and ground crew but tended to concentrate on preparing ISS resident crews and ground support staff. Even those astronauts who had experiencein operating the Canadarm required further training to successfully operate the space station's MSS using the newly-designed RWS. Training at the MOTS was never perceived as exclusive, however, and instead was considered an important part of the intensive preparation required for any ISS expedition.

Yet another facility was devoted just to verifying all SPDM tasks. The STVF simulated and verified all procedures used to employ the SPDM on ISS missions. Using a full-scale working model of the SPDM, tasks proposed for the manipulator were first tested in the simulator and then adjusted as required to meet the aim of the job. Astronauts and ground staff routinely carried out simulated tests in near-real conditions to evaluate the success or failure of any maneuver. The simulator was designed to provide the most realistic conditions possible. For example, the system incorporates force feedback into the computer simulation. This allowed the computer-simulated arm to react to any resistance that the SPDM might encounter (e.g., pushing a piece of equipment into its assigned slot) just as the real SSRMS would have encountered. Also, special lighting was arranged to simulate the harsh and sometimes blinding conditions of the sun or the deep shadows that occur in space

---

[32] More recently, these suborganizations were restructured; the SOSC was renamed the Remote Multi-Purpose Support Room (RMPSR) and the ESC was renamed the Operations and Engineering Center (OEC).

as the station moves around the Earth. For a final touch of reality, the SPDM Task Verification Facility's SPDM model was fully integrated into the MOTS, allowing the two to be synchronized for training simulations. Therefore, when a trainee performed a task in the MOTS, the SPDM mock-up at the STVF could react to the same instructions accordingly.

The deployed SSRMS and its follow-on components, and the MSS Operations Complex on Earth that supported it, soon became an integral part of all subsequent ISS expeditions, helping each crew and mission as they endeavored to construct the space station. To the interested observer it also provided yet another example of the increasing maturity of the Canadian space program over the course of the 1990s, its evolving expertise in space robotics, the level of cooperation and interoperability that existed between Canadian and American space programs, and the central role that Canada was destined to play in the exploration of space during the first decades of the twenty-first century.

Space Science and the 'Humble' Space Telescope

Though the human presence in space program was often considered the darling of the Canadian space efforts at the turn of the century, it was by no means the only activity then under way under the aegis of the CSA. As well as human spaceflight operations, the agency's space science branch sponsored two other missions in addition to its ongoing investment in the development of the RADARSAT program. The first of these other projects was known as the Microvariability and Oscillations of Stars (MOST), an ambitious cooperative effort to design and construct the world's smallest astronomical space telescope. The projected was initiated in 1996 by a group of researchers including Slavek Rucinski of Ontario's Centre for Research in Earth and Space Technology, Jaymie Matthews of the University of British Colombia, Tony Moffat of the University of Montreal, and Kieran Carroll of Dynacon Enterprises. Essentially, the group proposed a series of questions: was it possible to understand the Sun in the context of other stars? By placing a birthdate on the oldest stars in the solar neighborhood, was it possible to set a limit on the age of the universe? How do strong magnetic fields affect the physics of other stars and our own Sun? What are mysterious planets around other stars really like? How did the atoms that make up our planet and our very bodies escape from stars in the first place? These questions could potentially be answered through the space-based observation of nearby stars.

Specifically, the team sought to capture data on the asteroseismology of acoustic and gravity-mode oscillations in Sun-like stars, magnetic (rapidly oscillating Ap or roAp) stars, cool giants, pre-main-sequence delta Scuti pulsating stars, massive O and B stars, and other stellar classes, to probe uniquely their internal structures and evolutionary states Second, the MOST space telescope would provide data for the analysis of the transits and eclipses of exoplanets around Sun-like stars and red dwarf stars, to reveal their sizes, atmospheric compositions, magnetic fields and other properties. Third, the satellite could measure the turbulent variations in massive evolved (Wolf-Rayet) stars to understand how they add gas to the interstellar medium, and lastly, measure structure and variability of gas and dust disks around pre-main-sequence stars.[33]

---

[33] University of British Columbia, 'MOST: Canada's First Space Telescope: Mission at a Glance'. Accessed online on December 20, 2015 at http://most.astro.ubc.ca/page2.html

Fig. 6.48 MOST project insignia

In 1997, the CSA agreed to finance the design, development, launch, and mission control to the sum of CDN$8.5 million, with another CDN$1.2 million being secured through the Ontario Research and Development Challenge Fund in support of the project's ongoing development work at the University of Toronto Institute for Aerospace Studies (UTIAS) Space Flight Laboratory. At the same time, Jaymie Matthews was named principal investigator and mission scientist. Design and construction of the satellite began soon after with the end result living up to its assigned nickname 'the Humble Telescope' – a playful reference to NASA's Hubble Space Telescope. Weighing 54kg, the suitcase-sized micro satellite was only 65cm × 65cm × 30cm in diameter, powered by solar panels for just 35 watts of power. Its high-precision optical telescope with a collecting mirror was no bigger than a household pie plate, only 15cm in diameter. Despite its diminutive appearance, however, the MOST space telescope was a powerhouse. The optical telescope fed a CCD camera with a Marconi 47-20 frame-transfer device for collecting science measurements and for tracking guide stars for satellite attitude control. The instrument also contained a single broadband filter that selected light in the wavelength range 350–700 nanometers (nm). The camera itself was equipped with an array of Fabry micro-lenses that projected a large stable image of the telescope pupil illuminated by target starlight, which was key to the highest photometric precision of MOST. For low cost and high reliability, the instrument had no moving parts. The structure automatically maintained the same focus across a wide range of temperatures, and the rapid frame transfer of the CCDs controlled exposure times. The CCDs were cooled by a passive radiator system. All in all, it was a very compact yet formidable spacecraft.[34]

---

[34] Ibid.

The MOST satellite was launched from the Plesetsk Cosmodrome on June 30, 2003 aboard a refurbished Russian SS-19 ICBM Rockot/Breexe-KM as part of a multiple pay-load contract with Eurockot. The satellite reached its low-Earth polar orbit without inci-dent and soon afterwards provided substantial data that allowed the MOST team to report a number of discoveries. Designed to last just a year on station, the microsatellite contin-ued to function for more than a decade after its launch and at the time of writing continues to function in space on an as-needed basis.

The second project was the Science Satellite-1 (SCISAT-1), designed to make obser-vations of the Earth's atmosphere, specifically the depletion of the ozone layer and its impact on Canada and the Arctic. Manufactured by Bristol Aerospace of Winnipeg,

Fig. 6.49 The SCISAT logo

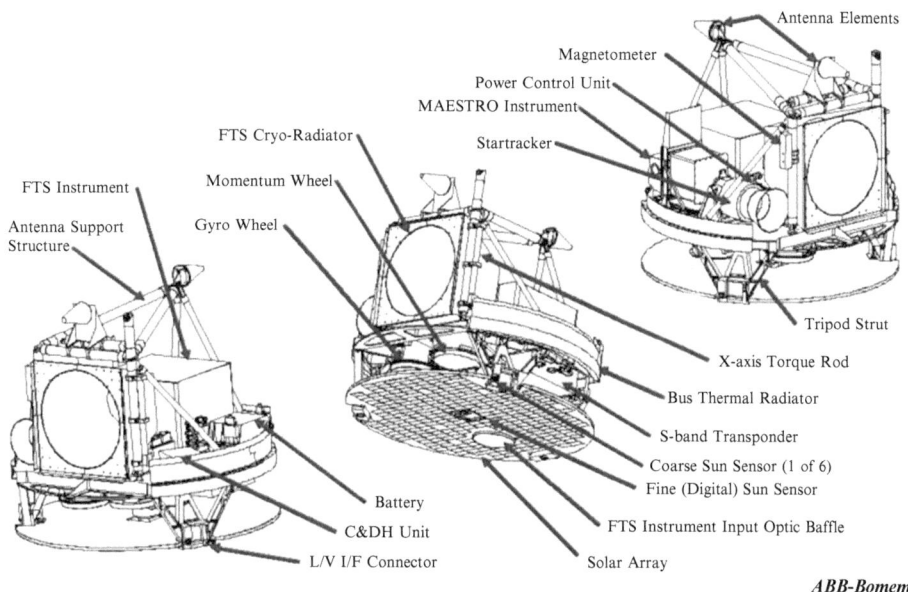

Fig. 6.50 Diagram showing the SCISAT payload

Manitoba, at a cost of approximately CDN$60 million the circular-shaped spacecraft was somewhat bigger than the MOST space telescope, weighing 150kg and requiring about twice as much power – 70 watts. On February 4, 1999, the government of Canada announced the selection of the Atmospheric Chemistry Experiment (ACE), which measured the chemical processes that control the distribution of ozone in the Earth's atmosphere, as the main science mission of the SCISAT-1. Dr. Peter Bernath from the Department of Chemistry at the University of Waterloo was the mission scientist and headed an international team of researchers from Canada, the United States, Belgium, Japan, France, and Sweden.

The SCISAT-1 was launched from Vandenberg Air Force Base California on August 13, 2003 using an Orbital Sciences Corporation Pegasus-XL air-launched rocket. The mission was designed to last at least two years but, as was the case with many Canadian satellites, SCISAT-1 served well past its original mandate.[35] Both the MOST space telescope and the SCISAT-1 were great achievements for Canada, ending a 32-year drought in Canadian scientific satellite deployments. It was therefore very unfortunate that their announcements came in the immediate aftermath of a space tragedy, one that ultimately affected the Canadian space program greatly and put in jeopardy the prospect of launching further scientific satellites anytime soon.

---

[35] University of Waterloo, *ACE Mission Information for Public Data Release.* N.d. Accessed online on February 2, 2016 at http://www.ace.uwaterloo.ca/v22data/ACEFTSPublicReleaseDocumentation.pdf

A New Space Strategy

Despite the many successes enjoyed in spaceflight there was never any guarantee that it would occur without some risk. Not even two years after Hadfield's historic flight, space shuttle *Columbia* carrying the crew of STS-107 was lost on February 1, 2003 when it disintegrated during re-entry at the end of its fifteen-day mission. It was later determined by the Columbia Accident Investigation Board that a piece of foam had broken off the launcher shortly after liftoff and struck the wing of the orbiter, critically damaging its thermal protection system. During re-entry at the end of the mission, the wing slowly overheated and came apart, leading to loss of control and the destruction of the orbiter.[36] Canada's prime minister, the Right Honourable Jean Chretien, expressed in a statement of condolence that, "on behalf of all Canadians, in our expression of sympathy to the families. Our prayers are with you, and with the people of the United States and Israel during this tragic time."[37] Marc Garneau expressed similar feelings to reporters at a press conference. "This is the kind of day you dread", he told them, "but you know all too well that this can occur. It is part of the risk you accept."[38]

The tragic loss of the STS-107 crew and the space shuttle *Columbia* overshadowed the joy that should have come with the launch of Canada's first space telescope in June. The *Columbia* disaster had a very similar effect on Canada's space program as the loss of *Challenger* did in January 1986. The shuttle fleet was immediately grounded until NASA's designated accident investigation board could complete its work and then, depending on its findings, no one knew for certain when the space shuttle fleet might resume operations. In the interim, the Russian space agency was forced to step up and provide crew launch and return via a new variant of its venerable Soyuz-TM spacecraft. While these ships could carry a three-person crew to orbit and back, this was still less than half the capacity of a space shuttle with its typical crews of seven. The reduction in crew transportation capacity also meant smaller ISS expeditions (for a period of time the station only maintained a two-person caretaker crew) and thus much longer wait times for international astronauts to be assigned to various future crews.

The situation forced Canada's space program once again to look inward at its own priorities, and to reconsider what were its vision, strategy, and longer-term. Marc Garneau, a former veteran astronaut and by then serving as the President of the CSA, led the development of a new space agency vision and strategy that replaced the Long Term Space Plans (LTSP) as the main framework guiding the CSA in its role as leader of the national space program. After a series of extensive consultations over the summer and fall of 2003, the CSA announced the release of the new Canadian Space Strategy on

---

[36] For a full account see NASA, Columbia Accident Investigation Board Final Report Volume 1. August 2003.

[37] AFP, 'Canada's PM Chretien Expresses Deep Sadness Over Columbia Shuttle Tragedy", *Space Daily*. February 1, 2003.

[38] Ibid.

November 12. Though the essential elements of older policies and the agency's core competencies remained more-or-less the same, the vision for Canada's space program was clearly redefined as exploration-centric and government-compliant. Specifically, it was the vision of the CSA to "integrate space fully and completely in Government of Canada departments and agencies as an invaluable tool to help fulfill their mandates and reach our government's goals for Canadians".[39] From a bureaucratic perspective, this was a clever tactical move, as the agency could co-opt stakeholders and thus leverage their own mandates and budgets in support of the space agency's goals. With the new vision directing the CSA to initiate, innovate, and integrate, or as another report put it, "to serve and to inspire Canadians through excellence", one could expect that the agency would be lobbying the government hard for political and financial support.[40] As for its core mission, the Canadian Space Agency Act's original statement still held true: "to promote the peaceful use and development of space, to advance the knowledge of space through science and to ensure that space science and technology provide social and economic benefits for Canadians".[41]

The newly-published strategy informed its readers that Canada must continue to use space for the betterment of Canadians in four main ways:

1. To look down upon the Earth to observe, monitor, and protect life below;
2. To look out into the depths of space to explore, learn, and discover more about the universe, as well as our place in it;
3. To look upon space as a means of communicating with each other by relaying information via satellites; and,
4. To look to space as a source of inspiration to inform Canadians about their country's advances in science and technology, and increase scientific literacy among our citizens.

Earth observation and resource management understandably remained a core competency for the CSA, as did its commitment to space science and exploration, mainly through the American space shuttle and space station programs. The re-engagement in satellite communications marked a noted shift in agency effort, but the space awareness and learning agenda was in essence just good business sense. Canada's space program remained an expensive endeavor and its total cost was likely to continue to grow going into the future, so the agency leveraged its unique appeal to nurture and inspire Canadians about the benefits of space exploration and in doing so create new advocates for the program's continued support.

---

[39] CSA, *The Canadian Space Strategy: Serving and Inspiring the Nation.* St. Hubert: The Canadian Space Agency, 2003,p.7.

[40] CSA, *2003–2004 Estimates: Report on Plans and Priorities.* Ottawa: Department of Industry, 2003, p.6.

[41] Op. cit., *The Canadian Space Strategy*, p.3.

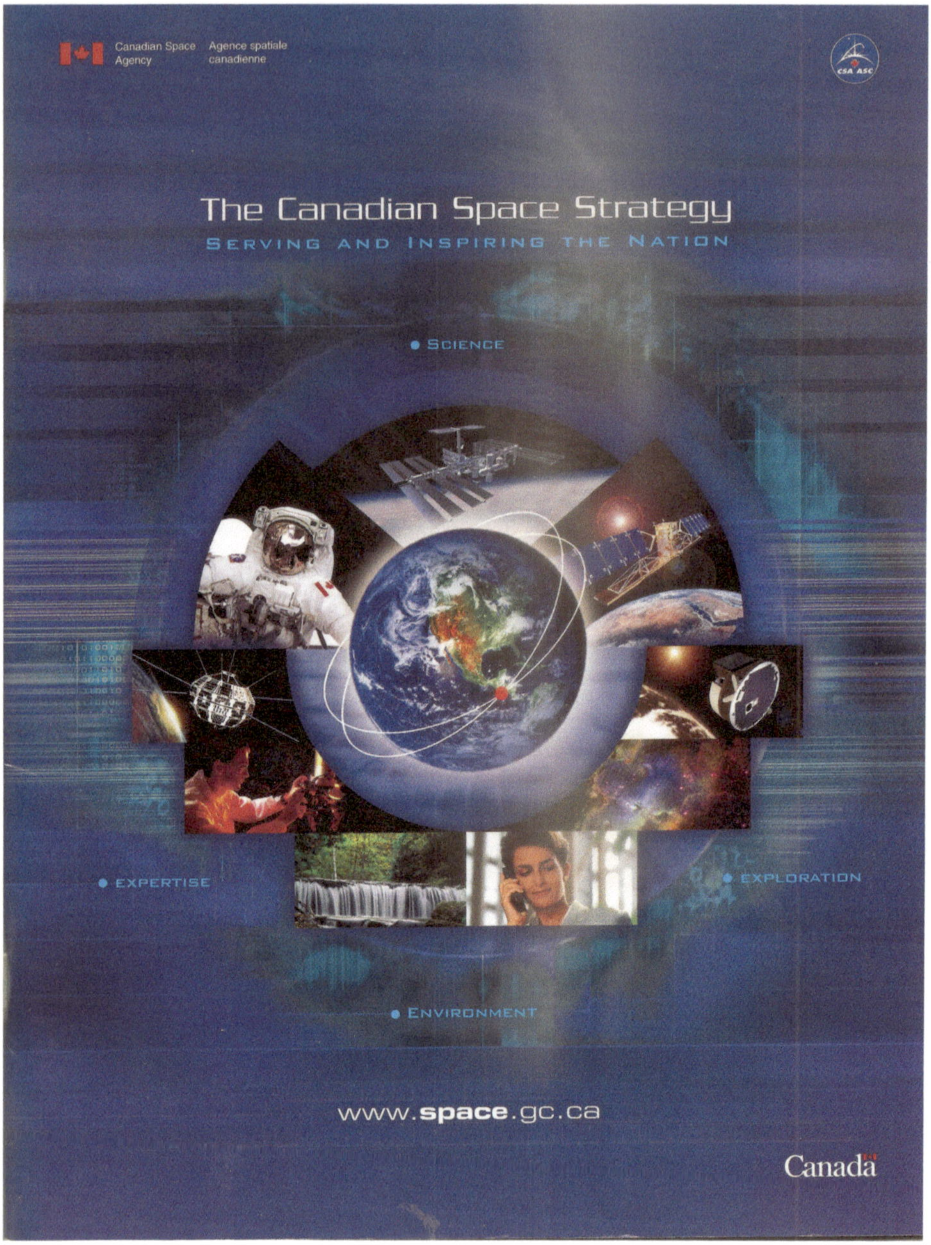

Fig. 6.51 *The Canadian Space Strategy* was published in 2003

Garneau was to be congratulated for his efforts. The CSA was still suffering organizational and financial challenges when he assumed the position of president in November 2001, and had been in the chair less than a eighteen months when the *Columbia* disaster further compounded those issues. Nevertheless, he found a way through, assessed the situation, and issued new "marching orders" that would successfully see the CSA through the next few years. He secured the return to flight for one of Canada's astronaut corps, and stretched every dollar of the agency's budget to support the main efforts of Earth remote sensing, space science, and human presence in space. Having accomplished all that he felt he could, Garneau subsequently announced his retirement from the CSA on November 28, 2005 to pursue a third career, this time in federal politics. His departure was hard felt by the agency, and the newly-elected government seemed to have considerable trouble replacing him. Dr. Virendra Jha was first chosen to succeed Garneau, but in the end only served for three months as acting president while a search for other suitable candidates continued. Replaced by Carole Lacombe in February 2006, likewise she lasted only just over a year before leaving the post. Another successor was not named until April 12, 2007 when Larry J. Boisvert assumed the job, but he only lasted nine months in the office before being replaced by Guy Bujold, who held the post for just eight months. It was not until the appointment of another Canadian astronaut, Dr. Steve MacLean, in September 2008 that the agency finally enjoyed some stability in the president's office.

The Overdue Satellite: RADARSAT-2

In May 1994, while RADARSAT-1 was still undergoing testing in preparation for launch, an agreement for enhanced cooperation in space was reached between the CSA and NASA that ultimately led to a commitment to launch Canada's RADARSATs. When the first spacecraft was launched in November 1995, both countries soon reaped the reward for their efforts, as RADARSAT returned much valuable data for a wide range of Earth remote-sensing programs and projects. Such was the success of RADARSAT-1 that the government of Canada soon after committed to the construction and launch of a second more advanced RADARSAT. Due to the cost of such a project, however, the government chose to partner with the private sector to produce an industry-led follow-on successor. In its decision, Cabinet approved an initial investment envelope in the range of CDN$400 million towards the project in exchange for data, with the prime contractor, McDonald Dettwiler and Associates (MDA) Limited, investing a further CDN$90 million in the spacecraft's development. It was expected to take four years to design, build, and test the new spacecraft, with a launch date then set for sometime during the year 2001.

The RADARSAT-2 was conceived and designed to be a considerably more advanced spacecraft than its predecessor. Weighing less than RADARSAT-1 (2200kg vs. 2750kg), its mission life was planned for seven years, two more than RADARSAT-1 had been designed for. While its SAR antenna was the same size, its larger solar arrays provided greater power and capability for both its radar as well as its bandwith capacity. The second spacecraft also sported other important new features. Whereas RADARSAT-1 had only a right-looking direction, RADARSAT-2 had the ability to conduct routine left- and right-looking operations which allowed it an increased re-visit time over its target for improved monitoring efficiencies. The RADARSAT-2 would still transmit in the C-Band, but it carried a vastly larger range of beam modes with increased nominal resolution in nearly every category. Overall, RADARSAT-2 would be a vastly superior spacecraft, giving Canada much greater capability to carry out is remote sensing tasks.

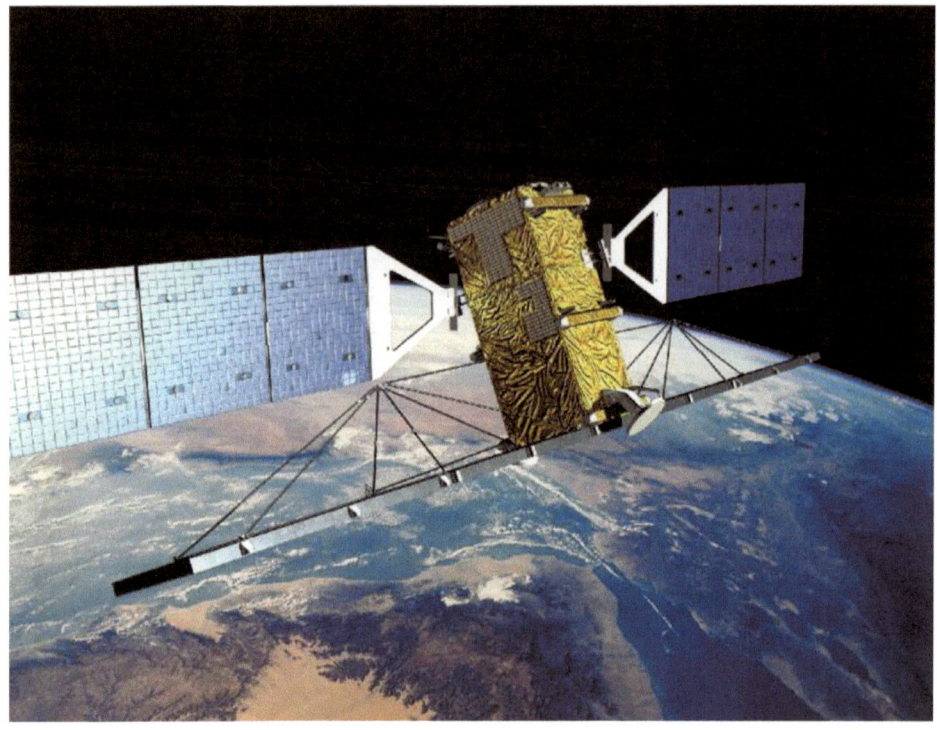

Fig. 6.52 An artist's concept of RADARSAT-2

Table 6.2 Comparison of RADARSAT-1 and 2 technical systems

Parameter	RADARSAT-1	RADARSAT-2
S/C mass at launch	2750kg	2200kg
Design life	5years	7years
Onboard data recording	Tape recorder (analog)	Soild-state recorder (384Gbit) and addressable data retrieval
Spacecraft location (tracking)	S/C ranging from ground	GPS receiver onboard
Imaging frequency	C-band at 5.3GHz	C-band at 5.405GHz
Spatial resolution of data	10–100m	3–100m
Polarization	HH	Fully polarimetric
Switching delay between imaging modes	about 14seconds	≤1second
Look direction of SAR antenna	Right	Left or Right (faster revisit times)
S/C attitude control	Sun sensors, magnetometers, and horizon scanners	Two star trackers for precision pointing
Downlink power transmitter	Standard ground antenna size of about 10m diameter is needed	Ground antenna size of 3m diameter is needed
On-board location accuracy device	None	GPS receivers (±60m real-time position information)
Yaw steering	None	Yaw steering for zero Doppler shift at beam center (facilitates image processing)

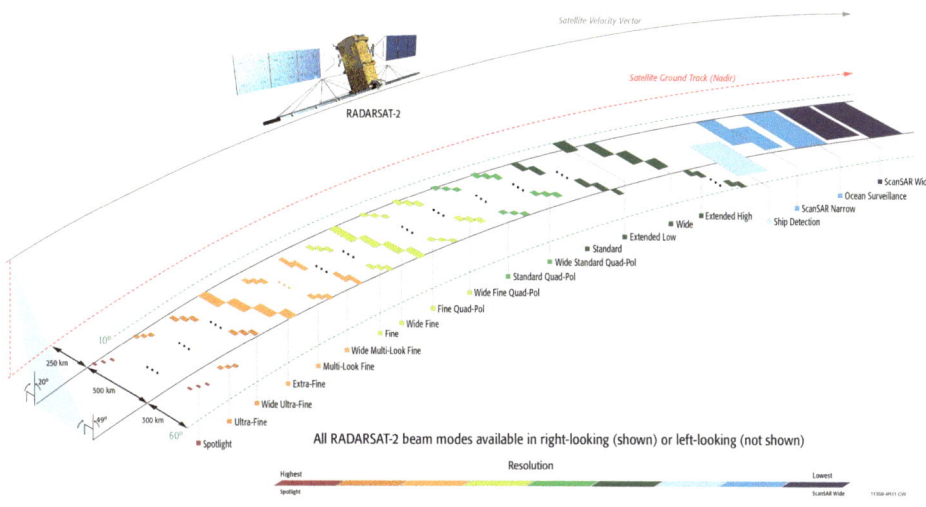

Fig. 6.53 RADARSAT-2 beam modes

The increased capabilities of RADARSAT-2 soon caused some concerns for the CSA's American partner, especially after it was announced in 1998 that the satellite would be eventually be owned and operated by MDA. NASA wrote to CSA President Mac Evans that Canada's decision to privatize the operation of RADARSAT-2 raised questions about the nature of the program, and made it difficult for NASA to justify supporting a foreign commercial SAR program. Amongst the American defence and technical communities, there were undoubtedly reservations about Canada owning and operating such a powerful remote sensing satellite. According to various government and media reports at the time, suggestions were made by the U.S. Department of Defense that RADARSAT-2 might even pose a security threat.[42] In November 1998, NASA reiterated its inability to launch RADARSAT-2 based on the concerns it raised earlier, and this forced the CSA, now responsible for the launch, to go shopping for another potential client. Meanwhile, during the same month the Treasury Board (TB) approved the RADARSAT-2 Major Crown Project with a funding envelope of $242.2 million. The CSA and MDA in turn signed a master agreement in December 1998 for the RADARSAT-2 mission, under a firm price agreement in which the government contribution was $225 million, in exchange for data. MDA was to invest $80 million in spacecraft development.[43]

The launch issue derailed the progress of RADARSAT-2 for some time. In early 1999 the CSA, clearly annoyed at NASA for backing out of providing launch support for RADARSAT-2, decided to explore the possibility of a launch by Russia. This idea immediately drew criticism from both American and other Canadian government officials, who

---

[42] Anon, 'Canadian satellite will take off despite U.S. protests: Manley', *Ottawa Citizen*, 19 February 1999, p.4.

[43] CSA, *Report on Plans and Priorities, 2010-11*. Ottawa: Minister of Industry, 2010, Annex 4 – Status Report on Major Crown Projects.

spent the remainder of the year trying to work out some agreement. Ultimately, the CSA saw a European option as their best way forward, and announced in December 1999 that an Italian company, Alenia Aerospazio, had been selected to supply the bus for the RADARSAT-2 spacecraft. It seemed very likely at the time that the ESA might also provide a launcher.

The Master Agreement between the CSA and MDA was updated in January 2000 to reflect changes in the schedule and the latest cost estimates. MDA remained responsible for spacecraft operations and business development, while the CSA was then responsible for arranging the launch and maintaining the long-term national archive of RADARSAT-2 data. The CSA also agreed to provide an additional "in-kind" contribution of certain assets, plus the services of its David Florida Laboratory and the NRC Institute of Aerospace Research Laboratory for spacecraft integration and testing.[44]

In March 2000, the Treasury Board approved an increase of $47.1 million to cover the cost of changing bus suppliers, required as a result of U.S. government restrictions imposed on Orbital Sciences Corporation – the proposed U.S. bus supplier at that time – and an increase of $12.3 million for upgrades to the then existing satellite ground station infrastructure. In June 2000, the Treasury Board approved another increase of $108 million to cover the cost of procuring a third-party commercial launch. In June 2001, the year that RADARSAT-2 was initially supposed to launch, the Treasury Board approved yet another increase of $6 million to cover the cost of critical modifications to be made to the RADARSAT-2 spacecraft in order to accommodate a potential future tandem mission with a RADARSAT-3. First, however, the spacecraft still needed to be built.[45]

As a result of the problems associated with sorting out the launch provider for RADARSAT-2, the development of the satellite was completed at a much slower pace than originally planned. The issue was further exacerbated by the fact that both the main contractor and the subcontractors experienced challenges with the production of some of the satellite components, which in turn caused significant delays in the assembly, integration and testing of the spacecraft. The Extendible Support Structure (ESS), one of the primary spacecraft subsystems, was only delivered to the Assembly, Integration and Test (AI&T) site at the DFL in October 2003. The Solar Arrays and the Bus were delivered to DFL in April and May 2004, respectively. The SAR antenna was delivered in September 2005. The assembly, integration and test of the RADARSAT-2 spacecraft at the DFL, along with the operations and preparations activities at the CSA in St. Hubert were not successfully completed until September 2007, over six years later than the satellite was originally supposed to be launched. After years of delay, the Euro-Russian corporation Starsem finally launched the RADARSAT-2 on December 14, 2007 from the Baikonur Cosmodrome, successfully sending it into orbit aboard a *Soyuz-FG/Fregat* rocket. After a short period of time checking out the spacecraft, it was declared commercially operational on April 27, 2008. Eight years later, the spacecraft continued to function and return data to clients worldwide.

---

[44] Ibid.

[45] Ibid. See also Government Consulting Services, *Evaluation of the RADARSAT-2 Major Crown Project*. Project Number 570-2782-3. September 2009, pp.1–3, 12.

Canada Returns to Spaceflight: STS-115

With the resumption of space shuttle operations in 2005 came the chance for Canadian astronauts to return to space as well. Next in line for a mission was the stoic astronaut Steve MacLean, the future space agency president, who must have appreciated the irony that once again, he had to wait patiently for NASA to recover from a tragic loss, successfully return to flight, and make shuttle operations safer for its crews. Despite being closer than other Canadian astronauts to the tragic events in terms of where he was in the flight rotation on both occasions, he nevertheless accepted the risk that went with being an astronaut and looked forward to his next flight assignment. Fortunately, he did not have to wait very long before making his second spaceflight.

Mission STS-115, also referred to as ISS 12/A, was the first space station assembly mission following the *Columbia* disaster. Launched on September 9, 2006 from the KSC's Pad 39B, it reached orbit without incident approximately nine minutes later to begin its chase of the ISS. Veteran NASA astronaut Brent Jett, making his fourth spaceflight, commanded the mission, with rookie Chris Ferguson serving as his pilot. The flight's four mission specialists included Joe Tanner, Dan Burbank, Heidemarie Stefanyshyn-Piper, and Steve MacLean.[46]

Fig. 6.54 Steven MacLean was the second Canadian to walk in space on mission STS-115

---

[46] NASA, *STS-115: Return to Assembly*. United Space Alliance, August 2006, p.15.

The main objective of STS-115 was the delivery and installation of the 15,900kg P3/P4 ITS and solar arrays to the port side of the ISS. The truss also contained a device called the Solar Alpha Rotary Joint (SARJ) which, once installed, would allow the P4 and P6 solar arrays to track the Sun as needed for electrical power generation. During the planned eleven-day mission MacLean was tasked to conduct the second of three planned EVAs alongside Dan Burbank, making him the second Canadian astronaut to walk in space. His other duties included being the prime operator of the space station Canadarm2 throughout the mission, overseeing its use during the P3/P4 truss segment installation, and also during the Orbiter Boom Sensor System handoffs to the shuttle arm, assisting Burbank and Ferguson with the inspection of *Atlantis'* thermal protection system. Along with Burbank, he was also the designated IVA member helping to choreograph EVA 1 and EVA 3. Lastly, MacLean was responsible for overseeing the opening of the shuttle's payload bay doors once *Atlantis* reached orbit.

As a result of the *Columbia* disaster, NASA had implemented a number of new mandatory safety procedures to be executed after crews had reached orbit. Crews now routinely photographed the space shuttle external tank and returned these images to Earth for analysis by ground teams. As well, space shuttles now carried an additional arm known as the Orbiter Boom Sensor System (OBSS), a crane extension for the Canadarm that was used to inspect the leading edges of the shuttle's wings and the nose cap for any sign of damage resulting from the launch. After MacLean had opened *Atlantis'* payload bay doors, he along with Ferguson and Burbank worked to grapple and unberth the OBSS for an inspection of the spacecraft. When their task was completed, the imagery was downloaded to mission control and the OBSS was returned to its starboard side sill in the payload bay for use again later in the mission.[47]

Fig. 6.55 Space Shuttle *Atlantis* closes with the ISS on flight day 3

[47] Ibid., p.6.

Flight day 3 was spent closing the distance with the ISS and docking. Before opening the hatches, Burbank and Ferguson used the Canadarm to lift the P3/P4 truss out of *Atlantis'* payload bay and prepared it to be snagged by the Canadarm2. After a secure connection was made, the hatches were opened between the shuttle and the station and STS-115 received a warm welcome from the ISS Expedition 13 crew. MacLean also had the privilege of officially joining the list of Canadian astronauts that had visited the ISS, as he transferred over to the station joining Expedition 13 science officer Jeff Williams in the *Destiny* module robotic workstation. From there, MacLean maneuvered the Canadarm2 to grapple the P3/P4 truss, preparing it for a handover the following day.

The first STS-115 EVA occurred on flight day 4. As mission specialists Tanner and Piper conducted their spacewalk, MacLean and Williams slowly moved the P3/P4 truss over into alignment with the P1 truss. Once they had it in place, the two spacewalkers continued their work to join the new truss to the station. On flight day 5, Steve MacLean and Daniel Burbank picked up where Tanner and Piper had left off the day before, conducting a lengthy EVA to complete the final preparations for the activation of the SARJ. During their spacewalk the pair removed six launch restraints, four thermal covers, and sixteen launch locks to set the stage for its activation. As they worked, MacLean and Burbank overcame several minor problems, including a malfunctioning helmet camera, a broken socket tool, a stubborn bolt, and a bolt that came loose from the mechanism designed to hold it captive. The stubborn bolt required the force of both spacewalkers to finally remove it.[48] Though tired after their seven hour and eleven minute EVA, both first-time spacewalkers could take pride in what they had achieved. Once back inside *Atlantis*, the NASA Mission Control Center (MCC) in Houston initiated a four-hour activation and checkout of the SARJ. A few software problems were encountered, but by the following day a workaround was achieved and the SARJ and P3/P4 truss were successfully checked out for operations.

Steve MacLean's spacewalk once again caught the attention of many back home, and he took time out from his busy schedule on 14 September to field a short congratulatory call from the new Canadian Prime Minister, the Right Honourable Stephen Harper, while other members of the crew and the ISS expedition also conducted various media interviews. With their public relations duties completed, the crew of STS-115 returned to other tasks, as there remained one more spacewalk to conduct before their main objectives were complete.

The next day, STS-115 successfully conducted its last spacewalk. The newly-installed solar arrays were also unfurled, giving the space station a new look as one end of it now appeared complete. With their primary objectives behind them, the crews were given a half day off duty on flight day 8, followed by the final transfer of items from *Atlantis* to the space station, including 41kg of oxygen supplies. The following day, the transfer work was completed and the two crews said their farewells before closing the hatches between the ISS and the space station.

---

[48] NASA, STS-115 MCC Status Report #9. 13 September 2006. Accessed online on January 10, 2016 at http://www.nasa.gov/mission_pages/shuttle/shuttlemissions/sts115/news/STS-115-09.html

ISS013E81053

Fig. 6.56 Canadian astronaut Steven MacLean performs a task to relocate articulating portable foot restraints (APFR) during the second of three scheduled spacewalks

Fig. 6.57 Steven MacLean and Daniel C. Burbank, both STS-115 mission specialists, participate in the second of three scheduled spacewalks as construction resumes on the orbital outpost. Here they are translating along the side of one of the station's trusses

Fig. 6.58 Space Shuttle *Atlantis* astronauts spread a second set of wings for the ISS. The new solar arrays were fully extended at 7:44am

After completing a fly around the ISS, space shuttle *Atlantis* backed away from the spacecraft and prepared for its own return to Earth. The decision was made to extend the mission by a day in order to give the crew sufficient time to complete its inspection of the shuttle's thermal protection system, after which *Atlantis* de-orbited and returned safely to Earth. For Canadian astronaut Steve MacLean, his 11 day, 19 hour mission to the ISS had come to an end. It was also to be his final spaceflight. After his mission he announced in the summer of 2008 that he was stepping down from the Canadian astronaut program to accept an appointment as president of the Canadian Space Agency, an office he would hold for four and a half years.

From Ocean Floor to Outer Space: Williams and STS-118

Since his last spaceflight in April 1998, Dafydd Williams had continued to serve as a senior manager in NASA, adding to his duties as Director of the Space and Life Sciences Directorate at the JSC a concurrent appointment as the first deputy associate administrator for crew health and safety in the Office of Spaceflight at NASA Headquarters in 2001. Beyond these appointments he also continued to take part in astronaut training to maintain and further develop his skills. In October 2001 he qualified as an aquanaut through his participation in the first joint NASA-NOAA NASA Extreme Environment Mission Operations (NEEMO-1) expedition, a space mission simulation and training session held

Fig. 6.59 CSA astronaut Dafydd Williams answers a question during a NASA press conference. He was Canada's third spacewalker on mission STS-118

at the Aquarius Laboratory anchored 19m underwater off the coast of Florida. He later returned to the Aquarius Laboratory in 2006 to lead the NEEMO-9 expedition as part of his preparation for his next spaceflight, now scheduled for the following year.

Mission STS-118 began on August 8, 2007 with the launch of space shuttle *Endeavour* from Pad 39A at the KSC. This was *Endeavour*'s first flight since November 2002, having flown the last mission prior to the *Columbia* disaster. STS-118, also referred to as the ISS-13A.1 assembly mission, was commanded by NASA astronaut Scott Kelly making his second spaceflight and piloted by NASA astronaut Charles Hobaugh, also making his second spaceflight. Five mission specialists were assigned to the flight, including Tracy Caldwell, Rick Mastracchio, Alvin Drew, Barbara Morgan, and CSA astronaut Dafydd Williams. *Endeavour* reached outer space approximately nine minutes after launch without incident, and was soon after given the go-ahead for on-orbit operations. The crew opened the payload bay doors, activated their Spacehab module, powered up the Canadarm, and performed a variety of other payload activation procedures before entering their first sleep shift. The following day, they spent most of their time completing an inspection of the thermal heat shield while *Endeavour* chased down the ISS for a rendezvous and docking on flight day 3.

Serving as mission specialist 2 on this flight, Williams was scheduled to take part in three of the four planned spacewalks that would deliver and assemble the starboard S5 truss segment to the ISS. Essentially, a "spacer" designed to extend the length of the ISS truss, the S5 was smaller than the other segments, only 3.37m in length and weighing just 1818kg. In addition, STS-118 carried three other payloads: the External Stowage Platform

3 (ESP-3), a key component of the ISS that held spare parts for the space station; a replacement Control Moment Gyroscope; and a Spacehab Logistics Single Module full of much-needed supplies. All these items would add increased capability to the space station.

After performing its backflip and completing the rendezvous pitch maneuver that brought the space shuttle upside down near the station so that photos of the underside could be taken, *Endeavour* docked with the ISS at approximately 2:02pm EDT as it passed over the Pacific Ocean northeast of Sydney, Australia. Shortly after the hatches were opened the crew activated the new Station to Shuttle Power Transfer System (SSPTS), a series of electrical converter units in the space shuttle that drew electricity from the station's power system to supplement the normal electrical output from the shuttle's three cryogenic fuel cells. This new capability was designed to extend the shuttle's stay at the ISS to up to twelve days if needed, without depleting the finite electrical power of the fuel cells themselves.[49] During the same day, the STS-118 lifted the S5 truss out of the payload bay and handed it off to the ISS using the Canadarm and Canadarm2, while Williams and Mastracchio prepared for their spacewalk the following day.

On August 11 (flight day 4), Dafydd Williams and Rick Mastracchio exited *Endeavour* to begin their six-and-a-half hour spacewalk to install "stubby", the nickname given to the S5 truss. During their spacewalk they also retracted the forward heat-ejecting radiator from the space station's P6 truss. This was the final step needed before the P6 truss could be relocated to its final position at the end of the port truss during the STS-120 mission later that year. On their second six-and-a-half hour spacewalk the pair completed the difficult task of replacing one of the space station's four gyroscopes that had failed in October of the previous year. Williams maneuvered the 272kg replacement to its new home on the Z1 truss segment, and stored the failed equipment outside the station to be returned home on a future shuttle mission. Williams did not participate in the mission's third EVA, but he was part of the team for the fourth spacewalk. Mission control decided to keep the last spacewalk short due to prevailing weather conditions on Earth that could have affected the space shuttle's safe return. A hurricane (Dean) had progressed into the Gulf of Mexico and it was expected to make landfall around the same time as *Endeavour*'s return. As a result, the last EVA lasted only five hours, during which both astronauts got a good look at Hurricane Dean, and were awed at the sight. "Holy smoke," was Anderson's initial comment. "Man, that's impressive", Williams replied. Anderson added, "They're only impressive when they're not coming towards you."[50] Dave Williams set two records during his third EVA. He officially became the Canadian with the most spacewalks – three in total – and he passed his colleague Canadian astronaut Chris Hadfield in total EVA time. Williams ended his last EVA with a total of 17 hours, 47 minutes of extravehicular time, another remarkable achievement for Canada's space program.

---

[49] NASA, *Press Kit – STS-118: Build the Station, Build the Future.* Florida: United Space Alliance, July 2007, p.6.

[50] Harwood, William, "Astronauts marvel at Hurricane Dean", CBS News, August 19, 2007. Accessed online on February 2, 2016 at http://www.cbsnews.com/network/news/space/118/STS-118_Archive.html

ISS015E21963

Fig. 6.60 As the mission's first planned session of EVA draws to a close, astronaut Dave Williams smiles for the camera in the Quest Airlock of the ISS while Space Shuttle *Endeavour* was docked with the station

ISS015E22371

Fig. 6.61 Astronaut Dave Williams participates in the mission's second planned session of EVA as construction and maintenance continue on the ISS

ISS015E22574

Fig. 6.62 Backdropped by a blue and white Earth and the blackness of space, Space Shuttle *Endeavour*, docked to the *Destiny* laboratory of the ISS, is photographed by a crewmember during the STS-118 mission's third planned session of EVA. The shuttle's Canadian-built Remote Manipulator System (RMS) robotic arm and station's Canadarm2 are also featured. The SPACEHAB pressurized logistics module is visible in *Endeavour*'s payload bay

ISS015E22730

Fig. 6.63 CSA astronaut Dafydd Williams floats next to his EMU helmet during STS-118

After completing its visit to the ISS and undocking, the crew of STS-118 headed for home. They returned safely on August 21, 2007, touching down on Runway 15 at the Kennedy Shuttle Landing Facility. After the crew left the space shuttle they were greeted by ground crew and took a moment to inspect their ship and the minor damage it had sustained during launch. When pressed by reporters about any concerns they might have had about their safety, Williams replied simply, "no". Every space mission carried risk, but it was a risk they all accepted.

Williams' successful spaceflight added yet another completed mission to the growing list of human spaceflight accomplishments for the Canadian space program, but for him it would be his last. Having accomplished all that he felt he could as an astronaut, he announced his retirement from the Canadian Astronaut Program effective March 1, 2008. He subsequently returned to the Canadian health sector, moving to Oakville, Ontario, where he became Chief Executive Officer of Newmarket's Southlake Regional Health Centre. With Williams' departure the Canadian astronaut corps had lost a key member, though there was still much ahead for those still flying.

Canada's Last Space Shuttle Mission: STS-127

Interestingly, the last Canadian mission on an American space shuttle overlapped with the first Canadian expedition aboard the ISS. CSA astronaut Robert Thirsk, who had flown on mission STS-78 in June 1996, was the first Canadian to travel into space aboard a Russian *Soyuz-TMA 15* rocket in May 2009 when he ventured to the ISS as part of Expedition 20. He was still serving aboard the space station in July when another Canadian astronaut, Julie Payette, visited the orbiting outpost as part of the NASA STS-127 mission.[51] This spaceflight, also referred to as the ISS assembly flight 2 J/A, was finally launched from the KSC's Pad 39A on 15 July 2009 after five abortive attempts to get into orbit. NASA astronaut Mark Polansky, making his third spaceflight, commanded the mission with rookie Doug Hurley serving as his pilot. *Endeavour* also carried five mission specialists on this flight, including Dave Wolf, Christopher Cassidy, Tom Marshburn, CSA astronaut Julie Payette, and Tim Kopra who was on his way 'up' to join ISS Expedition 20, replacing Japan Space Agency astronaut Koichi Wakata who would later return to Earth with the crew of *Endeavour*. STS-127 enjoyed an uneventful ascent into orbit, and after receiving the go-ahead for on orbit operations, prepared to chase down the ISS for a docking on flight day 3. The shuttle carried into orbit the final pieces of the Japan Space Agency's *Kibo* complex as well as a load of large spares to keep the space station well stocked after the space shuttle's impending retirement from operational service. A total of five spacewalks were needed to complete these tasks, but unlike the three previous Canadian missions, Payette was destined to remain inside the space shuttle and space station during this flight.

---

[51] Thirsk had a third Canadian visitor when Canadian entrepreneur and Cirque de Soleil founder Guy Laliberté travelled to the ISS as part of the *Soyuz TMA-16* crew in September 2009.

Fig. 6.64 Captured by a remote camera equipped with a special "fish-eye" lens, *Endeavour* and its seven-member STS-127 crew head toward Earth orbit and rendezvous with the ISS. Liftoff was on time at 6:03pm (EDT) on July 15, 2009 from launch pad 39A at NASA's Kennedy Space Center

As mission flight engineer and lead robotic operator, Payette was responsible for a range of duties including assisting with robotic and camera operations when *Endeavour* handed over the *Kibo* laboratory to the space station. She was also directly involved in the deployment and installation of the Japanese Exposed Facility – what the astronauts often referred to simply as the space station's "front porch", a module that allowed astronauts to expose scientific experiments directly to the extreme environment of space.

When the STS-127 crew closed with and docked at the ISS on flight day 3, the occasion was especially remarkable for the fact that, for the first time, a total of thirteen people were joined together in space, the largest population of astronauts to visit the ISS to date.

Fig. 6.65 CSA astronaut Julie Payette and Japanese Aerospace Exploration Agency astronaut Koichi Wakata are pictured onboard the orbital outpost

For Canada specifically, the large gathering also provided the first opportunity for two of its astronauts to be in space at the same time. Canadian astronauts Robert Thirsk and Julie Payette greeted each other warmly and took time for photos before resuming their duties. When, prior to her spaceflight, journalists in Houston asked about the meeting, she replied casually, "It's historic, but I see that as a normal evolution for a country which has always been an important player in the exploration of space."[52] Though still something of an

---

[52] Canadian Press, 'Payette to Join Fellow Canadian Thirsk in Space on Historic Mission', CBC News, May 28, 2009, accessed online on January 4, 2016 at http://www.cbc.ca/news/technology/payette-to-join-fellow-canadian-thirsk-in-space-on-historic-mission-1.800991

understatement, it was insightful of how both the astronauts and the agency perceived their involvement in human spaceflight twenty-five years after its first candidates had been selected. No longer did Canadians see themselves as interlopers, but rather, that Canada was firmly at the center of the action.

STS-127 spent nearly twelve days docked at the ISS completing the delivery of the *Kibo* complex, its exposed facility, and an integrated cargo carrier containing batteries and other essential supplies for the station. It must have often felt like a central station with so many people on board, but the comparative spaciousness of the ISS compared to its predecessors *Mir* and *Skylab* meant that many astronauts could move easily about the station and shuttle carrying out their various tasks. It was yet another small testament to the incredible design of the ISS and the potential it carried.

S127E011976

Fig. 6.66 CSA astronaut Julie Payette, STS-127 mission specialist, near the control panels and windows on the aft flight deck of the Earth-orbiting Space Shuttle *Endeavour*

After fifteen and half days in space the shuttle *Endeavour* returned safely to Earth, touching down on Runway 15 at the Kennedy Shuttle Landing Facility. For Payette, who has served as the CSA's Chief Astronaut from 2000 to 2007 and also as CAPCOM for many other missions, it was to be her final spaceflight. In 2013, she retired from the Canadian Space Agency to become Chief Operating Officer of the Montreal Science Centre, returning home to the city where she was born.

Conclusion

As the space shuttle era came to its end, so too did Canada's involvement in American space launches and ISS assembly. Robert Thirsk, who was on board the ISS as part of the Expedition 20/21 crew when Julie Payette and the crew of STS-127 visited them in July 2009, had traveled to the space station courtesy of a Russian-launched *Soyuz-FG*, which would become the norm for all ISS visitors until a suitable successor to the American space shuttle was designed, built, tested, and certified for human spaceflight. As there was no clear timeline on when a new American human launch vehicle might be ready to take over from the shuttle, for the foreseeable future it became clear that Canada's only access to the ISS was to be through its former adversary and cautious new partner, Russia.

Fig. 6.67 Julie Payette and Robert Thirsk – two Canadian astronauts in space

Still, when Canada officially joined the international effort to construct the International Space Station in the mid-1990s few could appreciate at the time just how critical the country's involvement would be. Half of all Canadian space missions were to the ISS, and Canada's astronauts participated directly in the deployment of station-critical systems such as the Canadarm2 as well as two of the massive solar array segments that gave the spacecraft life. Similarly, participation in the ISS assembly had afforded Canada a near-endless string of historic and technological firsts. It was a record of achievement that the country could be proud of, especially given the conditions under which it was met.

Looking back, the evolution of Canada's space program throughout the 1990s must not be perceived as either strictly linear or always positive. The agency that oversaw Canada's space activities was born out of political frustration, and was itself politicized from day one. It suffered policy and decision-making woes, financial hardship, and also was routinely subjected to both political meddling and, at other times, indifference. All this made the accomplishment of any goal a significant challenge; therefore, one must appreciate the magnitude of achievement that was the Canadian space program during this era.

7

Conclusion: Towards Space Station Expeditions

Having delivered on its promise to be a central partner in the ISS project, during the next ten years Canada increased its direct participation in station operations that would eventually lead it to obtaining the most desired crew assignment – command of an ISS expedition. The first Canadian astronaut to serve as a permanent ISS crewmember was Dr. Robert Thirsk, who was part of ISS Expedition 20/21 that lasted from 27 May to 1 December 2009. He was followed by Chris Hadfield who, as ISS commander during Expedition 34/35, did an extraordinary job in getting those distracted by daily life down on Earth excited once more about all the wonders of space exploration. Serving aboard the ISS from 19 December 2012 to 13 May 2013, Hadfield went on to become an iconic astronaut, connecting with both celebrities and also a very attentive public back on Earth in a way few others before him had; he even recorded a music video in orbit, singing a new rendition of David Bowie's famous song 'Space Oddity'. Through social media – including tweeting from orbit with Canadian actor William Shatner – and music, Hadfield did truly yeoman service to all space agencies and programs by demonstrating a common touch and remarkable ability to spark every person's imagination and wonder about space exploration. His impact on the International Space Station program was, in fact, great enough that at the time of this writing Canadians are still digesting it through his own published memoir as well as other social media reflections.

While the Canadian space station expeditions from 2009 onwards are still under way, and therefore a subject beyond the scope of this history, they undoubtedly will be the central focus of Canadian human spaceflight investment for many years to come. At the time of writing, Canadian astronaut David Saint-Jacques has just recently been assigned to travel to the ISS in November 2018 as a member of Expedition 58/59, and it is forecast that we will also travel there aboard a Russian *Soyuz-FG* spacecraft. Similarly, other newly-selected astronauts from the July–August 2016 CSA recruiting campaign will soon join Saint-Jacques and fellow astronaut Jeremy Hansen as the agency prepares to expand the Canadian astronaut corps once more.

© Springer International Publishing AG 2017
A.B. Godefroy, *The Canadian Space Program*, Springer Praxis Books,
DOI 10.1007/978-3-319-40105-8_7

Taking a broader perspective, from ancient civilization to the modern day, the idea of reaching the heavens beyond the Earth has held universal appeal for humankind. For Canada, the dream of becoming a space-faring nation was born out of both the necessity for defence and the desire for discovery. As one of the first countries to commit to an official rocketry and space program after the Second World War, Canada played an important, if not central, role in space exploration from its earliest beginnings to the modern era. From this benchmark, a remarkable story of scientific and technological achievement also evolved, one that is only now being told in greater detail by historians.

Throughout its history, Canada's space program has accomplished much because of – or perhaps at times in spite of – government direction and intervention that affected its scientific and technological development. The defence establishment was able to carry space development through the first three decades of the Cold War to the point where control of the space program's future was then assumed and sustained by the civilian bureaucracy. Throughout the whole time in question, science and engineering played a role, large at first but later limited to supporting the needs of commercial space application. Throughout the endeavor as well the government applied its typically shrewd methods and got what it needed: international prestige, a domestic communications system, a remote sensing capability, a human presence in space, a strong science and technology relationship with the United States, and an invitation to future space activities as they arose.

Interestingly, the traditions formed in the first decades of Canada's space program survived through to the present day. The successful deployment of the *Canadarm 2* to the ISS by Canadian astronaut Chris Hadfield and the crew of the space shuttle Endeavour in April 2001, for example, publicly re-emphasized the high level of cooperation and integration that Canada achieved with both the United States and other international space partners. This was in spite of the fact that, at the same time, it was at odds with the United States over the launch of RADARSAT-2. At first glance, Canada's presence as one of the five key partners in the ISS suggested a remarkably adept and capable space program with significant resources to devote towards large-scale activities. While the former might have been true, the latter was most certainly not. With a comparatively small overall yearly program budget of approximately $350 million, Canada's prestigious membership within the ISS community was often more the result of a savvy tradition of space strategy, policy, and diplomacy rather than a major financial effort.

Examining how Canada's space program arrived at this point, exploiting strategy, policy, and technology in lieu of large financial assets or material resources is crucial to understanding its remarkable success in light of what might otherwise be considered by some as something of a lackluster space program. With very limited resources Canada's space program not only survived decade after decade, it prospered as a constant example of national achievement. Canada's space program of course did not evolve in a vacuum. It responded to the indigenous and international influences that fuelled first the Cold War era space race and, later, the desire to be part of humanity's greatest engineering project ever assembled in orbit. The United States played a major role in supporting and encouraging Canada's space program throughout this evolution, yet it did not exert the same controlling or coercive influence more often witnessed in bilateral defence or trade policy. To the present day, Canada's space program evolved largely as a result of its own efforts rather than the efforts of other countries, which at one time or another cooperated, supported it, or challenged it. It is a legacy of which the country may be justly proud but, more importantly, it must continue to be supported if it is to remain a part of the national identity and agenda.

There remains much to be learned about the history of the Canadian space program. The recording of history is always an iterative process, and this survey of Canada's national space program from its beginnings in the 1940s through to 2009 constitutes only the first of what will hopefully be in the fullness of time many detailed narratives and analyses of Canada's space exploration activities. Such historical endeavors are critical to the overall health of the Canadian space program. First, histories assist the Canadian space program in recording, documenting, and disseminating information about the country's space activities as widely as possible. Second, such histories help both internal and external audiences understand and thus benefit from the study of past challenges and successes, experiences and lessons. Third, in addition to serving the needs of both scholars and practitioners the publication of space history is of great interest to broader general audiences such as journalists, students, and the public who follow the space program with great interest and wish to learn from it and live vicariously through it. This history, rather simply, is a first step towards fulfilling that role and it is hoped that others will soon follow.

Appendix A

Canadian Space Agency Presidents, 1989–2013

The following persons have served as president of the Canadian Space Agency since its formation in 1989:

Dr. Larkin Kerwin	1 March 1989–29 February 1992
Laurent A. Bergeron (acting)	1 March 1992–3 May 1992
Roland Doré	4 May 1992–31 July 1994
Dr. Karl Doetsch	(acting) 1 August 1994–20 November 1994
Mac Evans	21 November 1994–21 November 2001
Dr. Marc Garneau	22 November 2001–28 November 2005
Dr. Virendra Jha (acting)	29 November 2005–25 February 2006
Carole Lacombe (acting)	26 February 2006–11 April 2007
Larry J. Boisvert	12 April 2007–1 January 2008
Guy Bujold (acting)	1 January 2008–2 September 2008
Dr. Steven MacLean	2 September 2008–1 February 2013

© Springer International Publishing AG 2017
A.B. Godefroy, *The Canadian Space Program*, Springer Praxis Books,
DOI 10.1007/978-3-319-40105-8

Appendix B

Biographical Notes on Canada's Astronauts

Since the inception of its program in the late 1970s, a total of thirteen Canadians have been selected for astronaut training and assignment to spaceflight. The first group of six astronauts (Bondar, Garneau, Money, Money, MacLean, Thirsk, and Tryggvason) was officially named in 1983; the second group of four astronauts (Hadfield, McKay, Payette, and Williams) was named in 1992; and a third group of just two astronauts (Hansen and Saint-Jacques) joined in 2008. One other, Stewart, was selected in competition, but ultimately did not join the Canadian Astronaut Program.

Of the twelve who entered service, eight have flown in space on a total of sixteen missions. Of these, two – Garneau and Hadfield – made three trips into space; four – MacLean, Payette, Thirsk, and Williams – flew twice; and two – Bondar and Tryggvason – made a single journey into orbit. Two other selected astronauts – McKay and Money – never flew before resigning from the program, and the two most recently-selected astronauts – Hansen and Saint-Jacques – are in the process of completing training. At the time of writing, Saint-Jacques has been assigned to ISS Expedition 58/59, currently scheduled for November 2018. The following are brief biographical sketches of the astronauts during their service in Canada's space program.

Bondar, Roberta. Born December 4, 1945 in Sault Ste. Marie, Ontario, she graduated with a BSc in zoology and agriculture from the University of Guelph in 1968, an MSc in Experimental Pathology from the University of Western Ontario in 1971, a Doctorate in Neurobiology from the University of Toronto in 1974, and an MD from McMaster University in 1977. Admitted to the Royal College of Physicians and Surgeons of Canada in 1981, in her spare time she had also qualified as a scuba diver and a civilian parachutist. Bondar was selected for the first Canadian astronaut group in 1983, and began her training in February 1984. She was named chairperson of the Canadian Life Sciences Subcommittee for the Space Station Freedom Project in 1985. While awaiting a flight assignment in the aftermath of the *Challenger* accident in 1986, she continued her research into blood flow in the brain during microgravity, lower body negative pressure and various pathological states. In early 1990, Bondar was designated the prime Payload Specialist 1 (PS-1) for the first International Microgravity Laboratory Mission (IML-1). Fellow Canadian astronaut

A.B. Godefroy, *The Canadian Space Program*, Springer Praxis Books,
DOI 10.1007/978-3-319-40105-8

Dr. Ken Money was designated as her backup. Dr. Bondar later flew as a crewmember on mission STS-42 aboard *Discovery*, which lasted from January 22–30 January, 1992. During the mission, Bondar completed the Canadian Space Physiology Experiments (SPE) included in the IML-1 that investigated human adaptation to weightlessness; the human vestibular and proprioceptive (sense of body position) systems; energy expenditure; cardiovascular adaptation; nystagmus (oscillating eye movement), and back pain in astronauts. Following her mission, Dr. Bondar left the Canadian Space Agency later the same year, resigning her post on September4, 1992 to pursue other professional opportunities. She subsequently led an international team of researchers at NASA examining the effects of space travel on astronauts, after which she served for two terms as Chancellor of Trent University. Retiring from this post in 2009, Bondar then went on to establish a not-for-profit foundation focused on environmental awareness.

Garneau, J.J.P. Marc. Born February 23, 1949 in Quebec City, Canada, Joseph Jean-Pierre Marc Garneau joined the Canadian Forces as an officer candidate and earned a Bachelor of Science degree in Engineering Physics from the Royal Military College of Canada in Kingston in 1970. Talented and motivated, he went on to earn a Doctorate in Electrical Engineering from the Imperial College of Science and Technology, London, England, in 1973. After graduation, Garneau was posted to the *HMCS Algonquin* where he served as a Combat Systems Engineer from 1974 to 1976. He went on to a number of varied naval engineering assignments in Halifax and Ottawa from 1977 to 1982, after which he was selected to attend the Canadian Forces Command and Staff College in Toronto. Promoted to the naval rank of Commander while at Staff College, he was subsequently transferred back to Ottawa in 1983 and became design authority for naval communications and electronic warfare equipment and systems. Garneau applied to the National Research Council's call for astronauts later that year and was selected for the first astronaut group in December. He was seconded from the Department of National Defence to the Canadian Astronaut Program in February 1984 to begin astronaut training, later becoming the first Canadian to fly in space when he served as a payload specialist aboard *Challenger* on mission STS-41G in October 1984. He was promoted to the naval rank of Captain in January 1986, but later retired from the Canadian Navy in 1989 to pursue a career with the newly-created Canadian Space Agency. Garneau was appointed Deputy Director of the Canadian Astronaut Program, providing technical and program support in the preparation of experiments to fly during future Canadian missions once the shuttle re-entered service. Maintaining his astronaut flight status, Garneau was selected for Mission Specialist training at the Johnson Space Center in July 1992 and completed this program the following year. He subsequently worked with the Astronaut Office Robotics Integration Team and later as a Capsule Communicator (CAPCOM) in Mission Control. Selected for his second mission shortly after, Garneau flew aboard *Endeavour* as Mission Specialist 4 on mission STS-77 in May 1996. Success on this mission firmly established Garneau as one of Canada's most experienced astronauts, and subsequently led to his selection for a third spaceflight assignment. In December 2000, Garneau flew into space one last time on mission STS-97 as Mission Specialist 2 aboard *Endeavour*. Having logged 677 hours in spaceflight, Garneau left active astronaut status in February 2001 to accept the appointment as Executive Vice President at the Canadian Space Agency. He was subsequently appointed President of the agency on November 22. He eventually resigned from the CSA in November 2005 to pursue a successful career in federal politics.

Hadfield, Chris. Born August 29, 1959, in Sarnia, Ontario, Chris Hadfield became interested in flying from a young age. As an Air Cadet, he won a glider pilot scholarship at age 15 and a powered pilot scholarship at age 16. He graduated as an Ontario Scholar from Milton District High School in 1977 and joined the Canadian Forces the following year, going on to earn a Bachelor's Degree in Mechanical Engineering (with honors) from the Royal Military College, Kingston, Ontario, Canada, in 1982. Hadfield underwent basic flight training in Portage La Prairie, Manitoba, from which he graduated as top pilot in 1980. In 1983, he also took honors as the overall top graduate from Basic Jet Training in Moose Jaw, Saskatchewan. Hadfield trained as a fighter pilot in Cold Lake, Alberta on the CF-116 *Freedom Fighter* and CF-188 *Hornet* between 1984 and 1985, after which he was assigned to 425 Tactical Fighter Squadron, a unit that was at the time serving as part of the North American Aerospace Defence Command (NORAD). During his tour of duty with 425 Sqn., Hadfield flew the first CF-188 *Hornet* intercept of a Soviet Tupolev TU-95 "Bear" strategic bomber. In 1988 Hadfield was selected to attend the United States Air Force (USAF) Test Pilot School at Edwards Air Force Base, in California, and graduated top pilot earning the prestigious Liethen-Tittle Award. Afterwards he served as a Canadian exchange officer with the U.S. Navy at Strike Test Directorate at the Patuxent River Naval Air Station. His accomplishments from 1989 to 1992 included testing the F/A-18 and A-7 aircraft; performing research work with NASA on pitch control margin simulation and flight; completing the first military flight of F/A-18 enhanced performance engines; piloting the first flight test of the National Aerospace Plane external burning hydrogen propulsion engine; developing a new handling qualities rating scale for high angle-of-attack test; and participating in the F/A-18 out-of-control recovery test program. In 1991, Hadfield was named U.S. Navy Test Pilot of the Year, and he earned a Master of Science in aviation systems from the University of Tennessee in 1992. This same year, Hadfield applied for and was selected to join the second group of Canadian astronauts from a field of 5330 applicants. He was assigned by the Canadian Space Agency (CSA) to the NASA Johnson Space Center in Houston, Texas in August of the same year, where he began astronaut training; addressed technical and safety issues for Shuttle Operations Development; contributed to the development of the glass shuttle cockpit; and supported shuttle launches at the Kennedy Space Center, Florida. In November 1995, Hadfield served as Mission Specialist 1 on STS-74, NASA's second space shuttle mission to rendezvous and dock with the Russian Space Station *Mir*. During the flight, the crew of Space Shuttle *Atlantis* attached a five-tonne docking module to *Mir* and transferred over 1,000 kg of food, water, and scientific supplies to the Russian cosmonauts. Hadfield flew as the first Canadian mission specialist, the first Canadian to operate the Canadarm in orbit, and the only Canadian to ever board the *Mir* space station. Back on Earth, Hadfield represented CSA astronauts and coordinated their activities as the Chief Astronaut for the CSA from 1996 to 2000. In April 2001 Hadfield returned to orbit, serving aboard *Endeavour* as Mission Specialist 1 on STS-100 ISS assembly Flight 6A. His main task was the delivery and installation of Canadarm2, the new Canadian-built robotic arm, as well as the Italian-made resupply module *Raffaello*. During the 11-day flight, Hadfield performed two spacewalks, making him the first Canadian to ever leave a spacecraft and float freely in outer space. In total, Hadfield spent 14 hours 54 minutes outside, traveling 10 times around the world as he completed his tasks. From 2001–2003, Hadfield was the Director of Operations for NASA at the Yuri Gagarin Cosmonaut Training Centre (GCTC) in Star City, Russia. His work included coordination and direction of all ISS crew activities in Russia, oversight of training and crew support staff, as well as policy negotiation with the Russian space program and other international partners. He also trained and became fully qualified to be a flight engineer cosmonaut in the *Soyuz TMA* spacecraft, and to perform spacewalks in the Russian *Orlan* spacesuit. He retired from the Canadian Air Force in 2003, becoming a civilian astronaut with the CSA and moving on to the assignment of Chief of Robotics for the NASA Astronaut Office at the Johnson Space Center. In 2006, he was appointed Chief of International Space Station Operations, a position he held until 2008. He was then assigned as backup to Dr. Bob Thirsk who was then training for ISS Expedition 20/21, a long-duration spaceflight. Hadfield then went on to the ISS Operations Branch where he developed space station emergency procedures. In May 2010, Hadfield commanded the NASA NEEMO 14 undersea mission to test exploration concepts and advanced technologies on extraterrestrial bodies through living in an underwater facility off the Florida coast. In September, he was officially assigned to the ISS Expedition 34/35 and began training for what would become Canada's second long-duration spaceflight and first Canadian command of the space station. Hadfield flew to the ISS aboard the Russian *Soyuz TMA-07M* on December 19, 2012, serving as Flight Engineer 4 for Expedition 34. When Expedition 35 began on March 13, 2013, Hadfield remained onboard to serve as commander of the ISS. This mission generated considerable media attention and renewed interest in spaceflight thanks to his efforts. Hadfield returned to Earth at the end of Expedition 35 on May 13, 2013, and announced his retirement from the CSA the following month. Hadfield officially left the space agency on July 3, 2013 to pursue other interests.

Hansen, Jeremy R. Born January 27, 1976 in London, Ontario, and raised on a farm near Ailsa Craig, Ontario until moving to Ingersoll for his high-school years, Lt. Col. Hansen earned a Bachelor of Science in Space Science (first class honours) from the Royal Military College of Canada in 1999. He earned a Master of Science in Physics from the same institution in 2000, with a research focus on Wide Field of View Satellite Tracking. Prior to joining the Canadian Space Program, Hansen served as a CF-18 fighter pilot and held the position of Combat Operations Officer at 4 Wing in Cold Lake, Alberta. His responsibilities included ensuring the effectiveness of NORAD Air Defense Operations and the operability of Deployed and Forward Operating Locations (Bases). From 2004 to 2007, Hansen served as a CF-18 fighter pilot with 409 Tactical Fighter Squadron and 441 Tactical Fighter Squadron, where his responsibilities included NORAD Operations, Deployed Exercises and Arctic Flying Operations. He completed his CF-18 Fighter Pilot Training in 410 Tactical Fighter Operational Training Squadron from 2003 to 2004. Lt. Col Hansen's flight experience began at the age of 12, when he joined the Air Cadet Program. He obtained both glider and private pilot licences through this program by the age of 17. This training led to his acceptance to Collège Militaire Royal in Saint-Jean-sur-Richelieu, Quebec. Lt. Col Hansen was one of two recruits selected in May 2009 through the third Canadian Astronaut Recruitment Campaign, and was one of 14 members of the 20th National Aeronautics and Space Administration (NASA) astronaut class. In 2011, he graduated from Astronaut Candidate Training, which included scientific and technical briefings, intensive instruction in International Space Station (ISS) systems, Extravehicular Activities (EVAs, or spacewalks), robotics, physiological training, T-38 flight training, Russian language courses, and sea and wilderness survival training. While waiting for a flight assignment, Hansen has continued to work at NASA's Mission Control Center as CAPCOM for the ISS program. In addition to his day-to-day work, he remains committed to building and maintaining his skills; he is taking spacewalk and robotics training while continuing to learn Russian. He has taken part in geological expeditions in the High Arctic and participated in the European Space Agency's CAVES program, during which he lived underground for six days. He was also a crewmember for NEEMO 19 (NASA Extreme Environment Mission Operations), living and working on the ocean floor in the Aquarius habitat off Key Largo, Florida, for seven days simulating deep space exploration. Lt. Col Hansen also continues to fly the CF-18 fighter jet with the Royal Canadian Air Force.

MacKay, Michael John. Born May 10, 1963 in Bracebridge, Ontario, MacKay obtained his Bachelor of Science degree specializing in Engineering Physics in 1985, a Master of Science in Electrical and Computer Engineering from the Royal Military College in 1991. A commissioned air force officer, in 1986 he worked as a Software Support Officer with the Aircraft Maintenance and Engineering Division of the Canadian Forces Base, Cold Lake, Alberta, before being named the Mechanical Support Officer and later the Repair Officer, responsible for the periodic inspection of the Canadian Air Force's CF-5 fighter jet fleet, a position he held until 1991. After completing his postgraduate degree, Major McKay was employed as a lecturer with the Collège Militaire de Saint-Jean, teaching courses in Logic Circuit Design and Electronic Instrumentation and Microprocessor Systems Design. MacKay applied to the Canadian Astronaut Program in 1992 but initially did not make the final cut. He was subsequently selected, however, after astronaut candidate Rob Stewart withdrew from the program. He officially joined the Canadian astronaut program on June 9, 1992 as a payload specialist designate. From August 1993 until early February 1994, Major McKay was involved with a NASA shuttle payload that flew on February 3, 1994. He provided Space Vision System (SVS) support to the Wake Shield Facility, a free-flying payload that was released from the Shuttle during that mission. In February 1994, Major McKay was one of four astronauts who participated in the Canadian Astronaut Program Space Unit Life Simulation (CAPSULS), a seven-day simulated space mission which was preceded by several weeks of intensive training. MacKay was forced to leave the program in early 1995 for medical reasons, but remained in the employment of the Canadian Astronaut Program as an engineer supporting various projects. In October 1997, he officially left the CSA to return to active military service with the Directorate of Space Development at National Defence Headquarters in Ottawa.

MacLean, Steve. Born December 14, 1954 in Ottawa, Ontario, Steve MacLean received a Bachelor of Science (Honours) in Physics in 1977 and a Doctorate in Physics in 1983 from York University in Toronto, Ontario. From 1974 until 1976 he worked in sports administration and public relations at York University, where he also competed with the Canadian National Gymnastics Team. He taught part-time at York University from 1980 until 1983, and then became a visiting scholar at Stanford University under the renowned laser physicist and Nobel Laureate A.L. Shawlow. As a laser physicist himself, MacLean's research included work on electro-optics, laser-induced fluorescence of particles and crystals, and multiphoton laser spectroscopy. MacLean was selected in the first group of Canadian astronauts in December 1983 and began astronaut training in February 1984.

Originally scheduled to fly the second mission after Garneau sometime in 1986, his flight was affected by the *Challenger* disaster disaster and he did not go until many years later. From 1987 to 1993, MacLean was the Program Manager for the Advanced Space Vision System (ASVS), a computer-based camera system designed to provide guidance data that enhances the control of both Canadarm and Canadarm2. From 1988 to 1991 he also assumed the role of Astronaut Advisor to the Strategic Technologies in Automation and Robotics (STEAR) Program. MacLean flew onboard Space Shuttle Columbia as a Payload Specialist on STS-52 in October 1992. During this mission, he performed a set of seven experiments known as CANEX-2, which included an evaluation of the Space Vision System. MacLean was the Chief Science Advisor for the International Space Station from 1993 until 1994, when he was appointed Director General of the Canadian Astronaut Program for two years. In August 1996, he began training as a mission specialist at the Johnson Space Center in Houston, Texas. After successfully completing basic training in 1998, he continued with advanced training whilst fulfilling technical duties in the NASA Astronaut Office Robotics Branch. MacLean served as CapCom (Capsule Communicator) at the Johnson Space Center for both the International Space Station and the Shuttle programs. Selected for another space assignment he flew as a mission specialist aboard STS-115 in September 2006, becoming the second Canadian to walk in space during this mission. He subsequently retired from astronaut duties to be appointed president of the Canadian Space Agency in September 2008, a post he held until February 2013.

Money, Ken. Born January 4, 1935 in Toronto, Ontario, Money attended primary and secondary schools in Toronto and Noranda, Quebec. He received a Bachelor's Degree in Physiology and Biochemistry in 1958, a Master's Degree in Physiology in 1959 and a Doctorate in Physiology in 1961 from the University of Toronto. Concurrently, he enrolled in the Royal Canadian Air Force, attended flight training, and earned his Pilot's Wings at the Advanced Flying School in Portage La Prairie Manitoba in 1957. An accomplished pilot with thousands of logged hours in T-33, F-86, C-45, and Otter aircraft, he was also qualified to fly rotary aircraft such as the Kiowa Helicopter. Dr. Money was the Senior Scientist at the Defence and Civil Institute of Environmental Medicine (DCIEM) of the Department of National Defence (DND) in Toronto. During his tenure, which began in 1961, he has made numerous contributions to knowledge in vestibular physiology and subsequently wrote more than ninety scientific publications related to the field. He made advances in alleviating the problems of motion sickness and pilot disorientation in flight, and was a major contributor to the development of the Malcolm Horizon, an aircraft instrument that provided orientation information to peripheral vision. From 1962 onwards he worked frequently in a variety of capacities with NASA scientists in the area of space motion sickness and orientation. Money graduated from the National Defence College at Kingston in 1972, and was subsequently selected for the first Canadian astronaut group in December 1983. Assigned as a life sciences payload specialist, he was a co-investigator in the vestibular experiments conducted on the NASA Spacelab 1, D1, SLS-1, and STS-41-G missions. Awaiting a chance for spaceflight, he completed the Accelerated Free Fall parachute course at the Spaceland Parachute Center in Houston in 1988. Assigned as the backup to Dr. Roberta Bondar on mission STS-42 to support the IML-1 objectives, he later resigned from the astronaut corps after its successful completion in 1992.

Payette, Julie. Born October 20, 1963 in Montreal, Quebec, Payette received an International Baccalaureate from the United World College of the Atlantic in South Wales, UK, in 1982, a Bachelor of Engineering, – Electrical from McGill University, Montreal, in 1986 and Master of Applied Science - Computer Engineering from the University of Toronto in 1990. Before joining the space program, Payette conducted research in computer systems, natural language processing, automatic speech recognition and the application of interactive technologies in outer space. She was a system engineer with IBM Canada from 1986 to 1988; a research assistant at the University of Toronto from 1988 to 1990; and a visiting scientist at the IBM Research Laboratory, in Zurich, Switzerland in 1991. She was a research engineer for the Speech Research Group, Bell-Northern Research/Nortel, Montreal, when the Canadian Space Agency selected her to join the second astronaut class. After undergoing basic training in Canada, she worked as a technical advisor for the Mobile Servicing System to be installed on the ISS. In 1993, Payette established the Human-Computer Interaction Group at the Canadian Astronaut Program and served as a technical specialist on the NATO International Research Study Group (RSG-10) on speech processing from 1993 to 1996. In preparation for a space mission assignment, Payette obtained her commercial pilot licence, studied Russian, and logged 120 hours as a research operator on board reduced gravity aircraft. In April 1996, Payette was certified as a one-atmosphere, deep-sea diving suit operator. Payette also obtained her captaincy on the CT-114 military jet at the Canadian Forces Base Moose Jaw, Saskatchewan in February 1996. She obtained her military instrument rating in 1997. Payette subsequently logged more than 1100 hours of flight time, more than half of it on high-performance jet aircraft. Payette reported to the NASA Johnson Space Center in Houston, Texas in August 1996. She completed initial astronaut training in April 1998 and was assigned to work on technical issues in robotics for the Astronaut OfficeShe flew on Space Shuttle Discovery from 27 May to 6 June 1999 as a crewmember of STS-96. During the mission, the crew performed the first manual docking of the Shuttle to the International Space Station (ISS), and delivered four tons of supplies to the station. Payette served as a mission specialist, was responsible for the station systems and operated the Canadarm robotic arm while in orbit. She became the first Canadian to participate in an ISS assembly mission and to board the new space station. Following this flight, Payette represented the Canadian astronaut corps at the European and Russian space agencies from September 1999 to December 2002, where she supervised procedure development and equipment verification for the International Space Station Flight Program. Payette was appointed Chief Astronaut for the Canadian Space Agency in 2002. As a technical assignment, she also worked as a CAPCOM (Capsule Communicator) at the Mission Control Center in Houston. In July 2009, she flew as a Mission Specialist aboard STS-127, becoming the last Canadian to fly aboard the space shuttle. Payette retired from the CSA in July 2013 to pursue other careers in the civilian sector.

Saint-Jacques, David. Born January 6, 1970 in Quebec City, Canada, and raised in Saint-Lambert near Montreal, Saint-Jacques earned a Bachelor of Engineering in Engineering Physics from École Polytechnique de Montréal, Canada in 1993 and a Ph.D. in astrophysics from Cambridge University, UK, in 1998. He earned his M.D. from Université Laval in Quebec City, Canada in 2005, and completed his family medicine residency at McGill University in Montreal, Canada in 2007, where his training focused on first-line, isolated medical practice. Prior to joining the Canadian Space Program, Saint-Jacques was a medical doctor and the Co-chief of Medicine at Inuulitsivik Health Centre in Puvirnituq,

Nunavik, an Inuit community on Hudson Bay. He is an Adjunct Professor of Family Medicine at McGill University and has also worked as a Clinical Faculty Lecturer for the university's Faculty of Medicine, supervising medical trainees in Nunavik. Saint-Jacques began his career as a biomedical engineer, working on the design of radiological equipment for angiography at Lariboisière Hospital in Paris, France. During his graduate studies in astrophysics he developed adaptive optics and interferometry systems for the Cambridge Optical Aperture Synthesis Telescope and the William Hershel Telescope in the Canary Islands. His postdoctoral research included the development of the Mitaka infrared interferometer array in Japan and the Subaru telescope adaptive optics system in Hawaii (1999 to 2001), after which he joined the Astrophysics group at Université de Montréal. His international experience also includes engineering study and work in France and Hungary and medical training in Lebanon and Guatemala. Saint-Jacques was selected as an astronaut candidate by the CSA in May 2009 and moved to Houston to be one of fourteen members of the 20th NASA astronaut class. In 2011, he graduated from Astronaut Candidate Training, which included International Space Station (ISS) systems and procedures, spacewalks, robotics, T38-N flight training, field geology training, Russian language, and wilderness summer, winter and water survival training. Since graduation, he has been continuously training to maintain his skills as well as taking part in various geology, glaciology, speleology and other scientific expeditions such as the underwater NEEMO 15 mission and the underground CAVES 2013 mission. After completion of basic training, he was first assigned to the Robotics Branch of the NASA Astronaut Office, then successively acted as Support Astronaut for ISS Expedition 35/36; Lead Capsule Communicator for ISS Expedition 38; Deputy Capcom for Cygnus-1 and Cygnus-2 ISS resupply missions; and Lead Capcom for the Cygnus-3, Cygnus-4 and SpaceX-6 ISS resupply missions. He is currently a Capcom instructor and supports the Visiting Vehicle Operations in the ISS Integration Branch of the NASA Astronaut Office. At the time of writing he has been assignment to ISS Expedition 58/59, scheduled to fly to the ISS in November 2018.

Stewart, Robert. Born December 28m 1954, he graduated from the University of Toronto with a Bachelor of Science degree in Physics and Mathematics and, later, a Ph.D. in geophysics from the Massachusetts Institute of Technology. He obtained industry experience with Chevron, Arco's research centers, as well as Veritas Software Ltd. During this time, Stewart became one of the leading researchers in vertical seismic profiling, borehole geophysics, and tomography. . In 1987, he was chosen as the Chair in Exploration Geophysics at the University of Calgary. During his time as Chair, Stewart initiated the Consortium for Research in Elastic Wave Exploration Seismology (CREWES). He applied to join the second group of Canadian astronauts and was successful, being selected among a group of four new candidates in June 1992. Two weeks after being chosen, however, Stewart resigned because of uncertainty about his future career path and the possibility that he would not be assigned to a future shuttle flight. Despite his declining the appointment to the Canadian Space Agency, Stewart continued to research the connection between oil fields and meteorite impact, and he subsequently supported the Canadian Space Agency's Martian simulations project on Devon Island. He was replaced by Michael McKay.

Thirsk, Robert 'Bob'. Born August 17, 1953 in New Westminster, British Columbia, Thirsk received a Bachelor of Science degree in Mechanical Engineering from the University of Calgary in 1976, a Master of Science in Mechanical Engineering from the

Massachusetts Institute of Technology in 1978, a Doctorate of Medicine from McGill University in 1982, and a Master of Business Administration from the MIT Sloan School of Management in 1998. He was in the family medicine residency program at the Queen Elizabeth Hospital in Montreal when he was selected in December 1983 for the Canadian Astronaut Program. He began astronaut training in February 1984 and served as backup payload specialist to Marc Garneau for the October 1984 space shuttle mission STS-41G. He was subsequently involved in various CSA projects including parabolic flight campaigns and mission planning. He served as crew commander for two space mission simulations: the seven-day CAPSULS mission in 1994, at Defence Research and Development Canada in Toronto, and the eleven-day NEEMO 7 undersea mission in 2004 at the National Undersea Research Center in Key Largo, Florida. He led an international research team investigating the effect of weightlessness on the heart and blood vessels. He also worked with educational specialists in Canada to develop space-related curriculum for grade school students. In June and July 1996, Thirsk flew as a payload specialist aboard STS-78, which included the Life and Microgravity Spacelab (LMS). During this seventeen-day flight aboard *Columbia*, he and his six crewmates performed 43 international experiments devoted to the study of life and materials sciences. The life science experiments investigated changes in plants, animals, and humans under space flight conditions. The materials science experiments examined protein crystallization, fluid physics and high-temperature solidification of multiphase materials in a weightless environment. In 1998, Thirsk was assigned to NASA's Johnson Space Center in Houston to pursue mission specialist training. At the same time within the NASA Astronaut Office, Thirsk also served as a capsule communicator for the International Space Station (ISS) program. In 2004, Thirsk trained at the Yuri Gagarin Cosmonaut Training Centre near Moscow and became certified as a Flight Engineer for the Russian *Soyuz* spacecraft. He served as backup Flight Engineer to ESA astronaut Roberto Vittori for the *Soyuz-10S* taxi mission to the ISS in April 2005. During the ten-day mission, Thirsk worked as Crew Interface Coordinator at the Columbus Control Centre in Germany. Thirsk then returned to the Johnson Space Center in Houston to begin his own ISS Expedition crew training. Thirsk was subsequently selected as the first Canadian to join a permanent ISS crew, and flew aboard the space station from May to December 2009 as part of the ISS Expedition 20/21 crews. He later resigned from the astronaut corps and retired from the CSA in 2012 to pursue a new career in health research and academia.

Tryggvason, Bjarni. Born September 21, 1945 in Reykjavik, Iceland, he emigrated to Canada and later obtained a Bachelor of Applied Science in Engineering Physics from the University of British Columbia in 1972. He completed postgraduate work in engineering with a specialization in applied mathematics and fluid dynamics at the University of Western Ontario, later becoming a meteorologist with the cloud physics group at Meteorlogic Service Canada in Toronto in 1972. Later, he became a research associate in industrial aerodynamics at the Boundary Layer Wind Tunnel Laboratory at the University of Western Ontario from 1974 to 1979. He then went to be a guest research associate at Kyoto University, Japan, in 1979 and at James Cook University of North Queensland, in Townsville, Australia in 1980. He was a lecturer in Applied Mathematics at the University of Western Ontario from 1980 to 1982. In 1982 Tryggvason was appointed as Research Officer at the Low Speed Aerodynamics Laboratory at the National Research Council of Canada (NRC) and also lectured at the University of Ottawa and at Carleton University. Selected as one of the original six Canadian astronaut candidates in December 1983, he was assigned as a payload specialist and later served as backup to Steve MacLean on STS-52. Tryggvason served as project engineer for the Space Vision System (SVS) Canadian Target Assembly (CTA), which was also deployed during that mission. In addition to his mission assignment, he was the principal investigator for a number of other projects, including the development of the Large Motion Isolation Mount (LMIM) which flew numerous times on NASA KC-135 and DC-9 aircraft, and the Microgravity vibration Isolation Mount (MIM) which operated on the Russian space station *Mir* from April 1996 until January 1998 to support several Canadian and U.S. experiments in material science and fluid physics. He was the originator and technical director during the early development phase of the Microgravity Vibration Isolation Subsystem (MVIS) that the CSA developed for the European Space Agency Fluid Science Laboratory for the ISS. Assigned to his own spaceflight, Tryggvason served the payload specialist aboard *Discovery* on STS-85, which flew from August 7–19, 1997. His primary role on this mission was to test MIM-2 and perform fluid science experiments designed to examine sensitivity to spacecraft vibrations, in order to develop a better understanding of the need for systems such as the MIM on the ISS. In August 1998, he was invited to take part in NASA mission specialist training held at the Johnson Space Center. Following completion of mission specialist training, his NASA duties included serving as a crew representative for the Shuttle Avionics Integration Laboratory (SAIL), which was used to test shuttle flight software prior to onboard use. He also supported integrated simulations on the ISS Training Facility, and served as a CSA representative on the NASA Microgravity Measurement Working Group and on the ISS Microgravity Analytic Integration Team. From mid 2001 to 2003 he worked in the private sector while on leave from the CSA. He returned temporarily to work at the CSA in 2004, finally retiring from service in June 2008.

Williams, Dafydd (David). Born May 16, 1954 in Saskatoon, Saskatchewan, Dave Williams graduated from McGill University, Montréal, Quebec, with a Bachelor of Science, Major in Biology in 1976. He earned a Master of Science degree from the Physiology Department, and a Doctorate of Medicine and a Master of Surgery from the Faculty of Medicine, McGill University, in 1983. In 1985, he completed a residency in family practice in the Faculty of Medicine, University of Ottawa. He also obtained a fellowship in emergency medicine from the Royal College of Physicians and Surgeons of Canada, following completion of a residency in emergency medicine at the University of Toronto in 1988. He was by then a Fellow of the Royal College of Physicians and Surgeons and the College of Family Physicians of Canada. Williams pursued postgraduate studies in advanced invertebrate physiology at the Friday Harbour Laboratories at the University of Washington, Seattle, but his interests shifted to vertebrate neurophysiology when, for his Master's thesis, he became involved in basic science research on how adrenal steroid hormones modify the regulation of sleep–wake cycles. While working in the Neurophysiological Laboratories at the Allan Memorial Institute for Psychiatry, Williams assisted in clinical studies of slow wave potentials within the central nervous system. In 1988, Williams became an emergency physician with the Department of Emergency Services at Sunnybrook Health Science Centre, while also lecturing with the Department of Surgery at the University of Toronto. He served as a member of the Air Ambulance Utilization Committee with the Ministry of Health in Ontario, both as an academic emergency physician and later as a representative of community emergency physicians. From 1989 to 1990, Williams served as an emergency physician with the Emergency Associates of Kitchener, Waterloo and as the medical director of the Westmount Urgent Care Clinic. In 1990, he returned to Sunnybrook as medical director of the Advanced Cardiac Life Support Program and also as the coordinator of postgraduate training in emergency medicine. Subsequently, Williams became the Director of the Department of Emergency Services at Sunnybrook Health Science Centre and assistant professor of Surgery at the University of Toronto. In June 1992, the Canadian Space Agency selected Williams as part of its second class of astronauts. He completed basic training, and in May 1993 was appointed manager of the Missions and Space Medicine Group within the Canadian Astronaut Program. His assignments included supervising the implementation of operational space medicine activities for the Canadian Astronaut Program Space Unit Life Simulation (CAPSULS) Project. During this seven-day simulated space mission, which was conducted at the Defence R&D Canada, Toronto (formerly DCIEM), Williams was the principal investigator of a study to evaluate the initial training and retention of resuscitation skills by non-medical astronauts. He was also one of the crewmembers and the crew medical officer. In January 1995, Williams was selected to join the international class of NASA mission specialist astronaut candidates. He reported to the Johnson Space Center in March 1995 for a year of training and evaluation. Following his successful completion of this training in May 1996, he was assigned to the Payloads and Habitability Branch of the NASA Astronaut Office. As a representative of the Office, he participated in the JSC Institutional Review Board and Science Merit Review Committee, the Independent Advisory Team for the International Space Station Crew Health Care System (CHeCs), the JSC Radiation Constraints Panel and was involved in the development of the Human Research Facility. In April 1998, Dave Williams flew aboard Space Shuttle *Columbia* on STS-90. During the sixteen-day flight, called Neurolab, the seven-person crew served as both experiment subjects and operators for 26 individual life science experiments. These experiments, dedicated to the advancement of neuroscience research, focused on the effects of microgravity on the brain and the nervous system. Williams also functioned as the crew medical officer, the flight engineer during the ascent phase, and was trained to perform contingency spacewalks. From July 1998 until September 2002, Williams held the position of Director of the Space and Life Sciences Directorate at the Johnson Space Center in Houston. He concurrently held a six-month position as the first deputy associated administrator for crew health and safety in the Office of Space Flight at NASA Headquarters in 2001. In addition to these assignments, Williams continued to take part in astronaut training to maintain and further develop his skills. In October 2001, he became an aquanaut through his participation in the joint NASA-NOAA (National Oceanic and Atmospheric Administration) NEEMO-1 mission, a training exercise held in Aquarius, the world's only underwater research laboratory. During this seven-day exercise, Williams became the first Canadian to have lived and worked in space and in the ocean. In 2006, Dave Williams took the lead of the NEEMO-9 expedition as the crew commander, which was dedicated to assess new ways to deliver medical care to a remote location, as would be done in a long space flight. Following this he was selected as a mission specialist for flight STS-118/13A.1 and returned to space in August 2007. During the 11-day mission to add a truss segment and relocate solar arrays on the International Space Station, Williams performed three spacewalks, making him the Canadian record holder for time spent outside a spacecraft. Williams later retired from the CSA in March 2008 to return to a career a health research.

Select Bibliography

Archival Sources/Library and Archives Canada/Government Record Groups

Department of Industry (RG20)
National Defence (RG24)
Foreign Affairs (RG25)
Department of Communications (RG97)

Manuscript Groups

J. Alphonse Ouimet (MG 30 E481)
John H. Chapman Fonds (MG31 J43)

Canadian Museum of Science and Technology, Ottawa, Ontario

Files on space science, communications, and industries

Magellan Aerospace Ltd. (Formerly Bristol Aerospace)

Anon. *Black Brant Rocket Systems.* Winnipeg: Bristol Aerospace, 1968.

Other Online Sources

Friends of CRC/Les Amis du CRC [Communications Research Centre]. Accessed on January 21, 2015 at http://www.friendsofcrc.ca

Visual Documentary Sources

Buttignol, Rudy and Sobelman, David (directors). *Space Pioneers: A Canadian Story.* 52mins. 1988.

© Springer International Publishing AG 2017
A.B. Godefroy, *The Canadian Space Program*, Springer Praxis Books,
DOI 10.1007/978-3-319-40105-8

Canadian Broadcasting Corporation Archives

Marc Garneau interviews (several).

Newspapers and Journals

Canadian Aeronautics and Space Journal
Canadian Space Gazette
Globe & Mail (Toronto)
Journal of the British Interplanetary Society
National Post (Toronto)
Online Journal of Space Communication
Science Dimension
Spaceflight
Space News
Space Policy Journal
The Ottawa Citizen

Government of Canada Publications

Anon. *Annual Report 1976: Interdepartmental Committee on Space*, Ottawa: Queen's Printers, 1976.

Anon. *A Space Program for Canada.* Science Council of Canada Report No.1. Ottawa: Queen's Printer, July 1967.

Bennett, S.L. *Strategic Command, Control, and Communications: Capabilities, Doctrine, and Vulnerability.* Ottawa: Operational Research and Analysis Establishment, September 1983.

Canadian Armament Research and Development Establishment – General Information. CARDE Technical Note 1421/61 (Unclassified), September 1961.

Chapman, John H., Forsyth, P.A., Lapp, P.A., and Patterson, G.N. *Upper Atmosphere and Space Programs in Canada: Special Study No.1.* Ottawa: Science Council of Canada Secretariat, February 1967.

Department of Communications. *Alouette.* Ottawa: Queen's Printer, 1970.

---. *Alouette 1: Canada's First Venture into Space.* Ottawa: Information Services Booklet, June 1974.

Department of External Affairs. *Communications Satellites: The Canadian Experience.* Ottawa: Minister of Supply and Services Canada, 1980.

---. *Satellites: The Canadian Experience.* Ottawa: Ministry of Supply and Services, 1984.

Department of National Defence. *An Outline of the Nature and Organization of the Defence Research Board.* Ottawa: DRB, March 1961.

Department of Reconstruction and Supply. *Research and Scientific Activity, Canadian Federal Expenditures, 1938–1946.* Ottawa: King's Printer, 1947.

Demorest, C.P. *Black Brant IIA Preparation and Launching Data.* CARDE Technical Note 1728/66. Valcartier, Quebec, August 1966.

Drury, Hon. C.M. *White Paper on a Domestic Satellite Communications System for Canada*. Ottawa: Queen's Printer, 1968.

Florida, C.D. *The ISIS Satellites*. DRTE Technical Note No.619. Ottawa: Department of Communications – Communications Research Centre, April 1969.

Gelly, A. and Tardiff, H.P. *Defence Research Establishment Valcartier, 1945–1995: 50 Years of History and Scientific Progress*. Ottawa: Minister of Public Works and Government Services Canada, 1995.

Ghent, J. Canadian *Government Participation in International Science and Technology*. Ottawa: Norman Patterson School of International Affairs, 1979.

Godefroy, Andrew B. *Allies in Orbit: Canadian-American Defence Cooperation Through Space*. Ottawa: Department of National Defence, Directorate of Space Development, 2000.

Ministry of State, Science and Technology. *Canada in Space*. Ottawa: Minister of Supply and Services, 1978 and 1982.

National Research Council. *Annual Report*. 1943–44, 1944–45, 1945–46.

---. "Space and Upper Atmosphere Programs (Research) in Canada", *Annual Report, 1969–1980*.

Pennie, A.M. *Defence Research Board: Defence Research Northern Laboratory, 1947–1965*. Report No. DR179, Ottawa: Defence Research Board, April 1966.

Shaw, S.B. *The Mechanical Systems Design and Performance of Four DRTE Black Brant II Rockets*. Defence Research Telecommunications Establishment Report No. 1134, Ottawa: Defence Research Board, April 1964.

Standing Committee on Research, Science, and Technology. *Canada's Space Prgoram: A Voyage to the Future*. Ottawa: House of Commons, June 1987.

Canadian Space Agency Publications, St. Hubert, Quebec

International Space Station Commercial Utilization Policy. July 2001.

The Canadian Space Agency: Building on Our Past, Encompassing Our Future. Montreal: Canadian Space Agency, 1999.

The Canadian Space Program: A New Era for Canada in Space. St. Hubert: Canadian Space Agency, 2000.

European Space Agency Publications

European Space Agency and Canadian Space Agency, Europe and Canada: Partners in Space. ESA Publications Division, June 2000.

National Aeronautics and Space Administration (NASA) Publications

Space Shuttle Mission Press Kits
    STS-115: Return to Assembly
    STS-118: Build the Station, Build the Future

International Space Station Press Kits

> *Expedition 19 and 20: Full Partners*
> *Expedition 21 and 22: Assembling Science*
> *Expeditions 32-33-34: A Beehive of Activity*
> *Expedition 35–36 (no descriptor)*

Published Sources, Monographs, and Other Documents

Adams, J. *Bull's Eye: The Assassination and Life of Super Gun Inventor Gerald Bull*. New York: Times Books, 1992.

Almond, J, Franklin, C.A., and Warren, E.S. "A Perspective on the Canadian Satellite Program", *Canadian Electrical Engineering Journal*. 1:1 (1976); 47–60.

Alway, Peter. *Rockets of the World – Third Edition*. Ann Arbor: Saturn Press, 1999.

Anderson, Hugh J. "Ernest A. LeSueur: Pioneer Canadian Chemical Engineer", *Journal of Chemical Education*, 72 (May 1995); 390–393.

Anderson, Lars and Andersson, Johnny. "Development of the S19 Guidance System for the Reduction of Sounding Rocket Dispersion", *Proceedings, AIAA 4th Sounding Rocket Technology Conference*. Boston: AIAA, June 23–25, 1976.

Anon. "Big Eye in the Sky", *The Roundel*, (October 1962); 29.

Anon. "DRB Pioneers Missile Study". *Canadian Army Journal*, 15:3 (Summer 1961); 54.

Anon. "Defence Research Board Tests: Rockets Feature 1958 Experiments". *Canadian Army Journal*, 13:1 (January 1959); 17–24.

Anon. "Black Brant: Canadian Bristol Aerojet's Family of Sounding Rockets". *Flight International* (January7, 1965); 14–17.

Armstrong, A.W. 'The RCAF Approach to the Evaluation of a Supersonic Interceptor Armed with Air-to-Air Guided Missiles". *Canadian Aeronautical Journal*, 4:6 (June 1958); 194–199.

Aronsen, L. "Canada's Post War Rearmament". Canadian Historical Association (1981); 175–196.

Aronsen, L. "Planning Canada's Economic Mobilization for War". *American Review of Canadian Studies*, 15 (1985); 38–58.

Atkinson, G. and Watson, M.D. "Canadian Directions in Space Science", *Canadian Aeronautics and Space Journal*, 27:1 (1981); 3–10.

Avery, Donald H. *The Science of War: Canadian Scientists and Allied Military Technology During the Second World War*. Toronto: University of Toronto Press, 1998.

Babbit, J.D. *Science In Canada: Selections from the Speeches of E. W.R. Steacie*. Toronto: University of Toronto Press, 1965.

Babe, Robert E. *Telecommunications in Canada: Technology, Industry, and Government*. Toronto: University of Toronto Press, 1990.

Barrington, R.E. "Canadian Space Activities in the Past Quarter Century", *Canadian Aeronuatics and Space Journal*, 25:2 (1979); 153–169.

Bates, Jason. "Canada Funds Small Satellite Platform Demonstration". *Spacenews* (February 16, 2004); 8.

Beesley, A. et al. "Canada's Contribution to Outer Space Law and Arms Control in Outer Space." In *Space Strategy: Three Dimensions*. Toronto: Canadian Institute of Strategic Studies, 1987.

Bingham, A.L. "Problems Associated With Transpolar Airline Operations". *Canadian Aeronautics and Space Journal*, 8:2 (February 1962); 27–30.

Bland, David (ed.) *Canada's National Defence, Volume 1: Defence Policy*. Kingston: School of Policy Studies, Queen's University,1997.

Blevis, Bert. "The Pursuit of Equality: The Role of the Ionosphere and Satellite Communications in Canadian Development", in Butrica, Andrew J. (ed.). *Beyond the Ionosphere: Fifty Years of Satellite Communication*. Washington D.C.: National Aeronautics and Space Administration, 1997.

Blumle, L.J. et. al. "The National Aeronautics and Space Administration Topside Sounder Program". *Vol.II: Space Technology*. Washington: U.S. Government Printing Office, 1963.

Bondar, Barbara and Bondar, Roberta. *On the Shuttle: Eight Days in Space.* Toronto: Maple Tree Press, 1995.

Bondar, Roberta. *Touching the Earth.* Toronto: Key Porter Books, 1994.

Brown, J.J. *Ideas in Exile*. Toronto: McLelland and Stewart, 1967.

Bryden, J. *Deadly Allies: Canada's Secret War, 1937–1947*. Toronto: McClelland and Stewart, 1989.

Bulkeley, Rip. *The Sputnik Crisis and Early United States Space Policy: A Critique of the Historiography of Space*. Bloomington: Indiana University Press, 1991.

Bull, Gerald V. "Some Aerodynamic Studies in the CARDE Aeroballistics Range". *Canadian Aeronautical Journal*, 2:5 (May 1956); 154–157.

Bull, H. *The Control of the Arms Race: Disarmament and Arms Control in the Missile Age*. London: Weidenfeld and Nicolson, 1961.

Burrows, W.E. *This New Ocean: The Story of the First Space Age*. New York: Random House, 1998.

Cameron, I.R. "Manufacture and Testing of Black Brant Engines". *Canadian Aeronautical Journal*, 7:2 (February 1962); 61–66.

---. "Programmes in Rocket Propulsion at CARDE." *Canadian Aeronautics and Space Journal*, 9:12 (December 1963); 307–312.

Chapman, I. and Gibbons, M. "Innovation and the Senate Report on Science Policy." *Journal of Canadian Studies*, 13:1 (Spring, 1978); 30–37.

Chapman, Sidney. *IGY: Year of Discovery*. Ann Arbor, MI: University of Michigan Press, 1959.

Carruthers, D.A. "A Canadian Domestic Satellite Communication System". *Telecommunications Journal*, 35:8 (1968); 396–402.

Chapman, John. "Why Satellite Communications in Canada". *In Search/En Quête,* 6:2 (Ottawa: Department of Communications, Spring 1979); 10.

Chinnick, R.F. "Upper Atmosphere and Space Research in Canada." *Canadian Aeronautics and Space Journal*, 8:10 (October 1962); 198–213.

Clauser, F.H. "Flight Beyond the Earth." *The Roundel*, 1:1 (November 1948); 10–14.

Cooper, Henry S.F. *Before Lift-Off: The Making of a Space Shuttle Crew*. Baltimore: The Johns Hopkins University Press, 1987.

Coughlin, Flt. Lt. T.G. "Alouette II – Canada's Second Satellite Probes the Mysteries of Outer Space." *Sentinel*. 2:1 (January–February 1966); 8–13.

Currie, Balfour W. "Space Research – The Canadian Background". *Physics in Canada*, 16:2 (Summer 1960); 7–17.

Daniell, Robert. "The Space Station Remote Manipulator System". *SPAR Journal of Engineering and Technology*, 1:1 (May 1992); 1–6.

Davies, M.E. and Harris, W.R. *RAND's Role in the Evolution of Balloon and Satellite Observation Systems and Related U.S. Space Technology*. Santa Monica: RAND, 1988.

Debresson, C. "Have Canadians Failed to Innovate? The Brown Thesis Revisited." *History of Science and Technology in Canada Bulletin*, 6 (January 1982); 10–23.

Debresson, C. and Murray, B. *The Supply and Use of Technological Innovation in Canada*. Ottawa: Cooperative Research Unit on Science and Technology, 1983.

Dewitt, D.B. and Kirton, J.J. *Canada as a Principal Power*. Toronto: John Wiley and Sons, 1983.

Rt. Hon. Diefenbaker, John G. *One Canada: Memoirs of the Right Honourable John G. Diefenbaker*. 3 Vols. Toronto: MacMillan of Canada, 1976.

Doern, G. Bruce. *Science and Politics in Canada*. Montreal and London: McGill-Queen's University Press, 1972.

Doern, G. Bruce, and Brothers, James A.R. "Telesat Canada", in Tupper, Allan, and Doern, G.B. (eds.). *Public Corporations and Public Policy in Canada*. Montreal: The Institute for Research in Public Policy, 1981; 225–228.

Doetsch, K.H. and Middleton, J.A. "Canada's Space Station Program". *Canadian Aeronautics and Space Journal*, 33:4 (December 1987); 218–223.

Donaghy, Greg. *Tolerant Allies: Canada and the United States, 1963–1968*. Montreal and Kingston: McGill-Queen's University Press, 2002.

Dotto, Lydia. *A Heritage of Excellence: 25 Years at Spar Aerospace Limited*. Toronto: Spar Aerospace Ltd., 1992.

---. *Canada in Space*. Toronto: Irwin Publishing, 1987.

---. *The Astronauts: Canada's Voyageurs in Space*. Toronto: Stoddard Publishing Co., 1993.

Drury, C.M. "International Aspects of Possible Future Canadian Participation in Space Programs". *Canadian Aeronautics and Space Journal*, (February 1971); 33–35.

Edel, Howard et al. "Canadian RADARSAT Program", *Quest: The History of Spaceflight Quarterly*, 11:4 (2004); ?-?.

Eggleston, W. *National Research in Canada: The NRC 1916–1966*. Toronto: Clarke Irwin, 1978.

Evans, W.M. (Mac). 'The Canadian Space Program – Past, Present, and Future". *Canadian Aeronautics and Space Journal*, 50:1 (March 2004); 19–32.

Eyre, F.W. "The Development of Large Bore Gun Launched Rockets". *Canadian Aeronautics and Space Journal*, (April 1966); 143–149.

Fia, A.W. "Canadian Sounding Rockets: Their History and Future Prospects". *Canadian Aeronautics and Space Institute Journal*, 20:8 (October, 1974).

Fordyce, A.M. "How It All Started: The Goforth Paper". *Canadian Defence Quarterly*,1:4 (Spring 1972); 15–16.

Franklin, C.A. "John Herbert Chapman, 1921–1979". In *Proceedings of the Royal Society of Canada*. Series IV. Vol.18 (1980); 67–72.

---. Transcript of presentation made at the IEEE International Milestone in Engineering Ceremony. Shirley Bay, Ottawa, May 13, 1993.

Gainor, C. *Arrows to the Moon: AVRO's Engineers and the Space Race*. Burlington: Apogee Books, 2001.

---. *Canada in Space: The People & Stories Behind Canada's Role in the Exploration of Space*. Alberta: Folklore Publishing, 2006.

---. "The Chapman Report and the Development of the Canada's Space Program". *Quest: The History of Spaceflight Quarterly.*, 10:4 (2003); 6–22.

Galt, George. "Unscrambling the Future". *Saturday Night*, (October 1983); 15–26.

Garigue, Philippe. *Science Policy in Canada*. Montreal: The Private Planning Association of Canada, 1972.

Gingras, Y. *Physics and the Rise of Scientific Research in Canada*. Montreal & Kingston: McGill-Queen's University Press, 1991.

Gingras, Y. and Jarrell, R. "Building Canadian Science: The Role of the National Research Council". *Scientia Canadensis*, 15:2 (1991).

Godefroy, Andrew B. "Canada's Early Space Policy Development, 1958–1974." *Space Policy Journal*, 19 (2003); 137–141.

---. *Defence and Discovery: Canada's Military Space Program, 1945–1974*. Vancouver: University of British Columbia Press, 2004.

---. "From Alliance to Dependence: Canadian-American Cooperation Through Space, 1945–1999". MA thesis. Kingston: Royal Military College of Canada, 1999.

Goodspeed, Lieutenant Colonel D.J. *A History of the Defence Research Board of Canada*. Ottawa: Queen's Printer, 1958.

Granatstein, J.L. *A Man of Influence: Norman A. Robertson and Canadian Statecraft, 1929–1968*. Toronto: Deneau Publishers, 1981.

---. *Canada 1957–1967: The Years of Uncertainty and Innovation*. Toronto: McClelland and Stewart, 1986.

Grant, Dale. *Wilderness of Mirrors: The Life of Gerald Bull*. Scarborough: Prentice-Hall Canada Inc., 1991.

Green, J.J. "Science and Defence". *Canadian Aeronautical Journal*, (April 1955); 3–6.

Harland, David M. and Catchpole, John E. *Creating the International Space Station*. New York: Springer-Praxis, 2002.

Hart, John Edward. "Canada and North American Defence, 1940–1965". Masters thesis. University of Oregon, 1967.

Hartz, T.R and Paghis, I. *Spacebound*. Ottawa: Department of Communications, Minister of Supply and Services Canada, 1982.

Hendrick, Air Vice-Marshal M.M. "The RCAF – Technology on Active Service". *Canadian Aeronautical Journal*, 2:6 (June 1956); 181–186.

Heppenheimer, T.A. *Countdown: A History of Spaceflight*. New York: John Wiley & Sons, 1997.

Hilliker, Major Michael. 'The Impact of the Moon Agreement on Canadian Space Policy". POE450 course assignment paper, Royal Military College of Canada, 2003.

Hines, C.O., Paghis, I., Hartz, T.R., Fejer, J.A. *Physics of the Earth's Upper Atmosphere*. Englewood Cliffs N.J.: Prentice Hall, 1965.

Hyndman, J. "National Interest and the New Look". *International Journal*, 26 (Winter 1970–71).

Jackson, J.E., Knecht, R. and Russell, S. "First Results in the NASA Topside Sounder Program". *Volume II: Space Technology, 1959–1962*. (Washington: NASA HQ, 1963); 40–50.

Jackson, F. "Development of the 23 KS20000 Motor for the Black Brant IIB Vehicle". *Canadian Aeronautics and Space Journal*. 11:12 (December 1965); 377–383.

Jakhu, R. 'The Case for Enhanced India-Canada Space Cooperation", *Space Policy*, 25:1 (February 2009); 9–19.

Jarrell, Richard A. *The Cold Light of Dawn: A History of Canadian Astronomy*. Toronto: University of Toronto Press, 1988.

Jelly, Doris. *Canada: 25 Years in Space*. Ottawa: National Museum of Science and Technology, 1988.

Jockel, J. *No Boundaries Upstairs: Canada, the United States, and the Origins of North American Air Defence, 1945–1958*. Vancouver: University of British Columbia Press, 1987.

Jones, Greta. *Science, Politics, and the Cold War*. New York: Routledge, 1988.

Kuhn, M. "Telecommunication in Canada – A Century of Symbiotic Devlopment". *IEEE Communications Magazine*, 22:5 (May 1984); 104–114.

King, A., et. al. *Reviews of National Science Policy: Canada*. Paris: Organization for Economic Cooperation and Development, 1969.

Kirton, John (ed.) *Canada, The United States, and Space*. Toronto: Canadian Institute of International Affairs, 1986.

Koerner, Stephen. "Canada and the Post-War Reparations Programme". Unpublished paper presented at the 2000 Canadian Historical Association Conference.

Kofler, Linda L. "Reach Out and Touch, Canadian Style". *Space World* (September 1986); 21–23.

Lackner, Chris. "Canada's Rocket Man". *Globe & Mail* (July 3, 2004); A8.

Lagowski, R.G. "Attitude Determination for the Anik Satellites". *Canadian Aeronautics and Space Journal*, 23:2 (March–April 1977); 77–87.

Langille, Robert C. and Scott, James C.W. 'The Canadian Defence Research Board Topside Sounder Satellite". *Journal of British I.R.E.* (January 1962); 61–68.

Law, Cecil E., Lindsey, George R., and Grenville, David M. "Perspectives in Science and Technology: The Legacy of Ormond Solandt", In Proceedings of a Symposium held at the Donald Gordon Centre, Queen's University, Kingston, Ontario, 8–10 May 1994. Kingston: Queen's Quarterly, 1994.

Law, Cecil E. "Omond McKillop Solandt: Operational Research Pioneer". *INFOR*. 31:2 (May 1993).

Legacy, Denis. "More to Space Than Rocket Science: Canada's Civilian Space Program". *Policy Options* (May 2005); 64–69.

Leslie, Stuart W. *The Cold War and American Science: The Military-Industrial-Academic Complex at MIT and Stanford*. New York: Columbia University Press, 1993.

Lesueur, Ernest A. "Rocketeers". *Canadian Defence Quarterly*, 9:3 (April 1932); 374–382.

Lindsey, George R. *No Day Long Enough: Canadian Science in World War II*. Toronto: Canadian Institute of Strategic Studies, 1997.

Lister, Air Vice Marshal M.D. 'The Evolution and Current Status of NORAD". *The Roundel* (June 1962); 2–8.

Lithwick, N.H. *Canada's Science Policy and the Economy*. Toronto: Methuen, 1969.

Logsdon, J., et. al. (ed.) *Exploring the Unknown – Selected Documents in the History of the U.S. Civil Space Program, Vol II: External Relationships*. Washington: National Aeronautics and Space Administration, 1996.

Lowther, William. *Arms and the Man: Dr. Gerald Bull, Iraq, and the Supergun*. Toronto: Doubleday Canada Limited, 1991.

Maeda, Ken-Ichi and Silver, Samuel (eds). *Progress in Radio Science, 1960–1963*. Space Radio Science Vol. 8. Elsevier, 1965.

Mar, J. "Meteoroid Impact on the Topside Sounder Satellite". *Canadian Aeronautics and Space Journal* (November, 1962); 237–240.

Mar, J. and Warren, H.R. "Structural and Thermal Design of the Topside Sounder Satellite". *Canadian Aeronautics and Space Journal* (September 1962); 161–169.

Mayer, Roy. *Scientific Canadian: Invention and Innovation from Canada's National Research Council*. Vancouver: Raincoast Books, 1999.

McDougall, John N. *The Politics and Economics of Eric Kierans: A Man for All Canadas*. Montreal and Kingston: McGill-Queen's University Press, 1993.

McDougall, W. *The Heavens and Earth: A Political History of the Space Age*. New York: Basic Books, 1985.

McLin, Jon B. *Canada's Changing Defence Policy, 1957–1963: The Problems of a Middle Power in Alliance*. Baltimore: The Johns Hopkins Press, 1967.

McTaggart-Cowan, P.D. "The First Century of the Meteorological Service of Canada", in McIntyre, D.P. (ed.). *Meteorological Challenges: A History*. Ottawa: Information Canada, (1972); 313–319.

Middleton, W.E.K. *Physics at the National Research Council of Canada, 1929–1952*. Waterloo: Wilfrid Laurier University Press, 1979.

---. *Radar Development in Canada: The Radio Branch of the National Research Council of Canada, 1939–1946*. Waterloo: Wilfrid Laurier University Press, 1981.

Morrison, L.C. "South Atlantic Lookout: Canada Assists U.S. Missile Research Program with Infrared Detection Team on Ascension Island". *Roundel* (April 1960); 2–3.

Mowrey, D. and Rosenberg, N. *Paths of Innovation: Technological Change in 20th-Century America*. Cambridge: Cambridge University Press, 1998.

National Science Foundation. *Science, Technology, and Democracy in the Cold War and After: A Strategic Plan for Research in Science and Technology Studies*. New Mexico: Santa Fe: Institute Workshop, 18–20 September 1994.

Needell, Allan A. *Science, Cold War and the American State: Lloyd V. Berkner and the Balance of Professional Ideals*. Washington: National Air and Space Museum, 2000.

Newell, H.E. *Beyond the Atmosphere: Early Years of Space Science*. SP-4211. Washington: National Aeronautics and Space Administration, 1974.

Nicholls, R.V. and Onyszchuk, M. *Reminiscences of a Pioneer Canadian Chemical Engineer, 1890–1952: James Richardson Donald*. Montreal: Department of Chemistry, McGill University, 1989.

Nicolet, Marcel. "Historical Aspects of the IGY". *Eos Transactions*, 64:19 (10 May 1983).

Ohman, L.H. "A Satellite Launch Study Employing Black Brant VB Sounding Rocket Motors as Booster Building Blocks". *Canadian Aeronautics and Space Institute Journal*, 13 (November 1967); 427.

Rt. Hon. Pearson, L.B. *Mike: The Memoirs of the Rt. Hon. Lester B. Pearson*. 3 Vols. Toronto: University of Toronto Press, 1975.

Pennie, A.M. "A Quarter Century of Achievement". *Canadian Defence Quarterly*, 1:4 (Spring 1972); 6–14.

Pugliese, David. "Canadian Report Issues Warning of Possible Space Threats". *Space News* (2000); 4.

---. "Military Wants Way to Attack Satellites". *The Ottawa Citizen* (21 September 2003); A3.

Raboy, Marc. *Missed Opportunities: The Story of Canada's Broadcasting Policy*. Montreal and Kingston: McGill-Queen's University Press, 1990.

Radford, K.J. "Studies of Orbital Rendezvous". *Canadian Aeronautics and Space Journal* (May 1962); 105–111.

Rettie, R.S. "The Churchill Research Range and Canada's Research Range Program". In Partel, G.A. (ed.). *Space Engineering: Proceedings of the Second International Conference on Space Engineering*. Vol. 15 Part IX (Netherlands: Springer, 1970); 650–671.

Richmond, R.D. and Perrier J.F. "The Technology of Guided Missiles and its Effect on Industry". *Canadian Aeronautical Journal*, 3:4 (April 1957); 113–120.

Richter, Andrew. *Avoiding Armageddon: Canadian Military Strategy and Nuclear Weapons, 1950–1963*. Vancouver: University of British Columbia Press, 2002.

---. *The Evolution of Strategic Thinking at the Canadian Department of National Defence, 1950–1963*. Occasional Paper No.38/ CDISP Special Issue No.2. Toronto: Center for International and Strategic Studies, York University, August 1996.

Ridler, Jason Sean. *Maestro of Science: Omond McKillop Solandt and Government Science in War and Hostile Peace, 1939–1956*. Toronto: University of Toronto Press, 2015.

Roberts, A.K. "Development of the 9KS11000 Black Brant III Rocket Engine". *Canadian Aeronautics and Space Journal* (June 1962); 137–143.

Rodzinyak, Major Ken. "Like a Sapphire in the Sky: Canada's Surveillance of Space Project". Assignment paper for War Studies Program, Kingston: Royal Military College of Canada, 2002.

Roper, P. "Seeking a Clearer Channel: Canadian Ventures in Satellite Technology and Nation Building, 1958–1972". PhD dissertation. Ottawa: University of Ottawa, 2003.

Rose, Donald C. "Space Research in Canada". *Canadian Aeronautics and Space Journal*, 10:4 (April 1964); 89–92.

---. "Radio Observations on the Upper Ionized Layer at the Time of the Total Solar Eclipse of August 31, 1932". *Canadian Journal of Research*, 8:1 (January 1933); 15–28.

---. "The International Geophysical Year". *Engineering Journal* (August 1958); 45–60.

Sawyer, Robert. "Anik E: The Next Generation". *Aerospace Canada International* (Jan–Feb 1987); 14–16.

Shanko, Barry. "The MOST for the Least: Canada's Pioneering Space Telescope". *Spaceflight*, 48:1 (January 2006); 14–17.

Shepherd, Gordon and Kruchio, Agnes. *Canada's Fifty Years in Space: The COSPAR Anniversary.* Burlington: Apogee Books, 2008.

Siddiqi, Asif A. *Challenge to Apollo: The Soviet Union and the Space Race, 1945–1974.* Washington: National Aeronautics and Space Administration, 2000.

Simonelli, D. "Cooperation in Space". *European Community* (Jan–Feb 1978).

Slemon, Air Marshal C.R. "Meeting the Potential Soviet Threat From Space". An address to the USAF Association Convention. Las Vegas: Nevada, 1963. Reprinted in *Roundel.* (March, 1963); 2–4.

Smith, M.R. *Military Enterprise and Technological Change.* Cambridge: MIT Press, 1987.

Snow, C.P. *Science and Government.* Cambridge: Harvard University Press, 1961.

, Omond M. "Science and Industry in Canada". *Canadian Aeronautics and Space Journal*, 8:9 (September 1962); 157–159.

Spears, Tom. "Canada's Mars Mission Invites Input, Offers Cash". *National Post* (March 22, 2006); A1–2.

Spencer, Robert A. *Canada in World Affairs, 1946–1949.* Toronto: Macmillan, 1959.

Spruston, Lieutenant Colonel T.A. "Science and Politics: The Evolution of Canadian Space Policy". Masters thesis. Kingston: Royal Military College of Canada, 1976.

---. "The Defence Research Board's Untimely End: What it Means for Military Science". *Science Forum*, 8 (October 1975); 19–21.

Starnes, J. *Closely Guarded: A Life in Canadian Security and Intelligence.* Toronto: University of Toronto Press, 1998.

Stewart, I.A. "The Churchill Research Range and Canada's Satellite Program". *Canadian Aeronautics and Space Journal*,15 (October 1969); 307–313.

St. Germain. "Fire, Ice, and Politics: The Evolution of Domestic Satellite Communications in Canada". n.p.: University of Victoria, n.d. 24pp.

Stokes, Mark. "Canada and the Direct Broadcast Satellite: Issues in the Global Communications Flow". *Journal of Canadian Studies*, 27:2 (Summer 1992); 82–96.

Sutherland, R.J. "Canada's Long Term Strategic Situation". *International Journal*, 17:3 (Summer 1962); 180–199.

Thistle, M. *The Inner Ring: The Early History of the National Research Council of Canada.* Toronto: University of Toronto Press, 1966.

Thomson, Don W. *Skyview Canada.* Ottawa: Energy, Mines, and Resources, 1975.

Tutton, Michael. "NASA Lends Support to Canadian Spacecraft". *Globe & Mail* (February 2, 2007); A8.

Twomey, C. "The McNamara Line and the Turning Point for Civilian Scientist-Advisors in American Defence Policy, 1966–1968". *Minerva*, 37 (1999); 235–258.

Ulman, William A. "Russian Planes Are Raiding Canadian Skies: Special Report on U.S. Air Defenses". *Collier's* (October 16, 1953); 33–45.

Valentine, B. "Obstacles to Space Cooperation: Europe and the Post-Apollo Experience". *Research Policy I* (1971–1972).

VanKoughnett, A.L. and Kendall, D.J.W. "Canadian Directions in Space Science: An Update". *Canadian Aeronautics and Space Journal,* 33:4 (December 1987); 205–210.

Van Steenburg, R. "An Analysis of Canadian-American Defence Economic Cooperation". In Haglund, David (ed.) *Canada's Industrial Base.* Ronald Frye, 1988.

Vardalas, John. *The Computer Revolution in Canada: Building National Technological Competence.* Cambridge: MIT Press, 2001.

---. "From DATAR to the FP-6000 Computer: Technological Change in a Canadian Industrial Context". *IEEE Annals of the History of Computing,* 16:2 (1994), **pages?**.

Watson, G.D. "The Scientific Exploration of Space". *Canadian Aeronautical Journal,* 6:3 (March 1960); 81–89.

Watson, William. "We are Bringing Benefit to Canadians: An Interview with William Macdonald Evans, President of the Canadian Space Agency". *Policy Options* (Jan–Feb 2001); 7–14.

Wein, M. "Cameras on a Black Brant II Rocket". *Canadian Aeronautics and Space Journal* (April 1968); 147–151.

Whitaker, R. and Marcuse, G. *Cold War Canada: The Making of a National Insecurity State, 1945–1957.* Toronto: University of Toronto Press, 1994.

Wilkinson, R.. "Rocket Research in Canada". *Canadian Aeronautical Journal,* 5:4 (April 1959); 138–142.

Wooding, B. and Spruston, T.A. "The Canadian Armed Forces and the Space Mission". *Canadian Defence Quarterly,* 5:2 (Winter 1975); 15–20.

Zimmerman, A.H. "The Role of Science in Defence". ***Canadian Aeronautical Journal,*** **4:9** (September **1958**); **213–216**.

Index

A

Abrahamson, J. LGen., 156
ACE. *See* Atmospheric Chemistry Experiment (ACE)
ACSR. *See* Associate Committee on Space
    Research (ACSR)
ACTORS. *See* Atlantic Canada Thin Organic
    Semiconductors (ACTORS) experiment
Advanced Space Vision System (ASVS), 160,
    162, 211, 253, 299
Advisory Panel for Scientific Policy, 21, 25
Aerospace Industries Association of Canada
    (AIAC), 150, 179, 180
Aikenhead, B., 77, 153
Aircraft
    C-130 transport, 46
    CF-100 Canuck, 27, 35
    CF-105 Arrow, 24, 25
Alberta, 7, 56, 296–298
Alenia Aerospazio, 274
Alouette 2 (ISIS-X)
    II concept, 82, 83
    design, 82
    launch, 79, 85
    on-orbit, 84
Alouette Satellite (Project S-27)
    concept, 63, 65
    construction, 64
    design, 64, 65
    I, 63
    launch, 66, 70–72
    on-orbit, 66
American Carnegie Institution, 9
American National Research Council (ANRC), 14
Anik satellites
    A-1, 123, 126–128
    B, 140, 168
    C3, 169
    D, 140, 168
Aquatic Research Facility (ARF), 216, 217
Arcas rocket, 98
Associate Committee on Space Research (ACSR),
    55–58, 80
ASVS. *See* Advanced Space Vision System (ASVS)
Atlantic Canada Thin Organic Semiconductors
    (ACTORS) experiment, 216
Atmospheric Chemistry Experiment
    (ACE), 267
Auld, M., 44
Avcoat insulation, 53

B

Babbitt, J.D., 32
Baikonur Cosmodrome, 274
Ball Aerospace, 203
Barry, J.N., 132
Bergeron, Laurent A., 293
Berkner, L.V. Dr., 13, 59
Bernath, P. Dr., 267
Black Brant Rocket, 36
    II, 41, 42, 45
    IIA, 38, 41, 42
    III, 42, 45–48
    IIIA, 48
    IVA, 48–50, 52
    VA, 41, 48, 52
    VB, 52, 54, 91
    program (general), 46, 98
    Propulsion Test Vehicle (PTV-I), 41
Blevis, B., 125

© Springer International Publishing AG 2017
A.B. Godefroy, *The Canadian Space Program*, Springer Praxis Books,
DOI 10.1007/978-3-319-40105-8

Boisvert, L.J., 271, 293
Boland, E., 176
Boland Letter. *See* Boland, E.
Bondar, R., 159, 166, 184–186
Booker, H. Dr., 61
Breit, G., 8
Bristol Aircraft (Western) Limited, 38, 43, 44
Bujold, G., 271
Bull, G. Dr., 82, 90, 92

C
Cabinet War Commitee, 5
Cadieux, L. Hon., 108
Cameron, W.M., 32
Campbell, A.G., 31, 32
Canadarm
    RMS (*see* Remote Manipulation System
        (RMS)), 2, 252, 254, 255, 257, 259–263,
        276, 277, 281, 283, 288
    parts-special purpose dexterous manipulator
        (SPDM), 252, 253
Canadian Armament Research and Development
    Establishment (CARDE), 6, 16, 35–38,
    40–42, 45–47, 55, 58, 59, 90, 95, 96
Canadian Army, 6, 35, 44
Canadian Astronaut Program
    first candidates, 287
    initial advertisement 1983, 158–160
    second class, 233, 304
    selection process, 159, 160, 190
    third class, 157
Canadian Broadcasting Corporation (CBC), 106
Canadian Centre for Remote Sensing (CCRS),
    139, 199
Canadian Earth Resources Survey (CERES), 94, 95
Canadian Experiments (CANEX)
    CANEX-2, 191–194, 299
    first set, 166
Canadian Rocket Propulsion Program (CRPP), 35,
    36, 54
Canadian Space Agency (CSA)
    Act, 79, 181
    early attempts to create, 77
    headquarters location controversy, 183
Canadian Space Physiology Experiments (SPE),
    187, 295
Canadian space policy
    early policy (1958-74), 118
    the Five-Year Plan (1980), 167–169
    Interim Space Plan (1985-86), 171
    LTSP-I (1987), 171
    LTSP-II (1994), 239
    national space policy (1974), 121

Space Program Management Framework
    (SPMF)-1990, 238
Canadian space strategy
    1999, 273
    2003, 268, 270
Canadian Target Assembly (CTA), 183, 193, 303
Cape Canaveral, Florida. *See* Kennedy Space
    Center (KSC), Florida
Capsule Communicator (CAPCOM), 219, 287,
    295, 298–302
CARDE. *See* Canadian Armament Research and
    Development Establishment (CARDE)
CARDEPLEX propellant, 38, 40, 44
CBC. *See* Canadian Broadcasting Corporation (CBC)
CCRS. *See* Canadian Centre for Remote Sensing
    (CCRS)
CERES. *See* Canadian Earth Resources Survey
    (CERES)
Challenger disaster, 176, 178, 299
Chapman, J.H. Dr., 11, 60, 67, 70, 80, 83, 123,
    181, 182
Chapman Report, 82, 93, 96, 98, 102, 106, 109, 123
Chapman, S., 13
Cherwinski, W., 162
Chretien, J. Rt. Hon., 233, 268
Churchill Research Range
    establishment of site, 30
    flights, 39
    rocketry program, 19, 39, 96
Churchill, W. Hon., 3
Cold War, 2, 6, 7, 14, 18, 19, 28, 98, 106, 107,
    137, 139, 291
Collin, Art, 180
Columbia Accident Investigation Board, 268
COM DEV Ltd., 125, 203
Commercial Float Zone Furnace (CFZF), 216
Committee on Space Research (COSPAR), 55, 75
Commonwealth Telecommunications
    Organization (CTO), 168
Communications Research Center (CRC), 108,
    125, 128, 129, 131
Communications Technology Satellite (CTS)
    Hermes, 123
Constellations, Big Dipper, ix
Coons, D.O., 77
COSPAS-SARSAT. *See* Search and Rescue
    Satellite (SARSAT)
Cox, J.W., 11, 56
Crippen, R.B., 162, 163
CRPP. *See* Canadian Rocket Propulsion Program
    (CRPP)
Cryogenic Infrared Spectrometers and Telescopes
    for the Atmosphere-Shuttle Pallet
    Satellite-2 (CRISTA-SPAS-2), 230

CSA. *See* Canadian Space Agency (CSA)
CTA. *See* Canadian Target Assembly (CTA)
CTO. *See* Commonwealth Telecommunications
    Organization (CTO)
Cuban Missile Crisis, 76

D
DARA. *See* German Space Agency (DARA)
David Florida Laboratory (DFL), 129, 140, 203,
    206, 274
Davies, F.T., 9, 11, 72
Davies, G., 125
Davies, N.G., 132
Defence and Civil Institute of Environmental
    Medicine (DCIEM) Toronto, 160, 184, 299
Defence Research Board (DRB), 5–7, 9, 10,
    15–17, 19, 21, 23, 24, 26, 28, 29, 31–33,
    37, 41–44, 49, 51, 52, 54, 55, 57–60, 62,
    63, 72–74, 76–79, 86, 87, 90, 95, 96, 98,
    100, 108, 110, 117, 167
Defence Research Chemical Laboratories, 7
Defence Research Electronics Laboratory
    (DERL), 10
Defence Research Northern Laboratory (DRNL),
    7, 16, 39
Defence Research Telecommunications
    Establishment (DRTE), 10, 11, 45, 59–64,
    66, 69–72, 80, 82–84, 86, 88, 100, 108, 125
Department of Communications (DOC), 72, 100,
    108, 110, 122, 123, 125, 131, 149, 167,
    179, 180
Department of Defence Production (DDP), 24, 43,
    57, 83
Department of Defense Research and
    Development Office (DRDO), 39
Department of Energy, Mines, and Resources
    (EMR), 139, 167, 179, 201
Department of External Affairs (DEA), 29, 32, 95,
    109, 111
Department of National Defence (DND),
    108–110, 167, 295, 299
DFL. *See* David Florida Laboratory (DFL)
Diefenbaker, J.G. Rt. Hon., 19, 20, 22, 24, 27, 31, 76
Dimock, B.C. Maj., 143
Doern, B., 22
Doetch, K.-H. Dr., 150
Dominion Observatory, 15, 56
Doré, R. Dr., 194
Dotto, L., 160, 198
DRB. *See* Defence Research Board (DRB)
DRNL. *See* Defence Research Northern
    Laboratory (DRNL)

DRTE. *See* Defence Research Telecommunications
    Establishment (DRTE)
Drury, B. Hon. C.M.
    appointment to Task Force on Space, 103
    tabling of white paper on Domestic Satellite
      Communications, 103, 104, 123
Domestic Satellite Communications, 101,
    104–106, 109, 118, 120–122, 126, 127

E
Earth Observing System (EOS), 215
Earth Radiation Budget Satellite (ERBS), 164
Earth Resources Technology Satellite (ERTS). *See*
    Landsat
Edwards Air Force Base (AFB) California, 154,
    190, 261, 296
Electronics lab (EL), 13, 60, 72, 125
Engineering Support Center's (ESC), 263
Engle, J., 153
Erb, B., 77
European Space Agency (ESA)
    activities, 274
    cooperation with Canada, 113, 140
    creation, 140
Evans, W.M. Dr., 238
Ewart, D., 77
Extendible Support Structure (ESS), 274
Extravehicular Activity (EVA), 144, 198, 243,
    245, 249, 255, 258, 260, 276, 277, 281,
    283, 297

F
Far Ultra-Violet Spectroscopic Explore (FUSE),
    215
Fia, A., 44, 48
Florida, C.D. Dr., 125, 130, 131
Forsyth, P. Dr., 45, 56, 60, 80, 99
Forward, F.A., 22
Fort Churchill. *See* Churchill Research Range
Foulkes, General Charles, 5
Franklin, C., 86, 125

G
Gagarin, Y., 31, 297, 302
Galezowski, S., 77
Garneau, M., 159, 160, 163, 165, 166, 169, 170,
    197, 198, 206, 216–218, 221, 246, 248,
    250, 251, 268, 270, 271, 293–295, 302
Gemini Program
    Gemini Launch Vehicle Directorate, 142

German Space Agency (DARA), 184, 230
Glassco Commission, 76, 78
Glassco, J.G.. *See* Glassco Commission
Global Commercial Communications Satellite
    System (GCCSS), 74
Goddard Space Flight Center (GSFC), 61–64, 71,
    72, 88
Green, H. Hon., 20
Gruno, B., 125
Guided Missile Advisory Committee (GMAC), 35

H
Hadfield, C., 196, 197, 206, 208–210, 213, 219,
    255–258, 260, 261, 281, 290, 291, 296
Haggett, S., 44
Hansen, J. LCol., 290, 294, 297, 298
Harris, T.M. Flt .Lt., 143
Hartz, T.R. Dr., 86
Heaviside, Oliver, 8
Hellyer, P. Hon., 84
Helms, L., 61
Henry, J. Sqn. Ldr., 142
High Altitude Research Project (HARP),
    90, 92, 96
High Energy Project, 21
Hiltz, M., 212
Hinchey, M., 158
Hines, C. Dr., 60, 72
Howe, C.D. Hon., 3, 5
Hubble Space Telescope, 265
Hughes Aircraft, 123, 126, 127

I
IMAX films, 166
Inflatable Antenna Experiment (IAE), 216, 220
Integrated Service and Test Facility (ISTF), 172,
    175, 178, 252
INTELSAT, 74, 109, 126
Interdepartmental Committee on Space (ICS)
    activities, 111, 118
    creation, 115
    predecessor, 117, 239
Interim Canadian Space Council (ICSC).
    See Interdepartmental Committee
    on Space (ICS)
International Extreme Ultraviolet Hitchhiker-02
    (IEH-02), 230
International Geophysical Year, 1957-58 (IGY)
    general, 14
    special committee, 14

International Microgravity Laboratory (IML-1),
    184, 186–188, 190, 294, 299
International Polar Year (IPY)
    1882, 1
    1932, 1, 13
International Satellites for Ionospheric Studies
    (ISIS), 82
    ISIS-A, 82, 87
    ISIS-B, 82
    ISIS-C, 82, 90, 123
    ISIS-X (*see* Alouette 2)
International Space Flight Development Station
    (ISFDS), 28, 30
International Space Station (ISS)
    creation, 280
    Control Moment Gyroscope, 281
    External Stowage Platform 3 (ESP-3),
        280–281
    Japanese Exposed Facility, 285
    Kibo Module, 284, 285, 287
    Microgravity Analysis and Integration Team,
        226
    P3/P4 ITS, 276
    P4 and P6 soalr arrays, 276
    P6 Integrated Truss Segment (ITS), 248
    P6 Truss, 250
    S5 Truss, 280
    unity, 242
    Z1 Truss, 248, 249
    Zarya, 242, 247
    Zvezda, 241, 242
International Telecommunications Union (ITU),
    74, 168
Ionosonde, 8, 9, 84
Ionosphere, 1, 8, 16, 39, 59, 60, 62–64, 71, 72, 83,
    84, 119
Ionospheric research, 8, 9, 59, 74, 82, 121
ISS expeditions
    2, 255, 260
    13, 277
    20/21, 288, 290, 297, 302
    34/35, 290, 297
    58/59, 290, 294, 301
ISTF. *See* Integrated Service and Test Facility
    (ISTF)

J
Jackson, J.E. Dr., 86
Japanese Space Agency (NASDA), 230
Javelin rocket (U.S.), 69
Jha, V. Dr., 271

Johnson Space Center (JSC), Houston, 162, 194, 279, 304

K

Kármán Line, 16
Kennedy Space Center (KSC), Florida, 164, 166, 187, 192, 193, 209, 213, 216, 221, 224, 229, 230, 235, 242, 246, 248, 255, 275, 280, 284, 285, 296
Kennelly, A., 8
Kerwin, L. Dr., 152, 161, 182, 194, 195
Keyston, J.E., 29, 32, 57, 58
Klein, G.J., 68
Kuehn, L., 158

L

Lacombe, C., 271, 293
Landsat, 139, 140
Lapp, A. Dr., 80
Large Motion Isolation Mount (LMIM), 226, 303
Latching End Effector (LEE), 252, 253
Lathey, J.H. Flt. Lt., 142, 143
LePan, D.V., 27–29, 31, 32
Lesueur, E.A., 34
Life and Microgravity Spacelab Module (LM2), 221
Lindemann, F. Dr., 3
Lines, C.S. Capt., 143
Low, D., 171

M

MacFarlane, Sqn. Ldr., 143
Mackenzie, C.J. Dr., 3–5, 23
MacLean, M.A., 86, 293
MacLean, S., 159, 160, 162, 166, 178, 183, 184, 191–194, 197, 198, 211, 260, 271, 275, 277–279
MacNaughton, J.D., 150
Make or Buy policy, 119
Manipulator Flight Demonstration (MFD), 230
Mar, J., 86
Marchand, R., 158
Marconi, G., 8, 265
Martlet rocket, 91
Matthews, J., 264, 265
Maynard, O., 77, 235
McDonald Dettweiller Company/and Associates (MDA), 200, 203, 271, 273, 274
McGill Fence Early Warning Line, 12
McKay, M., 196, 198, 294, 298, 301
McNally, J., 203
MD Robotics, 252

Measurements of Pollution in the Troposphere (MOPITT), 215
Meek, J.H., 9
Microgravity Measurement Working Group, space shuttle program, 226, 303. *See also* Specific programs
Microgravity vibration Isolation Mount (MIM), 226, 228–230, 232, 233, 303
Microvariability and Oscillations of Stars (MOST), 264, 265, 267
Middleton, J.J.A., 173, 175, 177, 252
Millman, P.M. Dr., 15
Ministry of State for Science and Technology (MOSST), 115–118, 120, 122, 149, 167, 171, 173, 175, 179, 239
Mir Space Station
    docking module (DM), 208, 210–212, 216, 296
    Kristall module, 208
    Priroda module, 229
Mission Operations and Training Simulator (MOTS), 263, 264
Mobile Base System (MBS), 252, 253
Mobile Communications Satellite (MSAT), 171
Mobile Servicing Center (MSC). *See* Mobile Servicing System (MSS)
Mobile Servicing System (MSS), 177, 178, 198, 201, 215, 252, 253, 255, 263–264, 300
Mobile Transporter (MT), 252
Money, K. Dr., 159–161, 166, 183, 184, 190, 195, 197, 294, 295, 299
Montabetti, R. Dr., 12
Mortimer, A., 189
Mueller, G.E., 141

N

Nano crystal Getaway Special (NANO-GAS), 216
NASA Extreme Environment Mission Operations (NEEMO)
    1, 279, 304
    9, 280, 304
National Aeronautical Establishment (NAE), 91, 150
National Aeronautics and Space Administration (NASA)
    Apollo program, 77, 112, 141, 237
National Research Council (NRC)
    Act, 3
National Resources Mobilization Act, 2, 3
National Space Study. *See* Chapman Report
National Space Transportation System. *See* Space Shuttle
Naval Research Establishment, 7
Nelms, L. Dr., 125
Neurolab, 233–238, 304

Nike Family of Surface to Air Missiles
Ajax, 39
Cajun, 14, 40
Hercules, 30

O
Oberle, F. Hon., 177
Office of Space and Terrestrial Applications
(OSTA-3) payload, 164, 216
Orbital Refueling System (ORS) experiement,
164
Orbital Replacement Unit/Tool Change Out
Mechanism (OTCM), 253
Orbital Sciences Corporation (OSC), 267, 274
Orbiter Boom Sensor System (OBSS), 276

P
Paine, T. Dr., 149
Patterson, G.N., 80
Payette, J., 196, 198, 243–245, 247, 284–288, 294,
300
Pearson, L.B. Rt. Hon., 19, 22, 24, 102
Pegasus-XL air-launched rocket, 267
Petrie, W. Dr., 12
Photogrammetric Appendage Structural Dynamics
Experiment (PASDE), 212
Pickering, A. Sqn. Ldr., 142
Plesetsk Cosmodrome, 266
Point Mugu Test Range, 48
Power Data Grapple Fixtures (PDGF), 252, 253,
261
Privy Council Committee on Scientific and
Industrial Research, 21, 25, 109, 110
Project S-27. *See* Alouette Satellite
Propulsion Test Vehicle (PTV). *See* Black Brant
Pulse ionosonde, 8

Q
Queen's University Experiments in Liquid
Diffusion (QUELD) II, 230

R
RADARSAT
1, 203, 207, 215, 271, 272
design and construction, 203, 271
launch, 273, 274
on-orbit, 280
RADARSAT-2, 271–274, 291
RADARSAT-3 proposal, 274
testing, 203

Radio Propagation/Physics Laboratory (RPL), 7,
9–12, 16, 60
Raffaello Multi-Purpose Logistics Module
(MPLM), 261
Raine, H., 125, 128, 129
RCA Victor Ltd., 83, 119, 122
Remote Manipulation System (RMS).
See Canadarm
Risk Mitigation Experiment (RME 1328), 230
Roberts, J., 154
Robertson, N., 21, 27, 31, 32
Robotic Work Stations (RWS), 253, 263
Rose, D.C. Dr., 15, 32, 45, 55, 57, 77
Royal Canadian Air Force (RCAF), 15, 19, 27, 46,
57, 58, 73, 76–79, 86, 141, 143, 298, 299
Royal Canadian Navy (RCN), 9, 12
R-7 rocket, 31

S
Saint-Jacques, D., 290, 294, 300, 301
Salcudean, T. Dr., 226
SAR. *See* Synthetic Aperture Radar (SAR)
SARJ. *See* Solar Alpha Rotary Joint (SARJ)
SAS. *See* Space Adaptation Syndrome (SAS)
experiment
Satellite Assembly and Test Facility (SATF), 125,
128, 129
Sauvé, J. Rt. Hon. Mme., 121
Science Council of Canada, 22, 79, 81
Science Satellite-1 (SCISAT-1), 266, 267
Science Secretariat, 22, 79, 82, 102
Scott, J.C.W. Dr., 11, 12, 60, 61, 72
Scout rocket
Canadian option for use, 59, 93, 94
Scully-Power, P.D., 165
Search and Rescue Satellite (SARSAT), 137–140,
168, 169
SEASAT, 140, 199, 200
Second World War, 1, 3, 8, 9, 11, 15, 17, 21,
34–36, 68, 98, 99, 102, 116, 141, 167, 291
Secord, L., 144
SED Systems Limited, 125, 203
SES. *See* Suffield Experimental Station (SES)
Shuttle Glow Experiment (GLO-4), 212
Shuttle-Mir Program, 207, 208, 211, 212
Siddon, T. Hon., 171
Siksika (Blackfoot) Nation, viii
Skylab, 112, 115, 141, 287
Solandt, O. Dr., 5, 6, 10, 24, 36
Solar Alpha Rotary Joint (SARJ), 276, 277
Soyuz
FG, 274, 288, 290
TMA 15, 284

Space Adaptation Syndrome (SAS) experiment, 166, 189
Space Development Plan/Space Defence Program, 19, 73, 79, 80
Space Policy Framework (SPF), 194
A Space Program for Canada Report No.1. *See* Chapman Report
Space shuttle
    missions, 145, 153, 161, 162, 166, 178, 184, 208, 216, 284–287, 296, 302
    STS-100/A6, 252–261
    STS-107, 268
    STS-115 (ISS 12/A), 275–279, 299
    STS-118 (ISS 13A.1), 280, 281, 283, 284
    STS-120, 281
    STS-127 (ISS 2J/A), 284–287, 300
    STS-2, 153
    STS-41G, 163–166, 295, 302
    STS-42, 184–191, 295, 299
    STS-52, 184, 191–194, 211, 299, 303
    STS-63, 208
    STS-71, 208
    STS-74, 208, 212, 213, 216, 219, 296
    STS-77, 215–226, 295
    STS-78, 215–226, 284, 302
    STS-85, 226–233, 255, 303
    STS-90, 233–237, 241, 304
    STS-96, 241–252, 255, 300
    STS-97, 241–252, 295
    STS-98, 253, 255
Space shuttle obiters
    Atlantis, 210, 211, 276, 279, 296
    challenger, 161, 164–166, 176, 178, 191, 194, 197, 199, 226, 268, 294, 295, 299
    Columbia, 153, 154, 166, 169, 191, 192, 221, 223, 224, 226, 233–236, 268, 299, 304
    discovery, 166, 184, 186–188, 190, 230–232, 243, 246, 247, 295, 300, 303
    endeavour, 216, 219, 221, 242, 249–251, 255, 257–259, 261, 262, 280–283, 285, 287, 291, 295, 296
    enterprise, 264
Space Shuttle Remote Manipulation System (SSRMS). *See* Canadarm
Space Station Freedom, 170, 172–178, 252, 294
Space Systems Operations Complex (SSOC), 263
Space Task Group (STG), 112
Space Transportation System (STS). *See* Space Shuttle
Space Vision System (SVS), 155, 157, 160, 165, 166, 183, 193, 245, 298, 299, 303
SPAR Aerospace Ltd., 123, 125, 144, 146, 149–151, 172, 173, 203

Spartan-207 spacecraft, 216, 220
SPDM Task Verification Facility (STVF), 263, 264
Special Products and Applied Research (SPAR) Ltd., 122. *See also* Spar Aerospace Ltd.
SPAR Aerospace Ltd., 123, 144, 146, 149, 150, 172, 173, 203
Springhill Meteor Observatory, 15
Sputnik, 24–27
Station to Shuttle Power Transfer System (SSPTS), 281
Steacie, E.W.R. "Ned" Dr., 23, 33, 55
Steadman, E.W., 5
Stehling, K., 34
Stewart, R., 196, 294, 298, 301
Storable Tubular Extendable Member (STEM) system, 67–69, 140, 150
Strategic Air Command (SAC), 26
Suez Crisis, 30
Suffield Experimental Station (SES), 7
Synthetic Aperture Radar (SAR), 140, 199–201, 203, 205, 271, 273, 274

T
Task Force on Space/Satellites (TFS), 102–104
Technology Applications and Science-01 (TAS-01), 230
Technology Experiments for Advancing Missions in Space (TEAMS), 216
Telesat Canada, 109, 123, 126, 149
Thirsk, R. Dr., 159, 160, 162, 166, 183, 191, 197, 198, 216, 222, 224, 284, 286, 288, 290, 294, 301, 302
Thoma, A. Flt. Lt., 142
Thor-Agena B rocket, 66, 70
Thor-Delta E1 rocket, 90
Topside Sounder Project. *See* Alouette Satellite
Topside Sounder Working Group (TSWG), 86, 87
Torso Rotation Experiment (TRE), 224
Trudeau, P.E. Rt. Hon., 102, 109, 113, 114, 116, 154
Truly, R., 153
Tryggvason, B. Dr., 159, 160, 162, 166, 183, 191, 194, 197, 198, 226–233, 294, 302, 303
Tuve, M.A., 8

U
Uffen, R. Dr., 83
Union of Soviet Socialist Republics (USSR), 26, 35, 56, 100
United Nations (UN), 27–31, 33, 55, 75, 139
United States (US), 2, 3, 6–10, 13, 14, 17–21, 24, 26–28, 31–33, 35, 36, 39, 41, 42, 48,

57–59, 61, 62, 71, 74, 77, 80, 82, 84, 95, 96, 99, 100, 106, 107, 109, 111–115, 119, 122, 125, 128, 129, 137, 141, 142, 149, 163, 166, 167, 170, 172, 206, 215, 267, 268, 291, 296

University of Toronto Institute for Aerospace Studies (UTIAS), 265

Upper Atmosphere and Space Programs in Canada, Special Study No.1. *See* Chapman Report

Upper Atmospheric Research Satellite (UARS), 183

US Air Force (USAF)
    USAF Headquarters Space Systems Division (SSD), 142

US Department of Defense (DoD), 18, 42, 61, 273

Ussher, T., 144, 150

V

Vandenberg Air Force Base (AFB), 70, 137, 203, 267

VanKoughnett, R., 178

Velvet Glove Air-to-air guided missile, 13

Video Signal Converter (VSC), 255

Viking satellite, 169

Virtual Operations Training Environment (VOTE), 263

V-2 Upper Atmosphere Research Panel (V-2 UARP), 14

W

Wallops Island launch site, 46, 48

Warren, E. Dr., 60–62, 86

Webster, J. Sqn. Ldr., 142

White, Sqn. Ldr. Robert, 142

Williams, D.D., 196, 198, 233–238, 279–284, 294, 303

Willis, C., 157, 158

Winegard, W., 195

Y

Young Space Scientists Program (YSSP), 224

Z

Zimmerman, A.H. Dr., 24, 29, 33, 36, 95